THE IRRIGATION AND WATER SUPPLY SYSTEMS OF VIJAYANAGARA

VIJAYANAGARA RESEARCH PROJECT MONOGRAPH SERIES

General Editors
JOHN M. FRITZ
GEORGE MICHELL
D.V. DEVARAJ

Vol. 1. *Pots and Palaces* by Carla M. Sinopoli

Vol. 2. *The Ramachandra Temple at Vijayanagara* by Anna L. Dallapiccola, John M. Fritz, George Michell and S. Rajasekhara

Vol. 3. *The Vijayanagara Courtly Style* by George Michell

Vol. 4. *Religious Traditions at Vijayanagara as Revealed through its Monuments* by Anila Verghese

Vol. 5. *The Irrigation and Water Supply Systems of Vijayanagara* by Dominic J. Davison-Jenkins

The General Editors of the Vijayanagara Research Project Monograph Series would like to acknowledge a generous grant from Mrs. Eleanor Schwartz which has made possible the publication of this monograph.

The American Institute of Indian Studies, especially its President, Professor Frederick Asher, and Director-General, Dr. Pradeep Mehendiratta, has continued to promote this Series.

We also extend our thanks to Mr. Ramesh Jain and the staff of Manohar Publishers & Distributors for maintaining high standards and overseeing this monograph through the press.

The plan of the 'Stepped Tank' shown on the jacket is by David Chanter.

The Irrigation and Water Supply Systems of Vijayanagara

DOMINIC J. DAVISON-JENKINS

MANOHAR
AMERICAN INSTITUTE OF INDIAN STUDIES
NEW DELHI
1997

ISBN 81-7304-031-1

First published 1997

Published by Ajay Kumar Jain
Manohar Publishers & Distributors
2/6 Ansari Road, Daryaganj, New Delhi 110 002 for
American Institute of Indian Studies
D 31 Defence Colony, New Delhi 110 024

Laserset by AJ Software Publishing Co. Pvt. Ltd.
305 Durga Chambers, 1333 D.B. Gupta Road
Karol Bagh, New Delhi 110 005
and printed at Raj Kamal Electric Press
Delhi 110 033

Contents

List of Illustrations vii

Abbreviations xvi

Preface xvii

Acknowledgements xviii

Introduction 1

CHAPTER 1: The Kingdom of Vijayanagara 3

1.1 The historiography of Vijayanagara 3
1.2 The historical background 5
1.3 Foreign and indigenous literary accounts 8
1.4 Inscriptions near and at Vijayanagara 9
1.5 Vijayanagara kingship and social structure 12

CHAPTER 2: The City of Vijayanagara 19

2.1 Introduction to the environment of Vijayanagara 19
2.2 General characteristics of Vijayanagara in the contemporary accounts 22
2.3 Physical description of Vijayanagara 23
2.4 The internal organisation of Vijayanagara 28

CHAPTER 3: Observations in the Field: 1. Domestic Supply 32

3.1 Introduction to the field observations 32
3.2 The domestic water features of the 'Royal Centre' 33
3.2.1 Features to the east of enclosure IV (enclosures XVIII, XIX, and XX) 33
3.2.2 Features within enclosure IV 36
3.2.3 Features to the south and southwest of enclosure IV (enclosures XXI and XXII) 41
3.2.4 Features within enclosures I, III, V and IX 44
3.2.5 Features within enclosures XIV and XV 44
3.2.6 Isolated domestic water features in the 'Royal Centre' 45
3.3 Domestic water features outside the 'Royal Centre' and temple tanks 45
3.3.1 Domestic and ritual tanks 45
3.3.2 Wells 46
3.3.3 Large temple tanks 46

CHAPTER 4: Observations in the Field: 2. Agricultural Supply 48

4.1 The Kamalāpuram kere 48
4.1.1 The outlets on the Kamalāpuram kere 49
4.1.2 The supply of water from the Kamalāpuram kere to the 'Royal Centre' 51
4.2 The Rāya and Basavanna canals 52
4.2.1 Postconstructional changes to the Rāya and Basavanna canals 52
4.2.2 The Rāya and Basavanna canals in their present condition 54
4.3 The Hosūru and Kālaghatta canals 56
4.4 The Hiriya anicuts and canal 57
4.5 Agricultural water features of the Irrigated Valley 60
4.5.1 The Hiriya kāluve from the Kamalāpuram–Hampi road bridge to the base of Matanga Parvatam 60
4.5.2 The Matanga Parvatam revetted earthwork: The Bhūpati bund 62
4.5.3 The bridges below Matanga Parvatam 62
4.5.4 The third bridge below Matanga Parvatam: The old spillway or waste weir 63
4.5.5 The old sluice on the Hiriya kāluve 64
4.5.6 An inscription on the right bank of the Hiriya kāluve, close to the Matanga Parvatam bridges 65
4.6 The Anegundi anicut and canal 66
4.7 The Rāya bund, southwest of Hospet 67
4.8 The Mallappannagudi bund 70

CHAPTER 5: Interpretation and Analysis 72

5.1 Analytical Methods Employed 72
5.2 Evidence of pre-Vijayanagara water management in Bellary and Raichur districts 73
5.3 Domestic water supply 76
5.3.1 How did the domestic supply system operate? 78
5.3.2 When and by whom was the domestic supply built? 81
5.3.3 Does the technology used represent a development of new ideas or merely a continuation of an existing tradition? 83
5.3.4 To what extent did the success of Vijayanagara as an urban centre depend upon the operation of its domestic supply system? 88
5.4 Agricultural water supply 89
5.4.1 How did the agricultural supply system operate? 91
5.4.2 When and by whom was the agricultural supply system built? 100
5.4.3 Does the technology used represent a development of new ideas or merely the continuation of an existing tradition? 102
5.4.4 To what extent did the success of Vijayanagara as an urban centre depend upon the operation of its agricultural supply system? 103

Conclusions 104

Bibliography 107

Index 121

Illustrations 123

List of Illustrations

Figures marked 'VRP' are reproduced with the kind permission of Drs. John M. Fritz and George Michell of the Vijayanagara Research Project. Figures marked 'after VRP' are based upon material supplied by the Vijayanagara Research Project.

2.1 Map of Vijayanagara its environs (VRP).

2.2 Internal organisation of the site (VRP).

2.3 Vīrūpākṣa temple and the Hemakutam hill (after VRP).

2.4 Achyuta Rāya's temple (after VRP).

3.1 Map of the 'Royal Centre' (VRP).

3.2 Map showing water features in enclosure IV and its environs.

3.3 Open channel XIX/25

3.4 Open channel XIX/25.

3.5. Chunam skirting on the west side of XIX/25.

3.6 Panorama of the disturbed area to the south of Chandraśekhara temple.

3.7 Ruined tank south of Chandraśekhara temple.

3.8 Close-up of the tank's lining.

3.9 Open channel XIX/25; 90° turn to the west, view to north.

3.10 Open channel XIX/25; 90° turn to the east, view to east.

3.11 View to the north along open channel XIX/25; disarticulated sections.

3.12 View to the north along open channel XIX/25; disarticulated sections.

3.13 View to the east along open channel XIX/25; disarticulated sections.

3.14 View to the east along open channel XIX/25; disarticulated sections.

3.15 XIX/25 acting as a boundary wall; view to east.

3.16 XIX/25 acting as a boundary wall; view to east.

3.17 XIX/25 running across toe of hill/outcrop; view to east.

3.18 XIX/25 running across toe of hill/outcrop; view to west.

3.19 Map of enclosure XVIII (after VRP).

3.20 Open channel XVIII/2; view of north.

3.21 Open channel XVIII/2; view of south.

3.22 Close-up of the chunam lining in XVIII/2.

3.23. XVIII/2 passing through the crevice in a boulder; view to north.

3.24 Close-up of the plaster lining in the crevice.

3.25 XVIII/2, supported by brickwork, running along terrace.

3.26 XVIII/2 in excavation; view to north.

3.27 Plan and section of open tank XVIII/1.

3.28 Open tank XVIII/1; view to northwest.

3.29 Open tank XVIII/1; view to northeast.

3.30 Close-up of the plaster lining in XVIII/1.

3.31 Open tank XVIII/1; the east face of the pipe outlet.

3.32 Open tank XVIII/1; close-up of the east face of the pipe outlet.

3.33 Open tank XVIII/1; the west face of the pipe outlet.

3.34 Open tank XVIII/1; close-up of the west face of the pipe outlet.

3.35 Pipeline leading from XVIII/1 passing through enclosure wall.

3.36 Close-up of the pipeline leading from XVIII/1.

3.37 Pipeline passing through alley between XVIII and IV.

3.38 Pipeline running along the outside of the enclosure wall of IV.

3.39 Map showing location of 'Queen's Bath' (XX/1) (after VRP).

3.40 Map showing location of 'Octagonal Bath' (after VRP).

3.41 Supply spout to 'Queen's Bath' (XX/1).

3.42 'Queen's Bath' (XX/1); view to northwest.

3.43 Queen's Bath' (XX/1); plan and section (VRP).

3.44 Moat around the 'Queen's Bath' (XX/1)

3.45 'Queen's Bath' (XX/1) outlet.

3.46 'Queen's Bath' (XX/1) outlet; elevation and section

3.47 'Octagonal Bath' plan and section (VRP).

3.48 'Octagonal Bath'; central fountain.

3.49 'Water Tower'.

3.50 Plaster-lined bath (XIX/25); view to west.

3.51 Plaster-lined bath (XIX/25); plan and section (VRP).

3.52 Raised channel in IV running over outcrop; view to west.

3.53 Raised channel in IV running over outcrop; plan, section and elevation (VRP).

3.54 Raised channel in IV; view to south.

3.55 Raised channel in IV running over outcrop; iron-stained holes.

3.56 Rised channel in IV; elevation and section (VRP).

3.57 Diagrammatic sketch-map showing location of icons on the piers of the raised channel.

3.58 Śiva and Parvatī, mounted on Nandi.

3.59 Viṣṇu on Garuda.

3.60 Sugrīva.

3.61 Hanuman.

3.62 Nandi in a shrine.

3.63 Feeder channel to the 'Great Tank' (IVc/1); view to north.

3.64 The 'Great Tank' (IVc/1); plan (VRP).

3.65 The 'Great Tank' (IVc/1); view to west.

3.66 The Great Tank' (IVc/1); plan, section and elevation of side wall (VRP).

3.67 The 'Great Tank' (IVc/1); outlet.

3.68 Supply chute to the 'Stepped Tank'

3.69 'Stepped Tank'; view to southwest.

3.70 'Stepped Tank'; view to northwest.

3.71 Kannada cypher.

3.72 Plan of enclosure IVa/b (northern sector) (VRP).

3.73 Panorama of enclosure IV; view to west.

3.74 Plaster-lined, stone tank (IVa/18); plan and section (VRP).

3.75 Plaster-lined tank (IVa/23); plan and section (VRP).

3.76 Stone tank (IVa/24); plan (VRP).

3.77 Plaster-lined tank (IVa/31b); plan and section (VRP).

3.78 Plaster-lined brick tank (IVa/32); plan and section (VRP).

3.79 Stone tank (IVb/2); plan and section (VRP).

3.80 Plaster-lined tank with central pavilion (IVb/3) plan and section (VRP).

3.81 Stone-lined tank (IVb/4); plan.

3.82 Rock-cut basin (IVb/5); plan and section (VRP).

3.83 Conduit of type 1.

3.84 Conduit of type 1; IVa/9 and 16, plan (VRP).

3.85 Conduit of type 2.

3.86 Conduit of type 3.

3.87 Plaster-lined tank with central pavilion (IVb/3); view to west.

3.88 Plaster-lined tank with central pavilion (IVb/3); outlet

3.89 Stone-lined tank (IVb/4); view to east.

3.90 Stone-lined tank (IVb/4); outlet.

3.91 Covered channel IVd/17; view to west.

3.92 Covered channel IVd/17; plan and section (YRP).

3.93 Covered channel IVd/17; plan and section (VRP).
3.94 Well (IVA/10b); plan and section (VRP).
3.95 Well (IVa/34): plan and section (VRP).
3.96 Well (IVc/3); plan and section (VRP).
3.97 Well (WA/1); plan and section (VRP).
3.98 Complex of water features in the northeast corner of XXII (XXII/2 and 3); plan (VRP).
3.99 East face of XXII/3.
3.100 South face of XXII/3.
3.101 View eastwards along XXII/2.
3.102 Fragment of the dismantled tank.
3.103 Fragments of the dismantled tank built into a dry-stone wall.
3.104 Traces of plaster covering visible on XXII/2.
3.105 XXII/3; east elevation (VRP).
3.106 XXII/3; plan (VRP).
3.107 XXII/3; close-up of the single pipe outlet, recessed in notch.
3.108 XXII/3; close-up of the double outlet.
3.109 XXII/3; close-up of a hole for the fitting of a pivotal rod.
3.110 XXII/3; raised tank before excavation.
3.111 XXII/3; raised tank after excavation (VRP).
3.112 XXII/3; raised tank, inlet with notch (VRP).
3.113 XXII/3; three pipe outlets in the raised tank (VRP).
3.114 XXII/3; recess behind the pair of pipe outlets in the raised tank (VRP).
3.115 XXII/3; impression of a fourth outlet in chunam in the raised tank.
3.116 XXII/3; plastered insert in the raised tank.
3.117 Panorama of the west end of XXII/2.
3.118 'Octagonal Fountain' (XXII/1).
3.119 'Octagonal Fountain' (XXII/1); plan, section and elevation (VRP).
3.120 'Bhojanśāla'; view to northwest.
3.121 'Bhojanśāla; plan and section (VRP).
3.122 'Bhojanśāla'; carved slabs flanking the channel.
3.123 'Bhojanśāla', close-up of carved slab.
3.124 Well (Ib/9); plan and section (VRP).
3.125 Stone-lined tank (Ic/12); plan (VRP).
3.126 Plaster-lined, stone tank (IIIa/5).
3.127 Plaster-lined, stone tank (IIIa/5); plan and section (VRP).

3.128 Monolithic, stone trough (IIIa/7).

3.129 Well (V/4); plan and section (VRP).

3.130 Plaster-lined tank (V/10); plan and section (VRP).

3.131 Rubble-lined tank (V/12); plan and section (VRP).

3.132 'Water Pavilion' (XIV/2); plan (VRP).

3.133 Stone-lined, rainwater collection tank (XIV/9); plan and section (VRP).

3.134 Step-well (XVa/8); plan and section (VRP).

3.135 Collapsed step-tank (XVb/12); plan (VRP).

3.136 Pipeline in XVb.

3.137 Step-tank (NQy/3); plan and section (VRP).

3.138 Hemakutam hill; first rainwater collection tank.

3.139 Hemakutam hill; second rainwater collection tank.

3.140 Hemakutam hill; third rainwater collection tank.

3.141 Hemakutam hill; fourth rainwater collection tank.

3.142 Hemakutam hill conduit; view to south.

3.143 Hemakutam hill conduit; tank at the southwest corner of the Vīrūpākṣa temple.

3.144 Hemakutam hill conduit passing through prākāra wall of temple.

3.145 Hemakutam hill conduit flowing across the northern toe of the hill.

3.146 Rainwater collection tank at the Parameśvara maṭha.

3.147 Mallappannagudi step-well; plan and section (VRP).

3.148 Manmatha tank; plan (after VRP).

3.149 Manmatha tank; view to northwest.

3.150 Manmatha tank; entrance pavilion.

3.151 Manmatha tank; outlet.

3.152 Viṭṭhala tank; plan and section.

3.153 Bālakṛṣṇa tank; view to east.

3.154 Lokapāvana tank; view to northeast.

4.1 Map of the Kamalāpuram tank.

4.2 Road running along the top of the Kamalāpuram bund.

4.3 Outcrop included in the bund.

4.4 Close-up of the upstream revetting at the east end of the bund.

4.5 The upstream revetting in the centre of the bund.

4.6 Three-sided, masonry structure.

4.7 Kamalāpuram tank; first outlet.

4.8 Irrigation channel taking off from the first outlet; view to northeast.

4.9 Kamalāpuram tank; second outlet.

4.10 Kamalāpuram tank; second outlet, plan of the open tank on downstream face of the bund.

4.11 Kamalāpuram tank; second outlet, open tank on downstream face of the bund.

4.12 Kamalāpuram tank; second outlet, close-up of interior of the open tank on downstream face of the bund.

4.13 Kamalāpuram tank; second outlet, water passing into an irrigation channel.

4.14 Kamalāpuram tank; third outlet.

4.15 Kamalāpuram tank; third oulet, plan of the open tank on downstream face of the bund.

4.16 Kamalāpuram tank; third outlet, the open tank on downstream face of bund; overview from the top of the bund.

4.17 Kamalāpuram tank; third outlet, the open tank on downstream face of the bund.

4.18 Kamalāpuram tank; fourth outlet.

4.19 Kamalāpuram tank; fourth outlet, downstream face of the bund.

4.20 Kamalāpuram tank; buried, masonry structure to the east of the fourth outlet.

4.21 Kamalāpuram tank; plan of spillway.

4.22 Kamalāpuram tank; spillway, view to west.

4.23 Kamalāpuram tank; spillway, close-up of needles.

4.24 Kamalāpuram tank, the escape channel taking off from the spillway.

4.25 Kamalāpuram tank; fragmentary remains of a structure.

4.26 Rock-cut channel passing between the Archaeological Office and the Inspection Bungalow.

4.27 Rock-cut channel flowing parallel to the main street of Kamalāpuram.

4.28 Rock-cut channel bifurcating.

4.29 Left branch of the channel flowing towards the 'Royal Centre'.

4.30 Channel raised up on supporting piers.

4.31 Channel flowing through gateway into the 'Urban Core'.

4.32 Map showing major canals after the construction of the Tungabhadrā Dam.

4.33 Map showing important irrigational features on the right bank of the river between Vallabhāpuram and Hospet.

4.34 Map showing major canals before 1900.

4.35 Map showing the route of the Basavanna and Rāya canals from Amarāya to Kamalāpuram.

4.36 Basavanna canal flowing east of Amarāya; view to west.

4.37 Basavanna canal flowing north after passing under the Hospet Kamalāpuram road; view to north.

4.38 Basavanna canal flowing north of Bhattarahalli; view to west.

4.39 Basavanna canal in deep excavation; view to south.

4.40 Basavanna canal; type 1 revetting.

4.41 Basavanna canal; type 2 revetting.

4.42 Rāya canal flowing close to Nāgenahalli (view to southwest).

4.43 Rāya canal flowing close to Nāgenahalli (view to northeast).

4.44 Rāya canal flowing around sheet rock.

4.45 Rāya canal flowing under the Hospet–Kamalāpuram road; view to north.

4.46 Rāya canal; rubble revetting.

4.47 Rāya canal; fragment of original block-lining.

4.48 Maps showing the Hosūru canal before (above) and after (below) the construction of the Tungabhadrā Dam.

4.49 The Hosūru anicut.

4.50 Map showing the route of the Kālaghatta canal.

4.51 Map showing the routes of the Hiriya and Anegundi canals.

4.52 Map showing the spatial relationship between the Hiriya and the other major canals.

4.53 Diagrammatic plan of the Hiriya anicuts.

4.54 Hiriya anicuts; one of the five, smaller anicuts; view to northeast.

4.55 Hiriya anicuts; close-up of holes drilled in slabs.

4.56 Main Hiriya anicut; view to north.

4.57 Main Hiriya anicut; view to southeast.

4.58 Main Hiriya anicut; fragments of the original weir.

4.59 Main Hiriya anicut; inscription.

4.60 Hiriya anicuts, holes for the fitting of a temporary weir crest.

4.61 Escape weir on the Hiriya canal.

4.62 Hiriya canal; an original water feature on the first outlet, view to east.

4.63 Hiriya canal; an original water feature on the first outlet, view to north.

4.64 Hiriya canal held in embankment against an outcrop of sheet rock; view to west.

4.65 Hiriya canal flowing north of Kṛṣṇāpuram; view to west.

4.66 Hiriya canal flowing in deep excavation towards the ruined temple; view to west.

4.67 Hiriya canal flowing through the ruined temple; view to west.

4.68 Hiriya canal flowing around right side of a boulder; view to west.

4.69 Hiriya canal flowing around left side of a boulder; view to west.

4.70 Hiriya canal flowing towards the Kamalāpuram–Hampi bridge; view to west.

4.71 Hiriya canal flowing under the Kamalāpuram–Hampi bridge; view to north, showing old revetting.

4.72 Hiriya canal discharging into the river at Talarigattu.

4.73 Hiriya canal discharging into the river at Talarigattu; close-up of temporary weir.

4.74 Map of the irrigated valley from the Kamalāpuram–Hampi bridge to Matanga Parvatam (after VRP).

4.75 Panorama of the irrigated valley from the top of Matanga Parvatam; view to southwest.

4.76 Hiriya canal; the first outlet, view to west.

4.77 Hiriya canal; the fourth outlet, view to west.

4.78 Overview of the valley bund; view to southwest.

4.79 Paved road on the right bank of the Hiriya canal beneath Matanga Parvatam; view to northeast.

4.80 Right bank of canal broken away to reveal an earlier structure; view to west.

4.81 Right bank of canal broken away to reveal an earlier structure; view to south.

4.82 Bund across the valley below Matanaga Parvatam; view to northwest.

4.83 Upstream face of the valley bund; view to north.

4.84 Panorama of the exposed core of the valley bund; view to west.

4.85 Plan of the south end of the valley and the Śiva temple.

4.86 Valley bund; close-up of the revetting on the upstream face.

4.87 Valley bund; close-up of the revetting on the downstream face.

4.88 Upstream face of the Hiriya bridge; view to east.

4.89 Downstream face of the Hiriya bridge; view to northwest.

4.90 Plan of the Hiriya bridge.

4.91 Śiva temple on the valley bund; view to northwest.

4.92 Paved ramp descending from the top of the bund of the valley floor; view to northwest.

4.93 Close-up of the inscription marking the boundary of Achyutāpura.

4.94 Second bridge below Matanga Parvatam; view to northwest.

4.95 Plan of the second bridge below Matanga Parvatam.

4.96 Plan of the waste weir (VRP).

4.97 Longitudinal section through the waste weir (VRP).

4.98 Waste weir; view to northeast.

4.99 Waste weir; view to southwest.

4.100 Outlet feeding the old sluice; view to north.

4.101 Plan of the three-bayed structure (VRP).

4.102 Sectional plan of three-bayed structure (VRP)

4.103 North elevation of three-bayed structure (VRP).

4.104 Three-bayed structure before clearance; view to west.

4.105 Three-bayed structure after clearance; view to west (VRP).

4.106 Ceiling beams laid on top of three-bayed structure, view to southwest (VRP).

4.107 Remains of channel running over three-bayed structure; view to east (VRP).

4.108 Hiriya canal; inscription on the inside of crevice, view to east.

4.109 Hiriya canal; close-up of inscription.

4.110 Diagrammatic plan of the Anegundi anicut.
4.111 Anegundi anicut; crest of main weir, view to north.
4.112 Anegundi anicut; downstream apron, view to north.
4.113 Anegundi anicut; vertical face of the downstream apron, view to north.
4.114 Anegundi anicut; close-up of constructional technique.
4.115 Anegundi canal flowing east; view to east.
4.116 Setu aqueduct; view to northeast.
4.117 Upstream face of the Rāya bund; view to north.
4.118 Rāya bund; first outlet, view to northwest.
4.119 Rāya bund; southeast elevation of the first outlet.
4.120 Rāya bund; second outlet, view to northwest.

Abbreviations

APRSC	*Annual Progress Report of the Archaeological Survey Department,* Southern Circle.
ARASI	*The Annual Report of the Archaeological Survey of India.* Government Press. Calcutta, 1903-1936.
ARASM	*The Annual Report of the Archaeological Survey of Mysore.* Government Press. Bangalore.
ARHAD	*The Annual Report of the Hyderabad Archaeological Department.* Government of Hyderabad, Baptist Mission Press. Calcutta.
ARMI	*Administration Reports of the Madras Public Works Department (Irrigation Branch).* Madras, 1876-1947.
ARSC	*Annual Report of the Archaeological Department,* Southern Circle.
ARSIE	*The Annual Report on South Indian Epigraphy.* Government of India, Archaeological Survey, 1894-1965.
ASMG	*Annual Progress Report of the Archaeological Survey of Madras and Coorg.*
EC	*Epigraphia Carnatica.* Government Press, 16 vols. Mysore, Bangalore, Madras, 1889-1955.
EI	*Epigraphia Indica,* 36 vols. Calcutta.
FRMC	First Report of the Madras Public Works Commissioners. 'History and present state of the Maramut Department of Public Works under the Board of Revenue', In *Return to the House of Lords.* Dated 6th May 1853, printed 2nd June 1853.
GMAS	*Government of Madras, Archaeological Survey.*
HAS	*Hyderabad Archaeological Series.* Government of Hyderabad, Baptist Mission Press, Calcutta.
IAR	*Indian Archaeology: A Review.* Archaeological Survey of India, Calcutta.
IGI	*Imperial Gazetteer of India,* 26 vols. Calcutta, 1907-1909.
PPC	*Parliamentary Papers. Returns to the House of Commons.*
RIIC	*Report of the Indian Irrigation Commission (1901-1903).* Published in four parts. London, 1903.
SRMG	*Selections from the Records of the Madras Government (New Series).* Madras.
VR	Rangacharya, V. (ed.), *A Topographical List of the Inscriptions of the Madras Presidency, Collected till 1915 with Notes and References.* Government Press, 3 vols. Madras, 1919.

Preface

As capital of a successful kingdom, the city of Vijayanagara flourished, dominating the political and cultural milieu of South India for 229 years. Vijayanagara could not have operated as a capital in the hostile, semi-arid environment of the Deccan without effective water supply systems to provide for the practical and ritual needs of its inhabitants.

The objective of this study was to examine the development and application of hydraulic technology at the city during the fourteenth to sixteenth centuries A.D. A multi-disciplinary approach was used to answer specific research questions relating to the design, operation and chronology of the water supply systems. This was achieved by integrating field data, documented in textual, diagrammatic and photographic form, with historical epigraphical and archival data.

In order to provide a social and political framework for the analysis of water supply at the site, Vijayanagara's history, historiography, and contemporary non-material evidence was analysed. In addition, an investigation into the organisation, environmental setting, operation, and layout of the city was carried out to better understand the spatial relationship between water supply features and areas of settlement.

The monograph presented here was originally submitted as a doctoral thesis to the University of Cambridge. It was written between 1985 and 1988 at Magdalene College under the supervision of Dr. Raymond Allchin.

New York
December 1994

DOMINIC J. DAVISON-JENKINS

Acknowledgements

First and foremost, I must thank Dr. Raymond Allchin for his patient guidance during my time at Cambridge. His advice proved of inestimable value. I am also indebted to Drs. George Michell and John Fritz of the Vijayanagara Research Project, who supported my work throughout, providing generous academic and practical assistance. In particular, I wish to thank them for permission to reproduce maps, plans, elevations, and sections of selected features within the 'Royal Centre' and 'Urban Core'. I would also like to acknowledge the co-operation and assistance of Dr. A. Sundara of the Department of Archaeology and Museums, Karnataka, and of Karnatak University, Dharwad.

Further, I would like to express my gratitude to the following for their practical and academic advice: Dr. Bridget Allchin, Mr. Balasubramanyam, Dr. Sunil Chander, Mr. Chandrashekhara, Dr. Howard Chase, Dr. David Clary, Dr. George Erdosy, Dr. Ian Glover, Mr. Don Manning, Dr. Parvathi Menon, Mr. Roy Norman, Dr. S. Rajasekhara, Dr. John Smith, and Prof. Burton Stein. I would also like to thank the artist, Mr. Roy Keegan, and his wife, Sarasvati, for their generous hospitality during my time at Hampi.

For financial support, I am grateful to the British Academy, Magdalene College, the Ancient India and Iran Trust, and the University of Cambridge.

Finally, I would like to thank my wife, Sarah. Without her help and encouragement this monograph would not have been possible.

Introduction

The city of Vijayanagara was founded in 1336 A.D. on the Tungabhadrā river, about 120 km. west of Bellary in Karnataka, south India. As capital of a successful kingdom, the city flourished, dominating the political and cultural milieu of South India for 229 years. It is clear that Vijayanagara could not have operated as a capital in the hostile, semi-arid environment of the Deccan without effective water supply systems to provide for the practical and ritual needs of its inhabitants.

In many ways it is remarkable that the evidence for water supply at Vijayanagara has received so little attention by scholars. As a conspicuous and durable class of physical remains they lend themselves well to an investigation of water management in the context of the urban centre. Prior to the construction of archaeological and architectural inventories by Drs. Fritz and Michell of the Vijayanagara Research Project, the water features at Vijayanagara received only scant attention in the literature. A comprehensive and integrated study of the urban settlement is now under way. The investigation presented here is intended as separate, though complementary, to the work of the Vijayanagara Research Project. It is hoped also that it may have a relevance beyond the scope of this study to the wider issues of the development of hydraulic technology in the context of urban settlement.

The history of water supply in India has largely been overlooked to date. Few specialised studies have been devoted to an investigation of the physical, literary and epigraphical evidence for irrigation and water supply. Notable exceptions include Rotzer's (1984) study of the complex water systems at Bījāpur, constructed in the sixteenth and seventeenth centuries A.D., and Venkayya's (1906) study of south Indian irrigation entirely in the light of inscriptions. The evidence for Sinhalese agricultural water management on the ground has been extensively documented by Brohier (1935) and Fernando (1980). The social implications of Sinhalese water management have been discussed by Leach (1959) in a critical analysis of Wittfogel's (1957) theory of 'hydraulic civilisation'.

The remains of the irrigation and water supply systems of the city of Vijayanagara provide the archaeologist with a unique opportunity to study the development of hydraulic technology during the fourteenth to sixteenth centuries A.D. An examination of the use of irrigational technology at the site is essential to a greater understanding of the operation of the medieval urban centre in south India, whose growth and existence depended on a successful manipulation of the environment.

Chapter 1, entitled 'The Kingdom of Vijayanagara', provides a brief, historical background for the period and examines recent changes in the interpretation of Vijayanagara's history. The primary sources of non-material evidence for the period, literary records and inscriptions, are considered, in particular for evidence relating to water supply. The final section is devoted to an examination of Vijayanagara kingship and social structure, and attempts to define the responsibility of the Vijayanagara kings in relation to water management. Chapter 1 is intended as a social and political framework for the analysis of water supply.

In chapter 2, entitled 'The City of Vijayanagara', the organisation and environmental setting of the capital are explored. It begins with a brief introduction to the physical environment in which the city operated. This is followed by an examination of the contemporary accounts for information relating to its operation and layout. A physical description is provided which explains the spatial relationship between water supply features and areas of settlement. Finally, the internal organisation is discussed in order to determine the role of water supply in the operation and maintenance of the city.

Chapters 3 and 4 contain the core of this investigation, the field observations. The evidence on the ground has been divided into two categories, domestic and agricultural supply. Data was collected using the technique of surface archaeology. Visible evidence on the ground was extensively documented in textual, diagrammatic and photographic form. No excavation was undertaken. This technique was adopted for three major reasons. Firstly, the majority of water features survive on the ground. Secondly, the water features cover an extensive area. Thirdly, this technique was perfectly suited to the selective collection of data relating solely to water management. The field data is presented as a textual description, illustrated with photographs, drawings and maps.

In the fifth and final chapter, a multidisciplinary approach is used to integrate the field observations with historical, epigraphical and archival data. This methodological technique was adopted for three reasons. Firstly, to make better sense of the available evidence by placing it in the wider context of the social and economic developments of the period. Secondly, to facilitate the testing of hypotheses constructed using field observations against alternative data. Thirdly, to give the investigation a wider cultural and temporal setting by examining the antecedents of the irrigational technology witnessed at Vijayanagara. The evidence for pre-Vijayanagara water supply in Bellary and Raichur districts is discussed in order to establish whether the technology employed at Vijayanagara represents the continuation of a local tradition.

For the purposes of interpretation and analysis in chapter 5, the division between domestic and agricultural supply has been maintained. Each section begins with a brief literature review. This is followed by a discussion of the operation and design of the water systems, and an appraisal of the builders' understanding of the basic principles involved. The chronology and sponsorship of the systems is then examined. The penultimate sections seek to establish the antecedents and contemporary influences for the technology used. The final part of each section discusses how far the success of the city depended on the operation of its water supply systems.

In conclusion, we would like to state that this dissertation is intended as a preliminary investigation and that its objectives were from the outset limited to a number of specific research questions. These centred on the establishment of the design, operation and chronology of the water supply systems, and the relationship between royal responsibility and water management at the city.

CHAPTER 1

The Kingdom of Vijayanagara

1.1 The historiography of Vijayanagara

The following discussion provides a brief introduction to recent reinterpretations of the social and political events of the Vijayanagara period.

In the last ten years there has been a rigorous re-examination of the traditional views concerning the foundation, character and growth of Vijayanagra. A more complete, though polemic, historiography is now being constructed upon the work of Stein (1980), Breckenridge (1985), Kulke (1985) and Ludden (1985) in an attempt to devise a more satisfactory interpretation of the fragmentary literary and epigraphical evidence available for the period A.D. 1336-1565. Stein (1980) in particular has gone a long way towards explaining, or at least clarifying, some of the discrepancies which are contained in the traditional history of Vijayanagara. The conflict between the concept of dharmic ideology as a guiding movement for the Vijayanagara state and the presence of Muslims at the capital, the frequent use of Muslim mercenaries and the attacks on other Hindu kingdoms had received little attention prior to Stein's investigation into Vijayanagara statecraft (1980). It is, however, important not to underestimate the role of early writers in the establishment of facts.

According to the new interpreters, the early writers of Vijayanagara's history (Sewell 1900; Longhurst 1917) sought to find in the evidence available to them, social and political manifestations which mirrored those seen in contemporary Europe. It is not therefore surprising to find Vijayanagara described as a powerful, centralised empire with a heavily populated capital. These assumptions are essentially reliant on the back projection of European terminologies and social and political constructs. Once established, this view of Vijayanagara history was enshrined in the work of Sastri (1950) and others who propagated the notion of a 'Golden Age'. Ludden (1985:6) discusses this phenomenon in his study of Tirunelveli peasant history. The influence of nationalistic ideology is criticised by Stein in his essay, 'The State and Agrarian Order' (1975).

The work of Stein has initiated a re-appraisal of the accepted social and political history of Vijayanagara which recognises the relationship between the state and the economy. The rejection of the unitary, centralised state as a model for the political arrangements of medieval South India, has been of profound importance in this redefinition. The extant literary and epigraphical data relating to the Vijayanagara period makes better sense in the context of a social and political framework less rigid than that proposed by Sastri (1950), Mahalingam (1951, 1955) and Thapar (1981).

Perhaps Stein's most important contribution to the study of South Indian social and political organisation is the rejection of the widely held belief in the static and uniform nature of conditions (1975:68). Traditionally, the Vijayanagara period has been viewed as one of conservatism and stagnation. This is directly contradicted by the physical evidence which suggests religious and racial tolerance, eclecticism in art and building styles and political volatility. The acceptance of a more flexible approach to the problem of Vijayanagara's social and political history will undoubtedly lead to the construction of a more successful framework for a study of this period.

The redefinition of the state in the context of Vijayanagara has not, however, moved on from the 'segmentary model', culled by Stein from the work of Southall in East Africa. Stein (1980:264) states:

> In order to consider a better model of political arrangements in medieval South Indian states than the existing one, it is necessary to recognise that there are many kinds of states, of which the unitary, centralised state is but one. It is, however, precisely that one which most South Indian historians assume to have been normative during the medieval period, and except when the king was weak or the kingdoms troubled by natural disasters or invasions, these historians claim to have existed.

Stein correctly recognises the deficiencies of a rigid unitary model as a definition of political order in medieval South India, but his segmentary paradigm, whilst replacing the inflexible concept of the unitary state, is dependent on the operation of a strong centre. Ludden (1985:69-71) in his examination of Nāyaka political organisation follows Stein's association of a 'hierarchy of authority' with strong urban centres. Ludden connects the investment of large amounts of revenue in the construction of defensive precautions and the elaboration of cult temples with an increase in centralised authority. A political manifestation of the sort proposed by Ludden for the Nāyaka period would necessitate a complex bureaucratic structure (especially for the keeping of records and communication), for which there is no evidence.

Chander (1987:18-19) has found the segmentary state a convenient model from which to begin a new appraisal of political patterns in the Hyderabad region, but recognises its limitations as a rigid paradigm. He rejects the notion that a strong, or at least a dominant, centre was the necessary pivot of a regional political order, or that its centre was co-equal with the state (Chander 1987:48). In the development of his argument relating to the dual character of territorial sovereignty in the segmentary state, Stein fails to delineate the extent of the state domain. Stein accepts that political power is divided between the centre and peripheral foci of administration who each exercise actual political control over a part, or segment, of the political system encompassed by the state (1980: 274). However, no attempt is made to explain how a strong political centre, dominating the segmentary pyramid, successfully distinguishes itself through the manipulation of ritual supremacy from other peripheral units. This problem is addressed in greater detail in section 1.5.

Stein has argued that large urban centres in South India grew up around important cult temples which were already in existence (1980:246), and that such urban activity brought with it craft specialisation and extra-local trade (1980:481). At Vijayanagara, temple complexes of extra-local importance essentially postdate the foundation of the capital. The Virūpākṣa temple at Hampi was a site of some ritual importance before A.D. 1336, but achieved great pilgrimage status only after the massive investment of gifted cash by Vijayanagara kings and nāyakas. The association of a royal capital with a tīrtha may be interpreted as a conscious attempt by Vijayanagara kings to generate spatial and social identity.

This investigation considers the dynamics of resources a crucial determinant in the sphere of South Indian social and political development. Chander (1987:16-17) has discussed the political importance of resources for the pre-colonial Hyderabad region. The availability of population and agricultural land shaped the character and growth of Vijayanagara as a political entity, a factor which has been ignored by many scholars. Ludden (1985:18-26) provides an excellent appraisal of the resources of the Tirunelveli region, and includes a detailed discussion of population and settlement patterns. The analysis is flawed, however, by his failure to appreciate the scarcity of resources in the Medieval period.

Previous research suggests that the military campaigns of the Vijayanagara kings represented either wars of conquest intended to extend the spatial area under direct political control or raids into lucrative areas designed to finance military expenditure. These military invasions or raids may represent a more complex strategy than hitherto envisaged, related to the inherent shortage of resources discussed above.

The attraction of population densities to political centres remained the primary concern of South Indian kings. This issue has been examined by Richards (1933:235-237). The visible defence of urban centres and their hinterlands represents one way in which a king may depict his kingdom as 'strong' and thus lure population groups away from other centres of attraction. The coercion of demographic units was not a viable option in a period of high mobility and migration. Ludden (1985:13,72) suggests that legitimate state coercion developed in the colonial period and only reached fruition under the British colonial regime.

The defence of scarce resources against attacks from other political units remained a matter of considerable importance for South Indian kings. The destruction of other ritual/political centres would act as a temporary means of curbing the expansionist tendencies of rival kingdoms (Chander 1987:17,37).

Small revenue returns from the attenuated core zone and nominally subordinate local kings and nāyakas outside this area would be supplemented by the removal of portable wealth from the territory of other political centres (Ludden 1985:27). Military action served also to establish the concept of the 'domain' (Chander 1987:19). Through active campaigning outside the 'sarcar' land or core zone, Vijayanagara kings delimited the extent of their own territory and established within it 'domain homogeneity'.

It may be postulated that Vijayanagara as an urban centre was located in the only remaining area of agricultural and political potential not already defended by an existing political unit. This picture of political containment is diametrically opposed to the traditional view of Vijayanagara as an extensive and monolithic state standing against the southward movement of Muslim power. Although formulated for a later period, Chander's definition of political and economic capitals in the Hyderabad region correlates well with the picture of Vijayanagara as politically contained:

> Zamindars and nawabs were concerned, in practice, with the maintenance of a loyal community in a productive domain in the context of external hostility and pressure (1987:19).

The problem of defining the extent of Vijayanagara political power is addressed more thoroughly in section 1.5. In conclusion, the re-examination of the spatial distribution and relative chronologies of Vijayanagara epigraphs throughout South India for evidence of direct/sustained political control, remains as a line of profitable future inquiry. Two epigraphs of the early sixteenth century A.D. record the movement of population away from areas where tax demands were excessively high. In A.D. 1501 a group of Ramnad farmers sold their land and moved away in response to unreasonable taxation (Sewell 1932:234). An inscription of A.D. 1514 records the revision of taxes in South Arcot, following the abandonment of agricultural land by peasant farmers (Sewell 1932:239).

1.2 The historical background

This brief introductory background to the history of Vijayanagara, is intended as a framework into which more complex social and cultural data can be fitted, thus summarising available information, old and new, for the period A.D. 1336-1565.

In the early part of the fourteenth century A.D., the failure of the Delhi Sultanate to maintain authority in the area south of the Vindhya mountains resulted in the formation of a number of new kingdom. On the orders of Sultan Alā-ud-dīn Khalji (A.D. 1296-1316), Malik Kāfūr led his army deep into the south in A.D. 1310 and forcibly installed Muslim governors to administer these territories (Briggs 1908:i.373-375; Longhurst 1917:9; Sastri 1950:i.44-45). Thapar (1981:321,323) has argued that the power vacuum created by the return of Kāfūr and his army to Delhi in 1311 had two major effects. Firstly, a power struggle developed between the new Muslim governors, some of whom rebelled against the Sultanate's authority. Secondly, the established political order of the south was revitalised by the Sultanate's expansionist policies. This situation proved favourable for the emergence of new political entities.

By the time Sultan Muhammad-bin Tughlak (A.D. 1325-1351) ascended the throne, the threat to the Sultanate's authority in the south had become so serious that the Sultan himself was forced to lead an army into the Deccan. In A.D. 1326, Bahā-ud-dīn Gurshāsp, the Governor of Sagar, rebelled against the Sultan, his cousin, and was defeated in battle (Briggs 1908:i.418; Sastri 1950:ii.50; Sherwani 1953:18). Gurshāsp fled for protection to the king of Kampli, and Muhammad-bin Tughlak responded by laying siege to the fortified town of Kampli itself, which he proceeded to capture by force in A.D. 1327 (Sewell 1900:17; Briggs 1908:i.418-419; Sastri 1950:ii.51). The Governor of Sagar escaped with his life, but was handed over to the Sultan by the Hoysala King, Vīra Ballāla III, who feared that Muhammad might turn his attention towards his own kingdom further south.

The reduction of Kampli would not be important if it were not for two brothers, Harihara and Bukka, the sons of Saṅgama, who surrendered during the siege. Traditional accounts hold that these two Telugu noblemen returned to Delhi with the army of the Sultanate, where they were converted to the Muslim faith. When Malik Muhammad, the Muslim Governor of Kampli, found himself unable to control the area, Sultan Muhammad-bin Tughlak sent Harihara and Bukka to replace him (Sastri 1950:ii.122; Devakunjari 1970:6). In A.D. 1336, Harihara and Bukka renounced their adopted faith and rebelled against the Sultanate, founding their own kingdom. This traditional account of the foundation of Vijayanagara has been re-examined by Kulke (1985), who proposes, on the basis of epigraphical data, an indigenous origin for Harihara and Bukka.

Literary accounts suggest that the capital of this new kingdom was located at Hampi, 16 km. to the west of Kampli, on the south bank of the Tungabhadrā river. This site and the area directly to the south became known as Vijayanagara, the 'City of Victory', giving its name to the political entity Harihara and Bukka were to forge.

In A.D. 1336, with the help of the Madhavācarya Vidyāraṇya, a powerful local Brahman, Harihara was declared vice-regent of the Śaivite deity Virūpākṣa (Sewell 1900: 18-23; Longhurst 1917:10; Sastri 1950: ii.122; Devakunjari 1970:6-7). Madhavācarya Vidyāraṇya was a celebrated scholar, the brother of Sāyaṇa the commentator on the *Rigveda*, and himself author of several works on Vedānta. This association of a royal personage with a powerful cult deity must be compared to that of Anaṇgabhīmadeva III in Orissa (Panda 1985:89). the traditional interpretation of the literary and epigraphical evidences suggests that Harihara and Bukka, in a relatively short space of time, consolidated their authority over a large area of territory using both military and peaceful means. the extent of their actual power remains unknown. It is likely, however, that the disintegration of the old Hoysala kingdom further south and the general dissatisfaction with the imposition of a Muslim ruling class in South India considerably aided the two brothers in their task.

The inability of the Delhi Sultanate to control the Deccan during the latter part of Muhammad-bin Tughlak's reign encouraged the emergence of another powerful kingdom. In A.D. 1347 Abul Muzaffar Alā-ud-dīn was proclaimed King of the Deccan, thus founding the Baḥmanī kingdom (Briggs 1908:i.440; ii.290-291; Sastri 1950:ii.109; Sherwani 1953:36-37). The expansion of this new Muslim kingdom was checked, firstly, by the growth of Vijayanagara with whom it waged internecine war for possession of the Raichur Doab, and secondly, by the reorganisation of the kingdom of Warangal in Telingana (Sastri 1950:ii.122; Sherwani 1953:23; Panda 1985:88-96). Further south, Jalāl-ud-din Ahsan Shāh, the Governor of Ma'bar, established the Madura Sultanate in A.D. 1333 at considerable cost to the old Hoysala kingdom (Sastri 1950:ii.53; Sherwani 1953:222; Ludden 1985:44). During the reign of Bukka (A.D. 1343-1379), Kumāra Kampaṇa successfully annexed the Madura Sultanate for Vijayanagara in A.D. 1365 (Sastri 1950:ii.123; Watson 1964: 102-104; Devakunjari 1970:7; Rajasekhara 1985: 103-104). However, this part of the old Pandya kingdom was never brought under direct political control by Vijayanagara (Stein 1980:393; Ludden 1985:44).

On the death of Harihara in A.D. 1343, Bukka succeeded to the throne and during his reign the influence and prestige of Vijayanagara steadily grew. Indeed, it was during the reign of Bukka that the emergent Hindu kingdom first clashed with the Bahmanī Sultanate under Alā-ud-dīn Bahman Shah (Sewell 1900:30; Briggs 1908:ii.294; Sastri 1950:ii.123; Sherwani 1953:61). Thus a pattern of conflict between the two kingdoms was established, oscillations of power in the Raichur Doab characterising the period.

Harihara and Bukka were the first two kings of what has come to be known as the Saṅgama dynasty (Sewell 1900:404). From epigraphical sources it is clear that at least seven more kings followed them before the usurpation of Narasiṁha in A.D. 1490. This marked the beginning of the second, Saḷuva dynasty (Sewell 1900:107-113; Sastri 1950:ii.126-127). With the conquest of the Madura Sultanate in A.D. 1365, the kings of Vijayanagara laid claim to all the territory south of the Kṛṣṇa river, from coast to coast. However, the military campaigns of Vijayanagara were not always successful and the city itself was besieged by the Bahmanī Sultanate on four occasions; by Muhammad I in A.D. 1366 (Sewell 1900:38; Briggs 1908:ii.316); Mujāhid in A.D. 1376 (Sewell 1900:41-43; Briggs 1908:ii, 333-334); Fīrūz in A.D. 1406 (Sewell 1900:59; Briggs 1908:ii.383), and by Ahmad I in A.D. 1423 (Sewell 1900:69; Briggs 1908:ii.404).

The Kingdom of Vijayanagara reached its apogee during the reign of Kṛṣṇadeva Rāya (A.D. 1509-1530). The king waged war successfully in the Raichur Doab against Ismail 'Ādil Shāh of Bijāpur (Sewell 1900:138-160; Briggs 1908:iii.48-50; Sastri 1950:ii.129) and in the east against the kingdom of Orissa (Sewell 1900:130-132; Sastri 1950:ii.129). In addition to his military conquests, Kṛṣṇadeva Rāya also undertook a substantial programme of building and structural alteration in the city itself (Rajasekhara 1985:107-108). A number of public buildings and temples were initiated or modified. Major water supply features were also constructed to provide a supplementary water supply to the greater metropolitan area. The town of Hospet, situated to the southwest of the city, represents a suburban extension to the capital, called Nāgalāpura. This too was constructed by Kṛṣṇadeva Rāya, presumably to house additional population (Sewell 1900:162; Longhurst 1917:19; Devakunjari 1970:9, 68-69).

It was during the reign of Kṛṣṇadeva Rāya that the Portuguese first established themselves on the west coast of India, under Francisco de Almeida (Sastri 1950:ii.149). By skilful diplomacy and military force, they came to dominate the lucrative horse trade with Vijayanagara, providing cavalry mounts for the Hindu nobles (Stein 1980:400-402). The Vijayanagara armies also employed Portuguese mercenaries, using them to good effect at the siege of Raichur in A.D. 1520 (Sewell 1900:343-345; Sastri 1950:ii.129; Stein 1980:402-403).

Although the reign of Kṛṣṇadeva Rāya marked a golden period in the history of Vijayanagara, it is perhaps ironic that the seeds of decay were also planted at this time. The King's military success in the Raichur Doab can largely be attributed to the fragmentation of political power in the Deccan. At the end of the fifteenth century A.D., the Bahmanī Sultanate had broken up resulting in the formation of the five successor states; Ahmadnagar, Berar, Bīdar, Bījāpur and Golkoṇḍa (Sastri 1950:ii.120). Kṛṣṇadeva Rāya's crushing defeat of the state of Bījāpur in A.D. 1520 forced the Muslim kingdoms of the Deccan to consider the possibility of uniting to destroy a common enemy (Sewell 1900:155).

Kṛṣṇadeva Rāya died in A.D. 1529. A power struggle ensued, heralding a period of political unrest (Sewell 1900:165; Sastri 1950:ii.165). The throne of Vijayanagara eventually passed to Kṛṣṇadeva Rāya's cousin, Achyuta Rāya (A.D. 1530-1542), and subsequently to Rāma Rāya (A.D. 1542-1565), under whom on 23rd January 1565 the army of Vijayanagara was defeated by the combined forces of the Muslim succession states near Talikota (Sewell 1900:196-205; Briggs 1908:iii. 246-249; Longhurst 1917: 23-24; Sastri 1950:ii. 133-134; Devakunjari 1970:10). The rout of the Hindu army left the city of Vijayanagara without military protection, and it was not long before its inevitable destruction came.

Sewell (1900:208) rather romantically says of this:

> Never perhaps in the history of the world has such havoc been wrought, and wrought so suddenly, on so splendid a city; teeming with a wealthy and industrious population in the full plenitude of prosperity one day, and on the next seized, pillaged, and reduced to ruins, amid scenes of savage massacre and horrors beggaring description.

Sewell's description of this event does little to explain the immediate social and political consequences of Vijayanagara's destruction as a sacral-symbolic centre.

It is reported that the nobles who survived the battle fled to Penugoṇḍā in the east, where an attempt was made to resurrect what remained of the existing political order (Sastri 1950:ii.135). The city was abandoned as a royal capital, the population that it had contained dispersed, leaving communities in isolated parts of the settlement, such as Kamalāpuram and Hampi. The inhabitants of these villages continue to farm the agricultural zones in and around the city, including some areas within the walled enclosures of the 'royal centre'. It remains unclear how the political order of the areas outside the core zone of political domination was effected by the demise of the urban settlement. We suggest that the established political milieu was little changed by the passing of Vijayanagara as a sacral-symbolic centre.

1.3 Foreign and indigenous literary accounts

A summary glance through the corpus of available literature on Vijayanagara will indicate that archaeological and historical research has drawn heavily from contemporary written accounts and compiled histories. These works were collected together for the first time by Sewell who included extracts from them in *A Forgotten Empire* (1900). Sewell concentrated on the accounts of two Portuguese travellers, Domingos Paes and Fernao Nuniz, who provide lengthy and detailed descriptions of the city and its inhabitants. These accounts also contain records of daily life and dynastic histories compiled in the first half of the sixteenth century A.D.

References to Vijayanagara are to be found in the works of other contemporary European and Arab writers. Translations are provided by Major (1857), Winter Jones (1863) and Dames (1918). A chronology of the contemporary accounts may be constructed as follows:

	Date	Nationality
Ibn Batuta	*c.* 1340	*Moroccan*
Nicolo di Conti	*c.* 1420	*Italian*
Abd-er-Razzak	*c.* 1442	*Persian*
Ludvico di Varthema	*c.* 1504	*Italian*
Duarte Barbosa	*c.* 1510	*Portuguese*
Domingo Paes	*c.* 1520	*Portuguese*
Fernao Nuniz	*c.* 1535	*Portuguese*
Caesaro Federici	*c.* 1567	*Italian*

A very general account of the city is provided by the Russian traveller, Athanasius Nikitin, who was travelling in India during A.D. 1470. However, he probably came no nearer to Vijayanagara than Gulbarga.

The character and value of the accounts is mixed. Often recorded in written form after the traveller had returned to his native land, they are not without confusions and irrelevancies. However, these eclectic collections of recorded observations, reported knowledge, local customs, history and personal comment contain invaluable data for a study of the period A.D. 1336-1565.

Previous research has ignored the problems of political bias, exaggeration and poetic licence in the contemporary accounts. Passages relating to the size of Vijayanagara and its population have not been subjected to critical analysis. The use of select extracts from contemporary accounts by scholars has reaffirmed accepted assumptions regarding the city's extent, character and influence. We propose to make critical use of contemporary accounts as a source of data for water management at the site.

To date, the full potential of the literary accounts has remained unrealised, as the information contained in them has been utilised mainly for descriptive purposes. Sewell (1900), Longhurst (1917) and Devakunjari (1970) have attempted to identify structures at the site using contemporary descriptions of Vijayanagara with little success. Nevertheless, some attempts have been made to examine this material in a more imaginative and constructive manner. Stein (1980) has made a study of Vijayanagara social structure and kingship using contemporary references to the Mahānavamī festival. Following Stein, Fritz *et al.* (1984) have provided some interesting structuralist interpretations of the surviving remains at the site.

The history of the Bījāpur kings compiled from thirty-four standard contemporary histories by the Persian, Mahomed Kasim Ferishta (b. A.D. 1570), provides data relating to the political history of Vijayanagara. Briggs (1908) furnishes us with a translation of this work. The value of Ferishta's document, though politically biased against Vijayanagara, cannot be underestimated as many of the standard texts used in its compilation have now vanished.

In addition to the contemporary accounts

and later histories, there are other literary works which provide information pertaining to Vijayanagara. Several poetic compositions of the Medieval period are of importance in this respect. These include the Madhurāvijayam (Rajasekhara 1985:13) by Gaṅgadevī, wife of Kumāra Kampaṇa (fourteenth century A.D.) and the Āmuktamālyada (Rajasekhara 1985:29) ascribed to King Kṛṣṇadeva Rāya (sixteenth century A.D.).

The officials of the East India Company and later the British colonial administration have left us with detailed information concerning revenue, agriculture, population, geography and local history in South India in published and manuscript form. The Bellary Gazetteers of 1872 and 1904, compiled by Kelsall and Francis respectively, provide a wealth of diverse information. Much of the information contained in these works is of crucial importance to an understanding of pre-colonial conditions in the area. Colonial administrators had always sought to record information pertaining to the areas which were under their control, even if it was only to more efficiently extricate lucrative, agricultural revenue returns.

It is necessary here to indicate the importance of archival material to this study. The ancient irrigation systems of Bellary district fell under the jurisdiction of the Public Works Department of the Madras Presidency. This body methodically recorded the restoration and replacement of water features. The records of the Public Works Department have proved of invaluable assistance in understanding the complex changes made to medieval water features from 1846 to 1956. Parliamentary returns for the Public Works Department of the Madras Presidency have also yielded important information pertaining to the management of water in Bellary district.

1.4 Inscriptions near and at Vijayanagara

The published epigraphical record at the site may be divided into two categories: pre-Vijayanagara inscriptions and inscriptions of the period A.D. 1336-1565. The eleven pre-1336 epigraphs represent a confusing picture of settlement at the site prior to the foundation of Vijayanagara. They are located in only four areas:

1. The environs of the Virūpākṣa temple (2).
2. Hemakutam hill (1).
3. Anegundi (3).
4. Enclosure IV in the 'royal centre' (5).

Two inscriptions of the late twelfth and early thirteenth centuries A.D. indicate religious activity around the much elaborated Virūpākṣa temple complex which was formerly the site of a sacred tīrtha dedicated to Pampā (Rajasekhara 1985:101-102, 117). An undated epigraph in the Prasaṇṇa Virūpākṣa temple on Hemakutam hill refers to the erection of a temple by Vīra Kampilirāya, the chief of Kampli, and may be ascribed to the closing years of the pre-Vijayanagara period (Rajasekhara 1985:117). Three pre-Vijayanagara inscriptions are reported at Anegundi. Two Jain inscriptions date between the tenth and thirteenth centuries A.D., whilst the third records a land grant made by a Cālukya king in the tenth century A.D. (Shama Sastry 1917:285-291).

More interesting are the five pre-1336 inscriptions discovered in enclosure IV of the 'Royal Centre'. The oldest and most intriguing of these inscriptions is a first or second century A.D. fragment on white limestone recording a religious donation (probably Buddhist) in Prākrit (IAR 1975-6:20, 62). It reads:

Tarasa putasa dānam

The inscription may be associated with the five limestone Buddhist relief panels recently recovered by the Archaeological Survey of India from behind the 'Hundred-pillared Hall' (IVb/1). These fine panels, thought to have originated from Amarāvati or Nagārjunakoṇḍa, are tentatively dated by the present writer to the second or third century A.D. Two inscriptions of Kṛṣṇadeva Rāya record gifts to the Amareśvara temple at Amarāvati in A.D. 1515 (Sewell 1932:240; VR 1919:ii., Guntur,632,638; ARSIE 1892:266, 272). Although it is unlikely that the architectural fragments dating from the second or third century A.D. came to Vijayanagara before A.D. 1336, the removal of Buddhist pieces from Amarāvati by Kṛṣṇadeva Rāya must remain a possibility worthy of consideration. Two stele inscriptions were uncovered during the 1975/6 excavations (IAR 1975-6:20,62). Both were dated

A.D. 1076, corresponding to the reign of Vikramaditya Cālukya VI, and had been reused as building material during the Vijayanagara period. The more complete of the two referred to the donation of coins to two scholars by Someśvara Bhaṭṭopādhyāya, a daṇḍanāyaka of the king. The two remaining early epigraphs are dated to the second half of the eleventh century A.D. (Devakunjari 1970:31,62; Rajasekhara 1985:117) and indicate Hindu and Jain activity in this area of the site. No complete structures, of a date contemporary with these early inscriptions, survive. However, the numerous chlorite architectural fragments, which litter enclosure IV and were apparently reused in the Vijayanagara period, date to the pre-A.D. 1336 period of site occupation.

It is perhaps worth noting here that two Aśokan edicts in Brahmi have been found in the adjacent district of Raichur at Koppal and Maski (Allchin 1954:77,82), indicating that some Buddhist activity was concentrated in the general area of Vijayanagara during the closing centuries of the pre-Christian era. A third Aśokan edict, dated 258 B.C., is reported by Francis on the banks of the Chinna Hagari river (1904:26-27).

There are a small number of inscriptions which refer directly to the construction, modification and maintenance of water supply features. These inscriptions have received little attention to date outside specialised epigraphical publications.

At Munirabad, on the left bank of the Tungabhadrā river, 5 km. north of Hospet, an inscription of Vikramaditya Cālukya VI dated A.D. 1088, was found inside one of a group of early temples. The epigraph is written in Sanskrit and Kannada. It refers to the gift of the village of Huligi (modern name Munirabad) to Chauvedi-Āditya-Bhaṭṭa by Trailokyamalla-Āhavamalla Someśvara I, the father of Vikramaditya VI. The village and its lands were improved by the construction of an irrigation canal and a network of sub-canals taking off from the Tungabhadrā and paid for by Chauvedi-Āditya-Bhaṭṭa.

> Quickly surveying the banks of the Tungabhadrā river and seeing them charming, Chauvedi-Bhaṭṭa founded this canal to flourish as long as the sun, moon, earth, stars and the famous Meru (mountain) last, with its limbs (i.e., sub canals) like threads drawn out (after) cutting asunder the stalk of a lotus, and presented it (to the people); so that, lo, the mass of plantain trees growing there (has been) increased/satiated (with the supply of water) (HAS 1922:10).

The village was subsequently given to the Brahmins. The inscription is incised on a stele and celebrates this occasion. It is housed in the Someśvara temple, one of five small Śiva shrines located on the left bank of the river. The inscription is important for two reasons. Firstly, it associates the legendary settlement of Kishkindhā with the old fort at Anegundi. Secondly, it refers to the construction of the irrigation canal 1.5 km. to the south of the village during the eleventh century A.D. This canal is in current use and is known locally as the Premogal channel. It may be observed, therefore, that the use of irrigational technology was well established in this area in the eleventh century A.D. and was recognised as a method of realising agricultural potential in the context of permanent settlement. Further, this inscription indicates that landowners, in this case a man associated with the Cālukyan royal family, accepted a measure of responsibility in providing basic subsistence requirements for their dependent tenants.

There is considerable evidence for an occupation of this region during the Cālukya period. A Cālukya inscription of the tenth century A.D., which records the granting of land near Hampi, is located near Anegundi. It reads:

> Āhvanalladeva of the Cālukya dynasty, by Pampe on the southern bank of the Tungabhadrā river, ... made a land grant to his prime minister Maṇṇamayya (Shama Sastry 1917:290).

One of the minor capitals of the Cālukya monarch Jayasiṁha II (A.D. 1018-1042), called Pottalakere, seems be have been located on the site of the modern Danāyakanakere, southwest of Hospet (Francis 1904:29). The Kannada word 'kere' means tank and it is frequently added to the names of settlements which possess such devices. Today Danāyakanakere is provided with a large, modified, catchment drainage tank of Medieval design. The identification of Pottalakere with Danāyakanakere is, therefore, of considerable importance as it indicated that a bunded kere or catchment drainage tank existed at Danāyakanakere before the Vijayanagara period. An inscription at Kogali dated A.D. 1064 indicates that Kampli was the capital (nelevīdu) of the Cālukyan King, Viṣṇuvardhana Vijayāditya (Francis 1904:29). The Tamil poem Kalingattu Parani indicates that Vīra-Rājendra Coḷa set up a pillar of victory at Kampli to mark a successful military campaign during the eleventh century A.D. (Kanasabhai Pillai 1890:331,339). This evidence strongly suggests that the area which later formed the greater metropolitan region of Vijayanagara was also earlier the zone in which several Cālukyan capitals and settlements were located. More than this, these earlier settlements depended upon irrigational technology (river-fed canal systems and large catchment drainage tanks) to support their agricultural hinterlands.

During the 1987/8 field season, two unpublished epigraphs directly related to water management at the site were brought to the attention of the author. The first inscription, located about 1 km. east of the Matanga Parvatam bridges on the Turtha canal, refers to the breaking of a boulder by an engineer in order to facilitate the passage of water eastwards along the canal. The second inscription is carved on a boulder just upstream of the main Turtha anicut (one of six such devices, arranged in a line), 2 km. west of Hampi and contains a reference to the Śaivite deity Virūpākṣa. Both these inscriptions are discussed in greater detail in sections 4.4 and 4.5.6

Before the completion of the Tungabhadrā Dam in the 1950s, the existence of an epigraph dated A.D. 1521 and recording the construction of an anicut at Vallabhāpuram was reported by several authors (Kelsall 1872:231; Francis 1904:38; Venkayya 1906:210). The weir lay some 12 km. upstream of Mālāpuram and supplied the Basavanna canal. This inscription is now lost.

Epigraphs from the Vijayanagara period are predominantly religious in character, recording dedications and donations from royalty, powerful nāyakas, wealthy individuals and groups of individuals to temples and shrines. These records also contain political, dynastic and social information. It is important to note that to date, relatively few inscriptions, perhaps no more than one hundred, have been found at Vijayanagara. Those that survive, in Sanskrit, Kannada and Telugu, do not represent an impressive or consistent record of patronage to religious institutions. While many epigraphs remain undiscovered, one might expect a greater number and spatial distribution of inscriptions for a city that was occupied for 229 years. Ludden (1985:70) has discussed the dearth of inscriptions for the Nāyaka kings at Madurai.

Inscriptional material provides chronologies for most of the major religious complexes and a number of the smaller temples at Vijayanagara. Unfortunately, these epigraphs do not provide a means of dating secular structural remains at the site, as they are found on buildings which represent a totally separate architectural style to that found in secular constructions. Though separate and quite distinct, these two building styles are contemporaneous. However, there are a small number of inscriptions which make reference to secular structures. Examples include the bridge across the Tungabhadrā, built in A.D. 1383 (Devakunjari 1970:61), the octagonal well near Malpannagudi, built in A.D. 1412 (Devakunjari 1970:71; Rajasekhara 1985:117), and Achyutarāyapeṭe or Achyutāpura, the suburban extension to the city built by King Achyuta Rāya in A.D. 1534 (Devakunjari 1970:60; Rajasekhara 1985:119).

Recent field survey at the site has located inscriptions incised on large boulders within the urban settlement (Nagaraja Rao 1983:39-40). These short epigraphs record the names of watchtowers or 'bastions', which served particular quarters of the city as lookout posts. None of the inscriptions are dated, although they do provide contemporary labels for parts of the site that were previously unknown.

During the course of this investigation, reference will be made to the inscriptions cited above which provide information relating to water management during the Cālukya and Vijayanagara periods.

1.5 Vijayanagara kingship and social structure

The recent researches of Stein (1980), Breckenridge (1985) and Ludden (1985) have done much to dispel the notion of Vijayanagara social structure and conditions remaining in an unchanging stasis. The picture of social and cultural stagnation propagated in the work of Sastri (1950) and Thapar (1981) can no longer be maintained in the light of these recent studies (see section 1.1). The following is intended as a brief discussion of Vijayanagara kingship and social structure, with particular reference to social organisation and responsibility. This examination is considered essential by the author, for the more coherent understanding of the relationship between resources and management in the Vijayanagara period.

When studying political conditions at Vijayanagara the importance of water supply cannot be overestimated. Although temples and powerful individuals invested in the construction of tanks for villages, the extensive water supply systems necessary to supply the metropolitan area of Vijayanagara remained the responsibility of the king. In the semi-arid environment, the patronage of hydraulic construction projects was the key to increasing economic potential by the successful manipulation of resources.

As indicated by Stein (1980), the emergence of Vijayanagara represented a series of new developments in social and political organisation. These changes are most clearly visible in the sphere of agriculture. Although the nāḍu, an assembly representing several villages, remained the basic building block of agricultural organisation (Stein 1980:90-91; Ludden 1985:35), it was in a reduced or fragmented form. Throughout the period prior to the foundation of Vijayanagara, traditional village assemblies remained responsible for the collective management of land and land use, as well as irrigation, dispute settlement, taxation and temple gift allocation. Brahman settlements (brahmadeyas) regulated by their sabhas, or assemblies, became powerful agents of land control, acting as mediating agencies for temple investments (Ludden 1985:40). Stein (1980:413) has indicated that by the fifteenth century A.D., the donation of gifts to brahmans was replaced by the patronage of cult temples.

During the Vijayanagara period, ambitious individuals, in particular amaranāyakas, assumed the responsibility for agricultural management, replacing the autonomous village assemblies as the primary, or dominant, agents of land control (Breckenridge 1985:42-43; Stein 1985:81). The penetration of outsiders into the South Indian macro-region, Telugu warriors (or nāyakas) with dependent followers (Stein 1980:418), resulted in the reorganisation of resources. Further, there developed an increasing spatial complexity related to the breakdown of territoriality and localised ethnicity. To this picture may be added the complication of the two-fold division of non-brahman caste members into the valaṅgai (right-hand) and iḍaṅgai (left-hand) groupings, a process which began in the Coḷa period (Stein 1980:469-488).

Social reorganisation dictated changes in land tenure, which had previously been organised around the veḷḷānvagai (ordinary) and brahmadeya (special) distinction (Stein 1980:419). This period witnessed the development of a three-fold tenurial division based upon the amara (military service), bhaṇḍārvāda (crown) and mānya (tax free) grants (Stein 1980:420). A discussion of the controversy surrounding the ownership of land rather than the income derived from it, and the character of the amaram or military tenure, is beyond the scope of this investigation. It is sufficient to note that the epigraphical evidence from which information regarding agricultural organisation is derived, remains ambiguous in relation to both of the issues cited above.

The growth of a class of supra-local warriors, or nāyakas, in the fourteenth century A.D. (Stein 1980:397; Ludden 1985:45) marks the beginning of dramatic changes in the organisation of social and political order in South India. The integration and incorporation of micro-regions, related to an increase in the movement of population, the intensification/expansion of agriculture and the growth of intra- and extra-local trade, resulted in the demise of local institutions such as the nāḍu and brahmadeya (Stein 1980:408; Menon 1986:30).

The nāyakas of the Vijayanagara period may be seen to replace existing institutions as agents of social control. Nāyakas legitimised their right to power through the patronage of religious centres (Ludden 1985:46). Stein has argued that nāyakas became a new, intermediary level of authority within a changed, but nonetheless recognisable, segmentary state (1980:409). This position was based on the use of superior, military technology (i.e. the horse and firearms). It is assumed that the 'strong military centre', in this case Vijayanagara, was able to curb the aspirations of quasi-independent chiefs when they threatened the structure of the segmentary state. Opposition to usurpations of power by subordinate chiefs, or nāyakas, is seen by Stein as the response of a unit that is at the top of the segmentary pyramid. Although hegemony is maintained by manipulation of ritual supremacy, instability resulting from the decentralisation of political power was checked by direct military action. The economic implications of this inherent instability are not discussed. Stein's paradigm for South Indian statecraft demands that the 'centre' is always presumed to be militarily stronger than other subordinate political units and never threatened by internal disorder (1980:410).

We suggest a rather different explanation for the use of direct military action by Vijayanagara against local kings, or nāyakas, which depends on a constant state of political flux and dynamic economic activity. This situation would be characterised by independent political 'centres' vying for limited demographic and agricultural resources. A 'centre' in the context of this discussion may be defined as a basic political building block, perhaps synonymous with a kingdom. It would represent an agglomeration of demographic, mercantile and agricultural resources located in a large urban settlement and its surrounding hinterland. The urban settlement would house a king and his followers, though not a large bureaucracy. Political 'centres' would all be fairly equal in terms of military and political influence. Their location would coincide with productive agricultural zones, the order of which would be located in riverine and deltaic tracts. Vijayanagara would represent one such 'centre' and would be allied to other 'centres' by a complex network of alliances. Direct military action against another kingdom or 'centre' would then represent a more complex phenomenon than the checking of a subordinate kingdom's aspirations to hegemony (see section 1.1).

Stein (1980:410) has proposed that the tensions between the Vijayanagara state and the intermediary level of Telugu nāyakas were resolved by the intervention and active involvement of brahmans in administration and politics. Brahmans, as direct agents of the king, served as commanders of fortresses, which were economically supported by bhaṇḍāravāda or crown villages (Stein 1980:411-414). They also acted as military commanders, giving dignity, prestige and legitimation to political control. The tensions ascribed by Stein to the imposition of an intermediary level of political control, do not necessarily infer a condition of instability, requiring the participation of powerful brahmans in the workings of the state apparatus. Rather, they suggest, in our view, a positive stimulus to state formation.

Stein presumes brahmans to be loyal to the dominant 'centre' of the segmentary pyramid and its principal figure, the king himself. However, brahmans were powerful landowners in their own right. They received wealth as gifts, including land, from nāyakas and merchants as well as from the king. This suggests that brahmans could afford to be somewhat politic in their allegiances, a possibility not considered by Stein.

Temples legitimised the political dominance of patrons. However, the wealth collected and redistributed by temples, as cash or land, needed to be protected by an agent of political control. The relationship of powerful nāyakas, of which the Vijayanagara king was one, and religious institutions may be viewed as one of symbiosis. The patronage of temples by local magnates ensured:

1. the generation of legitimacy for the maintenance of political power;
2. the development of agricultural tracts through the reallocation of gifted cash to land improvement projects;
3. the attraction of scarce population.

In return, local rulers protected temples from military attack and furnished them with lavish gifts. Royal patronage undoubtedly increased the status of a temple, attracting pilgrimage. This provided a valuable source of revenue for temples. That large cult centres received patronage from more than one and sometimes many political leaders (including Vijayanagara kings) suggests that political dominance was achieved with differing measures of success by competing individuals. Although the

Vijayanagara period witnessed considerable agricultural expansion in spatial terms, the need to attract and protect scarce population remained the most pertinent problem for political units headed by ambitious individuals.

The expansion of agriculture during the Vijayanagara period has been discussed by Stein (1980), Breckenridge (1985), and Ludden (1985). It is clear that the movement of peoples in possession of new agricultural technologies and practices significantly increased the productive power of existing agricultural tracts (Ludden 1985:51), and made it possible to settle in areas away from the deltaic and riverine zones (Stein 1980:429; Breckenridge 1985:42-43; Ludden 1985:46). These developments are directly linked to the increased use of irrigational technology funded by wealthy individuals and the redistribution of gifted wealth by religious institutions. There is considerable evidence for preVijayanagara water management in Bellary and Raichur districts (see section 4.2). Direct investment in large scale irrigation projects by Vijayanagara kings, contributed to the realisation of the full potentialities of the 'productive agricultural core zone'.

The fourteenth to sixteenth centuries A.D. represented a period of increased extra-local movement and social mobility (Stein 1980:368; Breckenridge 1985:43-47; Ludden 1985:42). This development is closely related to changes in social organisation and land control. The increasing complexity of social stratification, which resulted from the movement of linguistic and ethnic groups into new domains, was complicated by the existence of communities not tied to settled agriculture. Chander (1987:25-26) has elucidated the important role that these pastoralist, mobile groups and hill and forest groups played in the social and economic organisation of the Hyderabad region.

The Vijayanagara period witnessed a steady increase of intra- and extra-local trade, culminating in the Portuguese intervention of the early sixteenth century A.D. This event marked the beginning of direct European involvement in the social, economic and political order of South Asia. The growth of trade in the period A.D. 1336-1565 involved two distinct developments. Firstly, the expansion of intra-local trade networks linked established productive core zones to mobile groups outside these areas. Mobile groups, particularly those from forest tracts, supplied rare goods in small quantities for internal consumption and export. Chander (1987:25) has suggested that groups outside settled agriculture, were at least as numerous, and economically as dominant as the settled agriculturalists. The political and economic influence of these mobile groups, in the light of this assertion, cannot be underestimated. Previous research has not considered the vital role played by these groups. Secondly, there is plentiful evidence to indicate that complex patterns of extra-local and foreign trade emerged in this period. The lucrative practice of taxing high value trade goods in transit is assumed to have provided a major revenue source for all kingdoms in South India, particularly Vijayanagara. The 'strong centre' of the unitary and segmentary state models has been viewed as essential in the attraction and protection of long distance trade. In our view, since trade remains in the hands of mobile traders, it would be attracted to wealthy 'centres' of consumption which offered favourable mercantile terms. The economic growth of a 'centre' would thus rely on the encouragement of trade.

We suggest that the accumulation of wealth by non-mobile, high status groups depended on the attraction of scarce population. This ensured that a workforce for an agricultural base, capable of supporting a defensible urban settlement was maintained. Wealth from agricultural revenue returns and military raids was concentrated in urban settlements. It attracted extra-local and foreign trade, providing an additional source of revenue. Thus, we suggest that important urban settlements existed in a finely balanced equilibrium, controlled by a number of primary and secondary factors.

Stein has suggested that the purchase of arms and payment of mercenary soldiers was financed by commerce (1980:79). The great augmentation in military activity, attributed by Stein (1985:79) to the increasing pressure exerted by the expansion of Hindu and particularly Muslim kingdoms, is here viewed as the result of an increased competition for scarce resources. The intrusion of a Muslim ruling class, eager to establish kingdoms in South India, changed the social and political order by increasing the demand for resources. Military growth may be considered wasteful in terms of the amount of people required to take part, who might otherwise be engaged in agricultural activities.

The importance of religious institutions as centres of socio-economic exchange has been mentioned above. An increase in temple building and expansion was made possible by patronage from groups and individuals. The predominantly royal gifts of the Coḷa and Pallava periods were replaced by mixed patronage. Kings, nobles, nāyakas and merchants all donated gifted wealth (Thomas1985:16-20).

Religious centres attracted population and established complex patterns of social and economic redistribution (Breckenridge 1985:55-63; Ludden 1985:33-34). By the fourteenth century A.D., temples had developed significant regional importance. They linked disjointed social groupings by investing gifted cash in agricultural projects and creating exchange networks (Stein 1980:428-429; Breckenridge 1985:54-55; Ludden 1985:31,33). Large temples employed hundreds of specialists and general service employees, who ensured the smooth running of the shrine as a centre of worship, pilgrimage, education, and socio-economic exchange. Considerable cash and land wealth was amassed by religious institutions, indicating that they were important agents of social and economic control in periods of resource scarcity.

Thomas (1985:16-20) has argued that an increase in temple building and modification is indicative of political instability. Considerable military activity and political reorganisation occurred in the Vijayanagara period. However, it was by no means an era of economic or cultural stagnation. Chander (1987:50) has suggested that a situation of conflict was indicative of dynamic economic activity in political core areas. Accepted assumptions regarding the nature of political conditions must be subjected to critical analysis. The replacement of the monolithic state by Stein's segmentary paradigm as an explanation of South Indian political order has left a number of problems unresolved. If political order was based on 'centres' with subordinate foci, how did a 'strong centre' maintain its monopoly of ritual supremacy? In addition to this, how did the dynamics of resources affect South Indian statecraft?

The fourteenth to sixteenth centuries A.D. can be viewed as a period of great change and adaption brought about by external and internal stimuli. An increased competition for scarce resources resulted from the incursion of powerful new outsiders, as both Muslim and Hindu were keen to assume the position of a ruling class. The attraction and maintenance of population, essential for establishment of an urban centre, outside the traditional 'productive core zones' (i.e., riverine and deltaic areas) necessitated the need for a more skilful presentation of the ruler's personality. Thus the importance of personal charisma cannot be underestimated.

Ludden (1985:68) has discussed nāyaka charisma for the period following the decline of Vijayanagara. He suggests that nāyaka charisma rose from the ritual supremacy of their capital and from the patronage of temples and brahmans (1985:68). The success of nāyaka power play depends partially on the use of personality and locality charisma. This hypothesis may go some way to explaining how ritual supremacy was conceded to only one 'centre', through character and site presentation and royal recognition of social and political responsibility.

However, the nature of the 'centre' requires further discussion. Stein has argued that ritual supremacy was held by a 'strong centre' which dominated the segmentary pyramid (1980:269). The 'centre' was universally recognised by peripheral foci which retained their own essential characteristics as separate units (1980:272-275). He writes:

> It is not an accident of preservation that the historical record of Vijayanagara kinship should exist in temple inscriptions. It was here that the civil, as against the military aspect of kingly rule was most clearly realised (1980:481).

The state apparatus of ritual supremacy is not elucidated. Religious institutions are seen as the means by which a disparate society was integrated. In one sense this assertion is correct.

Temple complexes, in their role as centres of redistribution, linked diverse linguistic and racial elements in exchange networks. However, these networks did not represent or correspond with political order, which may be presumed to have remained in a state of constant flux.

Temple patronage by kings is traditionally seen as a method of affirming royal authority outside the zone of direct influence. Yet, cult centres did not provide legitimation of the king alone. Large temples contain epigraphs which record donations from powerful nāyakas and local kings. It may be more useful to interpret royal patronage as an attempt to influence or manipulate the established pattern of economic organisation inside the 'political core zone'. The injection of cash into powerful 'centres' of attraction, ensuring the maintenance of social magnetism, benefited both king and temple. Such action served also to win the support of influential though politic brahmans, in whose hands lay the machinery of economic redistribution.

This enquiry does not accept that the temple epigraphs represent positive evidence for Vijayanagara sovereignty. In areas outside the 'productive agricultural core zone', which supported the urban settlement, inscriptional material records brief military occupations and religious gifts. It is not possible at this juncture to provide a satisfactory measurement of Vijayanagara's direct political influence in spatial terms. Previous research has taken epigraphical distribution as an accurate guide to the extent of Vijayanagara's domain. In this discussion, reference has been made to two separate, though linked, definitions of Vijayanagara's physical extent: 'political core zone' and 'productive agricultural core zone.' The former may be taken to represent the nuclear area from which the urban centre received at least nominal military and political support from local kings and nāyakas. The latter represents the agricultural lands under the direct political and military control of the urban centre, which provided revenue taxable by the king personally. Both these definitions describe areas far smaller than previous accounts have suggested.

The social networks forged by religious institutions are largely to be viewed as an apolitical phenomenon, by which limited resources were transported, transformed and organised. Vijayanagara kings vied with other kings and nāyakas for theoretical dominance of resources outside their direct jurisdiction by the injection of gifted cash into religious centres. Cash of this sort was derived from land tax, trade tax and military raids. The attraction of population to the 'productive agricultural core zones' remained the primary concern of South Indian kingdoms. We suggest that this was partly achieved through the presentation of the capital.

If one discards the concept of the 'strong centre' in favour of a weaker, more nebulous 'centre' or kingdom as the pivot of social and political order, it is possible to envisage its function as threefold; the attraction, maintenance and protection of resources. The *Ārthaśāstra* makes it quite clear that the king must recognise the importance of resources:

> Which is better of the two, the tract of land with forts or that which is thickly populated? The latter is better; for that which is thickly populated is a kingdom in all its senses. What can depopulated country like a barren cow be productive of (Shama Sastry 1951:325).

The manipulation of ritual supremacy may then be considered as one of several mechanisms that provide the king and his 'centre' with charisma.

The management of resources through a skilled presentation of charisma would require a number of other mechanisms, symbolic and practical:

1. Maintenance of dharma by the king.
2. The ritual association of the king with mythical figures and localities imbued with religious importance.
3. The buttressing of limited political power by the creation of elaborate and expensive power symbols. The greatest concentration of these power symbols would be located within a city in which the charismatic ruler would reside.
4. The provision of physical benefits for the attracted populace. these would include irrigation works, military protection and the support of cult centres. It was important for the king to remain closely

associated with temple complexes through patronage, particularly those which controlled extensive networks of redistribution. The provision of benefits for the settled population may be considered the primary responsibility of a charismatic king.

5. Displays of military strength, including military raids, military displays and elaborate fortifications (especially in the city).
6. Establishment and maintenance of 'domain' and 'community' by legitimation rituals, consensus building devices and incorporating legends (Chander 1987:6). The movement of the king and his entourage around the 'political core zone' was a method by which the 'domain' was delineated.

The hypothetical situation cited above is reliant on the existence of four general conditions. Firstly, a political order consisting of 'weak centres' of power dominating attenuated hinterlands of agricultural activity (the 'productive agricultural core zone'). Units would form loose alliances in complex patterns relating to resource distribution. The indefinite spatial growth of kingdoms would be restricted by the availability of population to defend the 'domain', though limited expansion would be possible through the exploitation of rivalries between local units. A monopoly of superior military means, including foreign mercenaries, firearms and cavalry, was the key to successful defence of the 'productive agricultural core zone', domination of the 'political core zone' and the attraction of population.

Secondly, agricultural revenue would represent a secondary source of state income. Trade tax and portable wealth derived from raids into the territory of other rival kingdoms would provide the primary source of income Agricultural wealth may be presumed to have remained in the control of the religious institutions, who continued to operate as centres of socio-economic exchange. A large body of population would remain outside the sphere of political control, following a pastoralist life-style.

Thirdly, social mobility and racial and religious tolerance amongst settled populations, especially in urban settlements. This condition is particularly evident in the physical remains at Vijayanagara, which are religiously and culturally eclectic. Duarte Barbosa, a sixteenth century A.D. visitor to the city, states:

> ... the king allows such freedom that every man may come and go and live according to his own creed without suffering any annoyance and without enquiry (Dames 1918:202).

Finally, a fragmentation of political power and the absence of a centralised bureaucratic 'centre' capable of collecting tax returns outside the 'productive agricultural core zone'. This condition suggests that the existence of charismatic kingship demands a decentralisation of power. The operation of a 'weak centre' does not necessitate a large bureaucracy, as the extent of direct political control remains localised. The important buildings of the 'royal centre' at Vijayanagara may be presumed not to have housed a large administrative staff responsible for the direction of government affairs, but rather the personal retinue of the king. Administrative power would remain delegated to nāyakas capable of enforcing tax demands and maintaining personal armies within the 'political core zone' (Chander 1987:18-19). A king dominating a successful 'productive agricultural core zone' would satisfy himself with the nominal allegiance of nāyakas outside this area and respectful epigraphical references made by them in important cult temples. The primary function of the urban centre remained the attraction of resources (especially population and trade) through the presentation of personality and site charisma. The agricultural self-sufficiency of Vijayanagara as an urban centre suggests that regional order and stability had very little to do with the involuted operation of the sacral-symbolic core. Barbosa states:

> In the city as well there are palaces after the same fashion, wherein dwell the great Lords and Governors thereof (Dames 1918:202).

It would appear that the urban centre housed the king and his immediate followers, on whom the ruler relied for personal support. This passage suggests that the king's authority did not extend indefinitely from the site, and probably no further than the 'productive agricultural core zone'.

In conclusion, the proposed paradigm for Vijayanagara statecraft, with a political order based on many attenuated 'centres' of power competing for scarce resources, would result in the creation of internal and external tensions. These tensions may be seen as inherently destructive. The most dangerous external stress factors would be the emergence of alternative 'centres' of attraction, capable of offensive and defensive military action. The destruction of a successful, rival 'centre' would satisfy a practical and an ideological function for Vijayanagara's competitors. Internal pressures would develop between the king and nāyakas within the 'political core zone', inevitably leading to usurpation and intrigue for which there is evidence in the historical record. Offensive or defensive military action would tend to distract attention away from internal disorder of this sort (Chander 1987:42).

CHAPTER 2

The City of Vijayanagara

2.1 Introduction to the environment of Vijayanagara

Before an examination of the water supply systems of Vijayanagara can be initiated, it remains to discuss the city and its environs in more general terms. The water supply systems were an essential part of the urban settlement. An assessment of this class of data cannot be separated from a study of the settlement and its day to day operation.

A basic understanding of the environment of Vijayanagara is essential for a study of irrigation an water supply. Human settlement on any scale would not have been possible in the semiarid environment without the development of a perennial water supply for agricultural and domestic needs. In the context of this investigation, the city of Vijayanagara may be taken as the greater metropolitan area stretching from Hospet in the southwest to Anegundi in the northeast (see fig. 2.1). Although the domestic water features are concentrated within the 'Urban Core', in which the bulk of the population dwelt, agricultural water features are spread throughout the greater metropolitan area. In addition to this, a number of the diversion weirs and catchment drainage tanks were located to the south and southwest of Hospet, outside the metropolitan area. All of the supply features that supplied water to the metropolitan area, along with those inside it, are discussed in this investigation.

The urban settlement covers an area of not less than 26 km^2 and is bounded on the north by the Tungabhadrā river and on the south by the Sandur hills, which are composed of two parallel ranges running southeast-northwest (see fig. 2.2). The area between the two ranges is known as the valley of Sandur. The Tungabhadrā river winds sinuously between and around granite ranges, forming a complex regime. Extensive deposits of shingle occur in several places. In the north of the site, a valley running southwest to northeast, and representing an earlier course of the Tungabhadrā, cuts the settlement into two separate and geographically distinct areas.

The zone to the north of the valley is characterized by a rocky and uneven landscape, whereas the southern zone is situated in a large plain, broken by occasional outcrops of igneous granite. It has been suggested that the valley is supplied with water by a natural watercourse which takes-off from the Tungabhadrā higher upstream (Fritz *et al.* 1984:10). However, an inspection of the valley has revealed that no natural connection with the river exists. An artificial irrigation canal, known locally as the Turtha kāluve, flows along the southern side of the valley. This water feature today has command over the cultivated land in the valley, drawing its supply from three distinct sources, which include a modified Vijayanagara period diversion weir, about 2 km. west of Hampi village and two other sources. The Turtha canal feeds two smaller channels, the Kiriyagava and Chika, which bifurcate from their parent canal in the irrigated valley. The granite range which makes up the northern side of the irrigated valley represents the highest part of the site. The lofty peak of Matanga Parvatam, located in the centre of the range, affords a spectacular view over the entire settlement.

The wild and beautiful granite formations which litter the site exemplify a complex erosional history in which exfoliation, or onion-peeling, has played an important part. The iron and manganese-rich granite formations are coarse-grained and fairly homogeneous. This mature topography is provided with a well established drainage pattern which is dendritic in character.

Agricultural production is concentrated in the flat, well-drained areas west and south of the urban settlement. The irrigated valley represents a major zone of agricultural production within the settlement. Today, deserted habitational areas within the 'Urban Core' are also used for dry farming by the local inhabitants of Hampi, Kamalāpuram and Kadirāmpuram.

The severity of the semi-arid environment in this location necessitates the use of artificial watering for almost any form of productive cultivation. Kelsall (1872:70-71) provides a good description of the climatic conditions of Bellary district:

> The climate of Bellary is characterised by extreme dryness in consequence of the air passing over such an extent of heated plains. Less rain is supposed to fall at Bellary than at any other place in South India. The quantity of moisture in the air as indicated by the hygrometer is exceedingly minute, though the air is remarkable for its transparency.

The average rainfall for Bellary district is under 59 cm. (Francis 1904:11). Rain falls mostly in June and July under the influence of the southwest monsoon. A second, less intense rainy period occurs in October and November, when the northeast monsoon visits the district. An idea of the temperature range can be gleaned from Kelsall's record for 1868 to 1870, which uses the average of the mean temperature of each day in the month (1872:72):

	1868	*1869*	*1870*
Jan.	N/A	80° F	80° F
Feb.	88° F	85° F	83° F
Mar.	86° F	91° F	89° F
Apr.	92° F	95° F	93° F
may	96° F	93° F	91° F
Jun.	84° F	89° F	83° F
Jul.	83° F	83° F	80° F
Aug.	81° F	79° F	82° F
Sep.	79° F	82° F	81° F
Oct.	82° F	82° F	78° F
Nov.	79° F	81° F	N/A
Dec.	79° F	81° F	N/A

The cool season lasts from November to February, after which the heat builds up to a peak in March, April and May. During this hot period almost all natural vegetation unwatered by perennial supply, withers to nothing. The supply of water in tanks evaporates. Some dry up completely, rendering them useless for perennial irrigation. Although river levels are considerably lower in the hot season, diversion weirs remain an effective method of procuring perennial water throughout the year. Kelsall (1872:2) says of this:

> The river at all time contains water, but in the dry season the channel being full of rocks will not admit of floats. In the rainy season it swells prodigiously, and is said to be eight or ten feet higher than the rocks. Its stream is then extremely rapid and muddy.

As a vital subsistence requirement for settled human habitation in the area, water is derived from four major sources:

1. Surface run-off or natural drainage water.
2. Sealed underground deposits or aquifers.
3. Diverted river flow.
4. Directly stored rainfall.

The local water supply systems used by the modern inhabitants remain dependent on these sources for their agricultural and domestic needs. The introduction of post-medieval technology has increased the efficiency of local water procurement techniques without making existing and indigenous systems, such as the anicut, catchment drainage tank and well, redundant.

However, the completion of the Tungabhadrā Dam at Hospet (see fig. 2.1) in the 1950s has significantly altered the agricultural regime of the area. This huge reinforced concrete device, 28 m. wide at the base and 50 m. high, traps the drainage from Dharwad, Bellary, Chitradurg, Shimoga, Chickmangalore and parts of Kanara. Its catchment area is estimated at 28,167 km^2. The construction of the dam on the Tungabhadrā at Mālāpuram has submerged four important Vijayanagara period diversion weirs. The canal system which was fed by these weirs survives in modified form as the Rāya and Basavanna canals. At present, these canals take off from the modern dam between the Right Bank Low Level and Right Bank High Level Canals.

The building of the modern Tungabhadrā Dam, which includes a hydro-electric plant in its Low Level System, has changed the natural level of the river, storing vast quantities of water upstream and regulating seasonal variations in flow downstream. Water is directed, via large modern canals, to Raichur, Bellary and Andhra. The building of the High and Low Level Systems has not made the Vijayanagara period canals downstream of the Tungabhadrā Dam superfluous. Rather they have been repaired, provided with concrete headworks and in two cases abutted onto the new dam headworks.

The major types of canal can be readily identified. The modified and extended canals, originally constructed in the Medieval period, are characterised by a sinuous form which takes advantage of the natural landscape, following the contours of outcrops and ranges, and maintaining a head of water for considerable distances. The canals built as part of the Tungabhadrā Dam Project are wider and carry a more substantial flow for greater distances. They predominantly follow straight lines and are constructed using mortared masonry with a cement facing. However, this simple dichotomy is complicated by extensions and modifications made to old canals from 1846 to the present day. This problem is addressed in section 4.2.1.

Hospet taluk, in which the metropolitan zone of Vijayanagara was firmly based, contains three basic soil types in the following proportions:

Black cotton-soil (or Regaḍa)	8%
Mixed soils (or Masab), esp. red	90%
Red ferruginous (or Lāl)	2%

It may be seen that the dominant soil type is the mixed red soil. Francis (1904:11) has provided a good definition of this type:

> The red and mixed soils vary widely in composition and quality, ranging from deep ferruginous loams down to poor varieties which appear at first sight to consist wholly and entirely of pebbles as big as hen's eggs but which nevertheless succeed in producing a crop if only the rainfall is sufficient.

Agricultural land use may be broken up into three categories:

1. Dry lands without irrigation (punjah or kūshki).
2. Wet lands with perennial irrigation (nunjah or tarri).
3. Garden lands (totacal).

The crops traditionally grown on these land categories in Bellary district can be listed as follows (Kelsall 1872:261-266; Francis 1904:78-87):

Dry land	Cholum, ragi, korra, cotton, indigo, wheat and flax.
Wet land	Paddy and sugar cane.
Garden lands	Coconut, betel, plantain, sugar banana, nut trees, turmeric, chillies, onion, hemp, wheat, coriander, tobacco, ragi and vegetables (aubergines, cucumbers, pumpkins and gourds).

Perennial watering was required for wet and garden crops. The construction of the Tungabhadrā Dam and its canals must have radically altered the crops which could be grown. It also changed the hydrological environment by raising the water table and allowing crops to be grown in areas where it had not previously been possible.

The agricultural regime of the area has changed significantly in the last 100 years, with the growth of intensive production of factory crops, especially sugar, banana and coconut, which were previously restricted to garden lands. This process was undoubtedly accelerated by the construction of the Tungabhadrā Dam. Some traditional garden crops have replaced staples on wet lands, whilst other wet crops (such as sugar cane) have increased in importance. There is some evidence to suggest that new crop varieties have been introduced to improve the agricultural yield. The indigenous red sugar cane was almost entirely superseded by the more productive white cane, an introduced species, by the time Kelsall examined the agriculture of Bellary district in the latter half of the nineteenth century (1872:275-276). It is, therefore, reasonable to assume that agriculture during the Vijayanagara period was very different in character.

2.2. General characteristics of Vijayanagara in the contemporary accounts

The most pertinent accounts of Vijayanagara are examined in this section to determine the way in which the city appeared to the contemporary visitor. The object is to extract information relating to the general organisation of the city and its water supply system. The contemporary accounts contain a variety of erroneous spellings of Vijayanagara. These include Bizenegalia, Bidjanagar, Bisinegar and Bisnaga.

The earliest contemporary description of the city is provided by the Venetian merchant, Nicolo di Conti, who visited Vijayanagara in A.D. 1420. Conti relates:

> The great city of Bizenegalia is situated near very steep mountains. The circumference of the city is sixty miles, its walls are carried up to the mountains and enclose the valleys at their foot, so that its extent is thereby increased. In this city there are estimated to be ninety thousand men fit to bear arms (Sewell 1900:82).

The visit of Conti coincided with the reign of Devaraya II (A.D. 1419-1444). Conti indicates that the defensive walls of the 'Urban Core' joined together the natural features. The date of his account approximates the movement of royal habitation to the flat plain (see section 2.3). One further account of the city in the fifteenth century A.D. is provided by Ab-der-Razzak, a Persian ambassador to the court of Shāh Rukh, who visited Vijayanagara in A.D. 1442. He states:

> The city of Bidjanagar is such that the pupil of the eye has never seen a place like it, and the ear of intelligence has never been informed that there existed anything to equal it in the whole world. It is built in such a manner that seven citadels and the same number of walls enclose each other (Sewell 1900:88).

This may be compared to Varthema's account of A.D. 1504. Varthema, a native of Bologna and of whom little is known, states:

> The said city of Bisinegar belongs to the King Narsinga, and is very large and strongly walled. It is situated on the side of a mountain, and is seven miles in circumference. It has a triple circle of walls. It is a place of great merchandise, is extremely fertile, and is endowed with all possible kinds of delicacies. It occupies the most beautiful site, and possesses the best air that were ever seen: with certain very beautiful places for hunting and the same for fowling, so that it appears to me to be a second paradise (Winter Jones 1863:125-126).

The accounts of both Razzak and Varthema refer to the elaborate defensive precautions. Varthema's description of a triple wall system correlates much better with the surviving physical evidence than Razzak's. However, Sewell preferred Razzak's description of the urban settlement and devised an ingenious explanation for the missing walls and 'fortalezas' or citadels (1900:88-90). This author strongly suspects that Razzak's account of the seven concentric walls results from exaggeration and a misunderstanding of the complex layout of the city.

A more detailed description is provided by Barbosa, who visited Vijayanagara in A.D. 1510. His description is as follows:

> Forty leagues (of this country) further inland there is a very great city called Bisnaga, wherein dwell folk without number; it is fenced about with strong ramparts and by a river as well, on the further side of a great chain of mountains. It stands on a level plain. Here dwells the king of Narsyngua, who is a Heathen and is called Rayen, and here he has great and far palaces,

> in which he always lodges, with many enclosed courts and great houses very well built .... (Dames 1918:i.200-202).

The account implies that the king stayed within the complex of enclosures housing the royal apartments, and goes on to describe the king's gardens and orchards in the 'Royal Centre':

> He also has gardens full of trees and sweet-scented herbs (Dames 1918:i.201).

Paes, whose account may be dated to A.D. 1520, also refers to an orchard of palms and fruit-bearing trees (Sewell 1900:256).

The contemporary accounts emphasise the distinction between the metropolis and the greater metropolitan area. The 'Urban Core' and the suburbs are referred to as 'cities' by the visitors. In our view, this confusion originates from the addition of the suffixes -pura and -nagara, which both literally mean 'city', to the contemporary names used to identify part of metropolitan area. Paes states:

> And on the northwest side (of Bisnaga) is another city called Crisnapor connected with Bisnagar, in which are all their pagodas (Sewell 1900:290).

Paes emphasises the distinction between the metropolis and the suburbs of the metropolitan area (see section 2.4).

An unusual feature of the general organisation of Vijayanagara was the maintenance of agricultural areas within the metropolitan area. Paes relates:

> Between all these enclosures are the plains and valleys where rice is grown, and there are gardens with many orange trees, limes, citrons and radishes, and other kinds of garden produce as in Portugal, only with lettuces or cabbages (Sewell 1900:243).

It is evident from Paes' account that both wet and garden crops were grown inside the metropolitan area. Nuniz also recorded this phenomenon (Sewell 1900:366). Paes indicates that the fields inside the city served the secondary function of providing grazing for livestock (Sewell 1900:259).

The supply of domestic water to the 'Royal Centre' attracted the attention of several visitors to the city. Barbosa states:

> The king has in his palace many women of position, daughters of great lords of the realm, and others as well, some concubines, and some as handmaids.... They bathe daily in the many tanks ... kept for that purpose (Dames 1918:i.208).

These accounts suggest that the numerous tanks and baths within the royal enclosures were used for ritual bathing by the king's entourage. Both Razzak and Paes recorded the presence of water conduits inside the 'Royal Centre'. Paes states:

> There are many groves of trees within it, in the gardens of the houses, and many conduits of water which flow into the midst of it, and in places there are lakes (Sewell 1900:256).

Elaborate precautions appear to have been taken to ensure the practical and ritual purity of the king's water. Nuniz reports:

> The king drinks water which they bring from a spring, which is kept enclosed under the hand of a man in whom the king has great confidence; and the vessels in which they draw the water come covered and sealed (Sewell 1900:375).

Such precautions were justified. The literary evidence suggests that intrigue and usurpation were common. In particular, Razzak recorded an attempt on the life of King Devarāya II by his treacherous brother in A.D. 1442.

2.3 Physical description of Vijayanagara

The metropolis and contiguous suburbs of Vijayanagara may be divided into two distinct zones, separated by the 'Irrigated Valley', which effectively runs parallel to the sinuous course of the Tungabhadrā river (see fig. 2.2). The rocky zone to the north of the valley is traditionally imbued with ritual and mythical importance and contains numerous temples and shrines. Four temple complexes, the Virūpākṣa, Bālakṛṣṇa, Achyuta Rāya and Viṭṭhala, are situated in prominent positions. This zone has been identified as the 'Sacred Centre' (Fritz *et al.* 1984). A ruined stone bridge, constructed in A.D. 1383 (Devakunjari 1970:61), connects the right bank of the river with the island of Virūpāpuragadda. Remains dating to the Vijayanagara period, including the Setu aqueduct, are found on the island.

Early shrines are located in the area of Hampi; to the north of the Virūpākṣa temple, within the walled precinct which encloses the Manmatha tank, and on Hemakutam hill, which overlooks the Virūpākṣa complex (see fig. 2.3). Hemakutam hill and the area to the north of the Virūpākṣa temple are protected by a defensive wall of imposing size. Inscriptional evidence suggests that the royal residence was located on Hemakutam hill during the fourteenth century A.D. and the opening years of the fifteenth century A.D.

An epigraph dated A.D. 1378 attributes the foundation of the city to King Bukka. It describes the city thus:

> Its fort walls were like arms stretching out to embrace the Hemakuta. The points of its battlements like its filaments, the suburbs like its blossom, the elephants like bees, the hills reflected like stems in the water of the moat, the whole city resembled the lotus on which Lakṣmī is ever seated. There with the Tungabhadrā as his foot-stool, and Hemakuta as his throne, the (Bukka) as seated like Virūpākṣa, for the protection of the people of the earth (Ramanayya 1933:52-53; EC,V, cn 256).

Another inscription of King Bukka II found in Hassan district dated A.D. 1406. states:

> ... his capital was the great nagari named Vijaya, situated in the Hemakuta, with the Tungabhadrā as its protector (Rajasekhara 1985:111; EC,V, hn 133).

The earliest Vijayanagara period inscriptions found in the environs of the 'Royal Centre' to the south are not earlier than the first quarter of the fifteenth century A.D. They are found in the Ramachandra temple and are ascribed to Devaraya I or II. In our view, the building of the Ramachandra temple at this time represents a movement of the royal residence from Hampi to the flat plain to the south. The construction of the bridge, which joined the right bank to Virūpāpuragadda in A.D. 1383, suggests that the island may have been a focus of early activity at Vijayanagara. On the left bank, the fortified settlement of Anegundi clearly predates and postdates the Vijayanagara period. Anegundi, which can be reached by coracle-ferry from Talarigattu, may also have served as an early citadel. The mythical associations of the area shall be discussed below.

Apart from the walls surrounding Hemakutam hill and the Manmatha tank, the zone identified as the 'Sacred Centre' does not appear to have been physically protected. However, individual temples are enclosed by prākāra walls, which may have been used for defensive purposes in times of danger (Stein 1980:404). This area, which is infused with great mythical importance and linked with the *Rāmāyaṇa* story, is considered a place of extreme sanctity. Today the Virūpākṣa temple remains an active Śaivite pilgrimage centre, attracting many thousands of devotees from all over Southern India every year. Although, the Viṭṭhala and Bālakṛṣṇa temples were abandoned in A.D. 1565, the Virūpākṣa temple has remained in use up to the present day.

Inscriptional evidence indicates that the environs of the Virūpākṣa temple were of religious importance during the Cālukya period (seventh to thirteenth centuries A.D.) (Rajasekhara 1985:101-102). There is also evidence for a Cālukyan presence in the form of architectural debris built into the fabric of the Virūpākṣa temple, and the large maṇḍapa at the west end of the Hampi bazaar. Hoysala associations with this area of the site furnish Kulke (1985:126) with evidence for his hypothesis concerning the origins of Harihara and Bukka, whom he believes were indigenous feudatories of the Hoysala kings. His theory is based on epigraphical evidence from sites elsewhere. As yet, no Hoysala inscriptions have been found at Vijayanagara. If the Hoysalas had held sway over this area, positive evidence in the form of inscriptions and architectural remains might be expected. References to Vijayanagara as Virūpākṣapura or Virūpākṣapaṭṭana by Hoysala kings of the thirteenth and fourteenth centuries A.D. (Ramanayya 1933:42-45) must surely represent territorial claims rather than evidence for direct military or political control.

The irrigated valley (see fig. 2.2) contains a single temple complex built during the reign of Achyuta Rāya (A.D. 1529-1542). It is dedicated to Viṣṇu as Tiruveṅgaḷanātha (see fig. 2.4). The shrine is popularly known as Achyuta Rāya's temple. It is situated in a gap in the northern range of the valley and faces north, towards the river. Its orientation contradicts the more usual pattern of facing to the east.

The valley is provided with water by the Turtha canal, which is held in embankment against the south side of the valley. A revetted earthwork situated below Matanga Parvatam, divides the valley into two. The earthwork is approximately 16-18 m. high and runs northwest-southeast across the valley. The water features of the 'Irrigated Valley' are described and discussed in greater detail in chapter 4. Apart from one Muslim-influenced structure (Fritz *et al.* 1984: 145), there is no physical evidence for habitation in the irrigated valley. The absence of pottery sherds suggests that it was used solely for agriculture in the Vijayanagara period. The irrigated valley is outside the defensive wall which encloses the 'Urban Core'.

The 'Urban Core', situated in the flat plain or natural basin to the south, contains a great density of habitational remains (see fig. 2.2). It is elliptical in shape and covers an area of 10 km^2. An isolated 1 km. section of defensive wall inside the southeast corner of the 'Urban Core' suggests that the 'Royal Centre' may have been surrounded by its own protective circuit. Two sections of defensive wall on Mackenzie's survey map of 1800 (Michell and Filliozat 1983: 2-3) indicate that the 'Urban Core' and some agricultural tracts were protected by a further concentric wall.

The wall which encloses the 'Urban Core' is constructed from large, granite blocks. They are laid in irregular courses without chunam or lime-plaster. Their thickness can be explained by the use of an earthen core which separates the two masonry skins. A more complete description of the constructional techniques used in defensive walls is provided by Fritz *et al.* (1984:43-45). The high, defensive walls make good use of natural landscape, joining together boulders and outcrops. Angled gateways of imposing strength, sometimes with small shrines, are located at strategic points. The variety of constructional styles employed in enclosure walls suggest that they were modified many times during the occupation of the site.

The 'Royal Centre' is located in the southwest corner of the 'Urban Core'. It constitutes an irregular grouping of rectilinear, walled enclosures. The enclosures contain standing buildings, remains of raised platforms and building foundations. The enclosures, labelled I to XXXII in a double clockwise spiral by Fritz *et al.* (1984), represent an agglomeration of building styles. The classification scheme devised by Fritz *et al.* has been employed in this investigation. In the discussion of the 'Royal Centre', the Fritz/Michell reference codes are placed in parentheses after the name by which the feature is commonly known.

The Archaeological Survey of India and the Karnataka State Department Archaeology Department, under the supervision of Dr. M.S. Nagaraja Rao, have investigated habitational areas within the 'Urban Core'. These studies have been augmented by an extensive programme of surface archaeology and architectural documentation, directed by Drs. Fritz and Michell. Detailed architectural and archaeological inventories of features on the ground have been assembled. These inventories include photographic and textual descriptions and drawings (plans, sections and elevations) of features within the 'Royal Centre'. A map series of the Vijayanagara metropolitan region (Scale 1:4,000) supplements this study.

The palimpsest nature of the site makes it extremely difficult to ascertain the function of individual structures. Excavations within the large walled enclosure labelled IV have revealed a complex arrangement of foundations adjoining two important features; the 'Hundred-pillared Hall' (IIIa/1) and the 'Throne Platform'

(IIIb/1). The 'Throne Platform' was enlarged on three occasions, attesting to a long period of use and modification (Fritz *et al.* 1984:99-102). Further investigation in the 'Royal Centre' will doubtless produce evidence for the rebuilding of other secular structures.

The 'Royal Centre' is dominated by the well-preserved Rāmachandra temple, located to the north of enclosure IV. It has been possible, on the basis of inscriptional evidence, to construct a chronology for the Rāmachandra temple. The temple was built in the early fifteenth century A.D. and extended several times thereafter. Fritz *et al.* (1984:149) have suggested that the temple may provide the urban settlement with a central axis.

Evidence for a complex water supply system is visible on the ground throughout the 'Royal Centre'. A supply system furnished baths, tanks, fountains and water pavilions. A network of wells and rainwater collection tanks was also provided. Enclosure IV contains the greatest density of water features. Recent excavations conducted by the Archaeological Survey of India, have revealed the 'Stepped Tank'. A finely constructed granite channel, raised above ground level on piers, runs east-west through enclosure IV. Numerous tanks, baths, basins and wells are located in and around enclosure IV. The water supply systems of enclosure IV were undoubtedly modified and enlarged several times. The domestic supply features of the 'Royal Centre' are discussed in detail in section 3.2.

Outside the 'Urban Core' there is considerable evidence for agricultural water management. To the southwest of Kamalāpuram village, a large earthwork forms the northern boundary of the Kamalāpuram tank. The ruins of an old fort are situated close to the north eastern corner of the tank, amongst the modern buildings of Kamalāpuram village. The fort's design is rectilinear, with circular towers at each corner and a large circular structure in the centre. Pharoach (1855:100) provides an early description of the town and the old fort:

> It has a large tank, a fort with a ditch and glacis, but the majority of the houses are outside the walls.

No trace of the ditch or glacis is visible today.

Devakunjari (1970:72) noted the existence of two bas-reliefs of the Gaṇḍabheruṇḍa (a double-headed eagle), an emblem associated with the early Vijayanagara kings, inside the fort. It is reasonable to assume that the fort was intended to protect Kamalāpuram village and its tank from attack. We have noticed defensive structures of a similar design and constructional style at several locations in the immediate area, including Munirabad, Gangāwati, Kampli, Ramsagar and Hosūru. Dry-stone forts are reported by Allchin at Śivapur (1954:138), Jamshed (1954:146,151,153), Billamrayan gudda (1954:179) and Mānvi (1954:196) in nearby Raichur district. Kelsall (1872:286) suggests that the use of such fortifications may have been widespread:

> Many villages in the district were originally within the walls of square forts with bastions at the corners but these forts have in time been dismantled and the stones removed for other purposes, and little now remains but the crumbling mud walls. Even these too have often been levelled and the earth used to fill up the moat which is generally a receptacle for rubbish and filth of all kinds.

Outside the defensive wall system of the 'Urban Core', there are scattered remains from the Vijayanagara period. These include religious and secular structures, inscriptions, images, tanks and wells. Some of the modern villages between Kamalāpuram and Hospet contain traces of Medieval settlement. Kadirāmpuram, located on the road from Gālammagudi to Hampi, is dominated by two Muslim tombs built in the Vijayanagara period. The isolated village of Hosūru, 4 km. northwest of Hospet, contains some interesting religious structures, including the Durgā-Hosūramma temple, an important pilgrimage place. The temple, which is built inside a dry-stone, rectilinear fort, was founded in the tenth century A.D. and reconstructed during the fifteenth century A.D. (an inscription dated A.D. 1415 is found within the temple). The fort is of the same design as those discussed above. An eclectic series of images and reliefs are contained within the temple compound. A diversion weir takes off from the Tungabhadrā just west of Hosūru, supplying the area to the north of the village with perennial water. The weir blocks the river flow between the right bank and a small island in the river, opposite the village. We found five Hindu shrines of modest size (orientated to the north, west and east), one Muslim structure and ruined walls on the island, indicating that it was occupied during the medieval period.

The villages on the modern road from Hospet to Kamalāpuram (Anantaśayanagudi, Kondanāyakanahalli, Mallappannagudi, Hosa Mallappannagudi and Gālammagudi) line the main route into the 'Urban Core' (see fig. 2.1). Entrance-ways, temples and small shrines indicate that these villages had their antecedents in the Vijayanagara period. The Anantapadmanabha temple at Anantaśayanagudi is particularly noteworthy for its unusual design. Its main shrine is surmounted by a large, barrel-vaulted śikhara. The temple was constructed in A.D. 1524 by Kṛṣṇadeva Rāya to serve the new town of Sale Mahārāyapura (Devakunjari 1970:69; ARSIE 1922:683).

The town of Hospet (the name is a corruption of Hossa petta or 'new town') has, since the 1950's, been overshadowed by the Tungabhadrā Dam. Hospet was constructed as a suburban extension to the city by Kṛṣṇadeva Rāya in the sixteenth century A.D. It was then known as Nāgalāpur. Surprisingly little physical evidence from the medieval period survives on the ground, excepting a few small shrines and images. Some structures of the Vijayanagara period were dismantled in the nineteenth century in the interests of hygiene. Kelsall (1872:18) states:

> The town was greatly improved in 1866 and 1867 by levelling the old fort wall and filling up the ditch, formerly a receptacle for all kinds of rubbish.

Francis (1904:278) describes the old remains of Hospet at the beginning of the twentieth century:

> To-day beyond a few fragments in the western portion of the town (still called 'the fort') and the fallen rampart which runs southwards from the Divisional Officer's bungalow, hardly a trace of these walls is to be seen. In the 1866 famine, workers on relief were employed in throwing down much of the fort wall into the ditch which then surrounded it, this latter having become a receptacle for all sorts of unsavoury rubbish.

We were unable to locate any trace of the wall or moat during the 1986 and 1987/8 field seasons.

Extensive evidence for agricultural water management during the Vijayanagara period survives outside the 'Urban Core'. Diversion weirs survive at two points on the right bank of the river. In addition, two old canals are now abutted to the Tungabhadrā Dam, and irrigate the agricultural land between Hospet and the 'Urban Core'. The remains of two bunds are found at Mallappannagudi village and on the Dharwad road, just southwest of Hospet.

There are strong mythical associations which link some events in the Rāmāyaṇa to the Hampi area. Evidence on the ground suggests that Anegundi was an important defensive and habitational area during the Vijayanagara period. Inscriptional evidence suggests that the site was settled from at least the tenth century A.D. (Shama Sastry 1917:285-291). Anegundi is believed locally to be Kishkindhā, the residence of Sugrīva and Vāli, the quarrelsome monkey-chiefs. Legend holds that on his eviction from Kishkindhā, Sugrīva took residence on Matanga hill with Hanumān. Francis (1904:260) writes:

> Some of the most dramatic scenes in the great epic of the *Rāmāyaṇa* occurred at a place called in the poem Kishkindhā, and it is asserted by local Brahmans and generally acknowledged by the learned in such matters that this Kishkindhā was close to Hampi.

The Munirabad inscription of A.D. 1088 (HAS 1922:10) provides documentary evidence for the identification of Anegundi as Kishkindhā.

Although events from the Rāmāyaṇa are identified with particular locations, only Kishkindhā is identified as a settlement. Kishkindhā's inhabitants, the vānaras, are recorded as having inhabited caves and rocky places, and using stone tools. This correlates well with the emerging picture of the Neolithic culture in the area (Allchin 1963:97). The association of Anegundi with Kishkindhā may be very ancient. Local tradition identifies the important ashmound at Nimbāpur, close to the city of Vijayanagara, as the site of Vali's cremation (Allchin 1963:52). Allchin suggests:

> ... whoever collected the source material or made the observations which Vālmīki used cannot have been aware of the original, mundane purpose of the ashmounds. Thus he may either have been no more than a casual visitor, or he visited the area at a time when the ashmound practice was no longer current. As we must expect stone tools to have disappeared reasonably fast after the beginning of the South Indian Iron Age, which we date from the fifth or fourth century B.C., we have reason to expect the Kishkindhā episode to reflect information from not later than that time, being after the ashmounds had been abandoned as cattle-pens and before stone tools entirely disappeared (1963:97).

To us, it seems likely that the process by which Anegundi and its environs were associated with mythical episodes from the *Rāmāyaṇa* began during the closing centuries of the pre-Christian era.

2.4 Internal organisation of Vijayanagara

The contemporary accounts and the evidence on the ground have been examined in sections 2.2 and 2.3 for information relating to the layout of the city. In this section, we discuss the organisation of Vijayanagara in more detail and thus determine the role of water supply in the operation and maintenance of the city. The complexity and scale of the water supply system largely resulted from the incorporation of agricultural zones inside the city (see section 2.2). This necessitated an extensive agricultural supply system, bringing water into the metropolitan area in great quantities, in addition to the domestic supply system.

Recent archaeological attention has been focused on the complex area known as the 'Royal Centre'. The Vijayanagara Research Project has undertaken a coordinated programme of excavation and surface archaeology. This includes the construction of detailed architectural and archaeological inventories of the evidence on the ground. The numerous buildings of the 'Royal Centre' represent an eclectic collection of ceremonial and residential structures which must be associated with the king and the royal household. Elaborate provisions were made for the supply of perennial water to a large number of ritual and ornamental water features associated with royal habitation and performance. This area was clearly the core of the city.

Two explanations are proposed for the existence of so many expensive and elaborate structures within enclosures of the 'Royal Centre'. Firstly, it is economically more viable to build afresh than to demolish an existing monument and build again on the same spot. This may, in part, explain why so many buildings are located in the same area. Secondly, in the sphere of temple architecture, modification and extension typified the Vijayanagara period. However, in secular architecture it was undoubtedly more prestigious to build new monuments than to embellish those constructed by another king or nāyaka. The cost of building monuments, in the light of a re-examination of kingship and social structure, may have been met through the mechanism of warfare and raids into lucrative areas which provided portable wealth for reinvestment in royal building projects.

The 'Royal Centre' was surrounded by the 'Urban Core', which, judging from the great quantity of habitational debris on the ground, must have been densely populated. Cottam suggests:

> The inner zone of Indian cities is characterized by high density of population and high income and high ranking castes, whereas Western cities exhibit a low density of population and low income and low status groups resident in the core area (1980:330).

This hypothesis correlates well with the evidence from Vijayanagara. However, the palace areas of other early medieval cities, such as Kathmandu and Patan in Nepal, have a much lower population density. The temple enclosures of some South Indian religious complexes, including Madurai and Śrīrangam, have similarly low population densities.

Previous research has viewed Vijayanagara as a capital, housing a population of enormous size. There are perhaps three major reasons for this. Firstly, contemporary accounts contain frequent references to the imposing size of the city and its population (see section 2.2.). However, these references refer only to the 'Urban Core' and not to the greater metropolitan area, which was composed of suburban areas and agricultural tracts. Secondly, Ferishta's history of the conflict between Vijayanagara and the state of Bījāpur gives unconvincingly large estimates for the size of South Indian armies. From these exaggerated estimates, it has been inferred that population of the capital was huge. Thirdly, the whole of the metropolitan area, including the 'Urban Core', is generally regarded as having been densely populated. Although the 'Urban Core' was densely inhabited, the population of Vijayanagara may not have been huge, perhaps no more than one hundred thousand people in total.

Early census figures compiled by the British of South India indicate that the population levels were relatively small (Chander 1987:16-17). It is likely that the conditions which the British recorded in South India mirrored those of the pre-colonial period. Demographic impoverishment and the dynamics of resources in medieval South India have been discussed in section 1.5 The economic self-sufficiency of Vijayanagara may reflect a need to ensure an effective agricultural base for the city in a climate of resource scarcity. Agricultural crops within the metropolitan area were defensible. It may be presumed that the 'Urban Core' did not house the agricultural workers. This central zone, or metropolis, was actually a centre of activity and residence, associated with economic exchange (intra and extra-local trade) and the demonstration of ritual power by the king and powerful nāyakas. It was to this area that the lucrative long-distance trade in high value goods was attracted, as indicated by the presence of Chinese pottery sherds.

In our view, the work force which maintained the agricultural tracts within the metropolitan area, dwelt in the suburbs outside the 'Urban Core'. The location of satellite, suburban areas delimited the extent of the metropolitan area. Nāgalāpur (modern-day Hospet) represents the westernmost extension of the city. The road from Nāgalāpur to the 'Urban Core' was lined with small suburban communities, the most important being Anantaśayanagudi. Kadirāmpuram must also be considered in this class. The northern flank of the metropolitan area was protected by the old fort of Anegundi. On the eastern side of the city, the fortified town of Kampli provided military protection, although there is no evidence to suggest that the metropolitan area of the city stretched much beyond the outer wall of the 'Urban Core' on this side. Inside the metropolitan area, agricultural production was concentrated in the irrigated valley, and between the 'Urban Core' and Nāgalāpur. The suburban zones may be presumed to have also included garden tracts, watered by a perennial supply. The inclusion of agricultural tracts within the metropolitan area is mentioned by several contemporary visitors to the city.

A number of suburbs cluster directly around the 'Urban Core'. The most important are Hampi, Bālakṛṣṇapura, Achyutāpura, and Viṭṭhalāpura. They are located immediately north of the 'Urban Core'. One further suburb, Kamalāpuram, dominated by the Paṭṭābhirāma temple, abutted the 'Urban Core' to the south. These units were each centred around an important cult temple and represented a separate community with an individual identity. Inscriptional and literary evidence reinforces the assumption that the city was divided up into a metropolis, the 'Urban Core', and suburban communities, which together represented the greater metropolitan area (see sections 2.2. and 2.3). It is worth mentioning at this point that, with the exception of Hampi, Kamalāpuram and Mallappannagudi, the suburbs of Vijayanagara are the result of the great expansion of the city during the sixteenth century A.D. under Kṛṣṇadeva Rāya and Achyuta Rāya. The division between the metropolis and suburbs appears in the accounts of the contemporary visitors to the city (see section 2.2).

The 'Urban Core' encloses a greater part of the flat plain south of the 'Irrigated Valley', incorporating several valleys and small ranges. It has been demonstrated above, that this area was dominated by the 'Royal Centre'. But how was the rest of the Urban Core organised? Fritz (1983:51-56) has provided evidence for a developed network of roads, which suggests that an effective system of communication and transport existed. Cultural debris is scattered throughout the 'Urban Core' providing evidence for complex pattern of habitation and economic production. Recent investigations have revealed an extensive area of high status dwellings in the open area labelled enclosures XXIV and XXV which about the northwest corner of the 'Royal Centre.' Numerous small temples and shrines are located throughout the 'Urban Core' and suggest that a wide range of cults were patronised. Wells, rainwater collection tanks and temple tanks provided for the domestic water needs of the inhabitants.

With reference to the organisation of the 'Urban Core' the sixteenth century A.D. Portuguese traveller, Barbosa writes:

> The other houses of the people are thatched, but nonetheless are very well built and arranged according to occupations, in long streets with many open places (Dames 1918:i.202).

This description correlates well with Dutt's analysis of the traditional Indian city. He states:

> Every city was a conglomeration of wards or muhallas, each muhalla being self contained and enclosed by walls. When extension became inevitable, it was easy to add some similar muhallas to the original number (1925:188).

The evidence contained in the contemporary accounts appears to reflect a division of the urban settlement into contiguous wards relating to profession, demonstrating a desire to maintain spatial and social distinctions. An examination of the evidence on the ground is required to test this hypothesis.

A preliminary study of the distributions of zones of economic activity (agricultural and non-agricultural) within the metropolitan region of Vijayanagara has been initiated by Morrison (1988). This study has investigated patterns of surface artefact distribution along exploratory transcepts, and has addressed the complex problem of postdepositional change, both cultural and non-cultural. The development of this project is likely to yield interesting new data on the internal organisation of the site.

An examination of information pertaining to the traditional economy of Hospet may indicate the location and types of economic activity within the metropolitan area. It is reported that in the nineteenth century, the traditional economy of the Hospet taluk rested upon the trade in jaggery (produced locally), weaving, iron founding and metalworking. Kelsall (1872:17-18) states:

> The chief industrial pursuits in the taluq are the weaving of silk and cotton cloths (known all over the district by their thickness); the manufacture of brass and copper vessels at Hospet, and the iron foundries Kamalapur where cauldrons used in sugar-boiling are made. Iron-ore is procured from the Ramandrug hills.

However, by the beginning of the twentieth century the traditional economy of the area had clearly undergone significant changes. Writing in 1904, Francis reports that:

> The chief industry of the place is cotton weaving ... Five or six families make brass toe-rings, cattle-bells, etc., but not brass vessels. The trade in jaggery (most of it goes by rail to the Bombay side) is still large, but the decline in prices—due, apparently to the competition of sugar refined by European processes—has affected it adversely. The jaggery is made from the cane grown under the Tungabhadrā canals (1904:281).

The economic activity recorded in the nineteenth century represents a long standing tradition of established centres of production. The origin of this tradition must be sought in the Vijayanagara period.

Iron production was noted at Kamalāpuram by several scholars in the nineteenth century (Pharoah 1855:100; Kelsall 1872:17-18). It had disappeared by the beginning of the twentieth century. Francis (1904:282) states:

> Until recently the manufacture of the huge shallow iron pans in which the cane-juice is boiled was a considerable industry in Kamalāpuram. The iron was brought back from Jambunath Koṇḍa—the noticeable dome-shaped hill at the Hospet end of the Sandur range—and was smelted and worked by men of the Kammara caste.

The industry was seriously affected by the importation of cheap iron from England. Fragments of iron and slag attest to the production of iron within the metropolitan area during the Vijayanagara period. Centres of production may well have been located in the suburbs, such as Kamalāpuram. It is quite probable therefore, that the production of iron at Kamalāpuram reflects similar activity in the Vijayanagara period. An investigation on the ground is required to test this hypothesis.

Agricultural production in the environs of the metropolis was concentrated in two areas, the irrigated valley and the large zone between Nāgalāpur and the 'Urban Core'. The intensive production of wet and garden crops in the metropolitan area was facilitated by the supply of perennial water from three sources:

1. The Basavanna, Rāya, Hosūru, Hiriya and Anegundi anicut-fed canals.
2. The Kamalāpuram and Mallappannagudi tanks.
3. Wells.

The implications of agricultural production inside the city are far-reaching. This is discussed in section 5.4.4.

CHAPTER 3

Observations in the Field: 1. Domestic Supply

3.1 Introduction to the field observations

The aim of the investigation in the field was the collection of data relating to the management of water during the Vijayanagara period. The investigation centres on four specific research questions:

1. How did the irrigation and water supply systems of Vijayanagara operate?
2. When and by whom were they built?
3. Did the technology used represent a development of new ideas or merely the continuation of an existing tradition?
4. To what extent did the success of Vijayanagara as an urban settlement depend upon the operation of its water supply systems?

The field observations were collected during two seasons at the site (1986 and 1987/8). The primary objective was the construction of a detailed written and photographic record of surviving features. The technique of data collection used was field-walking. Visible evidence on the ground was mapped and recorded in photographic and note form. Where possible, diagrammatic sketches and scale drawings were made. This methodological approach excluded any hidden or buried evidence that might otherwise be revealed by excavation. This limitation has not proved too serious when cataloguing agricultural supply features, as irrigational evidence is physically more robust and larger than its domestic counterpart. However, it must be noted that this survey in the 'Royal Centre' was subject to the fragmentary nature of the remains and the substantial changes of level which have occurred in this part of the site.

The field data is presented as a textual description. Plans, elevations and sections of selected features elucidated an extensive programme of photographic documentation. In the interests of standardisation, the comprehensive classification scheme devised by Drs. Fritz and Michell (Fritz *et al.* 1984) has been adopted. All Fritz/Michell reference codes are placed in parentheses after the name of the feature. It has proved necessary to include some epigraphical, archival and literary evidence in the field observation chapters in order to best explain their present condition. However, the analysis of the field data will be dealt with separately in chapter 5.

The evidence for water management on the ground is divided into two separate categories: domestic and agricultural supply. The evidence for domestic supply is discussed in this chapter, whilst agricultural supply is discussed separately in chapter 4. This division is based upon the functional differences between the two systems. It also serves three secondary purposes. Firstly, it differentiates between two seperate types of features with individual characteristics. Secondly, it allows the extant evidence to be conveniently broken up into more manageable form. Thirdly, it reflects the spatial distribution of the two classes of evidence, that for domestic supply being concentrated within a spatially restricted zone in and around the 'Royal Centre', whereas the agricultural supply features are spread over a wide geographical area, from Hospet in the west to Kampli in the east.

An examination of domestic supply features in the 'Royal Centre' must to some extent overlap with the work of the Vijayanagara Research Project (Fritz *et al.* 1984). The discussion of the 'Royal Centre' has built upon its pioneering work, and has integrated some of its general survey material with a large body of new field data specifically relating to water supply.

In discussing the evidence for agricultural water supply at Vijayanagara, there is one limitation that must be considered: the construction of the Tungabhadrā Dam. On completion of the Tungabhadrā Dam Project in the 1950s, four anicuts of the Vijayanagarā period were submerged upstream of Mālāpuram. Any analysis of these features is limited to an examination of pre-Tungabhadrā Dam Project literature, survey maps and archival material. Modern canals, which can be distinguished by their constructional style from Medieval and modified medieval canals, are mentioned only in passing. They remain outside the scope of this investigation.

On a terminological note, the following local names for water features have been used in this investigation:

Anicut — A diversion weir (a corruption of the Tamil word aṇaikkaṭṭu meaning dam).

Bund — The earthwork of a catchment drainage tank (a corruption of the Sanskrit word setubandha meaning bridge, causeway or dam).

Kāluve — The Kannada word for a canal or irrigation channel.

Kere — The Kannada word for a catchment drainage tank (eri in Tamil).

3.2 The domestic water features of the 'Royal Centre'

The surviving water features of the 'Royal Centre' have been divided into five groups according to their geographical location. The classification system for the 'Royal Centre' devised by Fritz *et al.* (1984) (see fig. 3.1) has been employed in the interests of standardisation. The division of the domestic water features of the 'Royal Centre' into six groups has been adopted in order to break up this complex and densely grouped series of remains into a more manageable form. The six groups are:

1. Features to the east of enclosure IV (enclosures XVIII, XIX and XX).
2. Features within enclosure IV.
3. Features to the south and southwest of enclosure IV (enclosures XXI and XXII).
4. Features within enclosures I, III, V and IX.
5. Features within enclosures XIV and XV.
6. Isolated domestic supply features in the 'Royal Centre'.

Although we shall point to inscriptions and other dating evidence, the main discussion of dating is contained in chapter 5.

3.2.1 Features to the east of enclosure IV (enclosures XVIII, XIX and XX)

An open channel, labelled XIX/25, runs south-north between the 'Queen's Bath' (XX/1) and the Chandraśekhara temple (see fig. 3.2). In our view, the channel represents the primary source of perennial supply for the 'Royal Centre'. The connection of the channel with the Kamalāpuram tank is discussed in section 4.1.2.

The channel is constructed from roughly-hewn stone blocks in two parallel rows set in think mortar (see fig. 3.3). The fragmentary remains of the channel are first visible in front of the Chandraśekhara temple. South of this point, there is no trace of the water-course on the ground. Much of the fabric of the channel has been reused for other purposes, such as dry-stone walling. At present, the channel is used as the boundary wall of a field. The design of the channel is interesting. Its vertical sides have been sealed with a lime-plaster coating, providing a friction-reduced passage (see fig. 3.4). In addition, part of the channel's western side are provided with an oblique skirting of lime-plaster (see fig. 3.5).

Approximately 100 m. of the channel were uncovered during the first quarter of this century by the Archaeology Department of the Southern Circle (ARSC 1912-3:49). A report states:

> A gravelled pathway has been made alongside of the trench in which the channel is situated, but the sides of trench will have to be provided with retaining walls to keep the earth and stone from falling into and blocking up the trench during rains. I pointed out to the Overseer on the spot what had better be done in order to complete this work and make it

> presentable to visitors. On no account shall any more of the ancient aquaduct be exhumed as the work is quite unnecessary, as we already have sufficient portions of it exposed to view (ARSC 1912-3:49).

Two important points emerge. Firstly, the channel was at this time below ground level and had been protected by a covering of earth. Secondly, further portions of it survived below ground and were not uncovered. It would appear from the present condition of the channel that the precautions suggested in the report were not implemented as approximately 150 m. of channel are visible. No evidence for the retaining walls is extant. To us, it seems likely that they were never built.

The damaged condition of the channel, which was clearly better preserved when uncovered, may be attributed to the clearing of scrub vegetation (ARSC 1919-20:16). This allowed the local inhabitants to dry farm the area. Cultivation of this sort is largely responsible for change of level in this part of the city. South of the Chandraśekhara temple, the channel has also suffered as a result of stone-robbing (see fig. 3.6). Pits of considerable size have been dug between the metalled road from Kamalāpuram to Hampi and the temple. Amongst the debris inside one of the pits are what appear to be the fragmentary remains of a plastered tank (3.5 m. by 3 m. and 80 cm. deep) (see figs. 3.7.and 3.8). This tank lies directly south of the supply channel. We suggest that the two may have been connected.

The channel remains consistent in its dimensions, 50-60 cm. deep and 50 cm. wide. It is supported by a raised core of rubble in areas where the ground level falls away and served to maintain the command of the channel. A modern path or track, leading from the 'Queen's Bath' (XX/1) to enclosure XIV and the Paṭṭananda Ellammā temple (XVb/5), runs parallel to the channel, crossing it in front of the Chandraśekhara temple.

Approximately 40 m. to the north of the point at which it first becomes visible, the channel makes a 90° bend to the west, continuing in that direction for 3 m. (see fig. 3.9). The conduit then turns 90° northwards and runs for 11 m. before making a third 90° turn to the east (see fig. 3.10). After approximately 3 m. the channel makes a fourth 90° turn, returning it to its original northwards route along a north-south axis. The function of the detour is unclear as there are no visible obstacles around which the channel has to pass. We suggest that this feature acted as a water-brake, transforming potential into kinetic energy by obstructing the rush in one direction. The channel continues northwards for 53 m. before making a sharp turn to the west. Its remains are fragmentary, with much of the heavy foundation lime-mortar lying in broken sections. We suggest that the channel, which rests on crude foundations that have been laid directly on the ground, was originally protected by earthen banks built up over the sides of its foundations. Dry cultivation and erosion appear to have destroyed the earthen revetting, exposing the brittle lime-mortar core on which the channel rests, causing it to fracture and fall away (see figs. 3.11 and 3.12).

The channel is more incompletely preserved after this last 90° turn to the west, although it runs along the surface of the ground for at least 100 m. Some sections have survived complete, whilst others have disintegrated (see figs. 3.13 and 3.14). Moving east-west, the supply channel heads towards the hill/outcrop in enclosure XVIII, to the immediate west of which lies the alley that separates enclosure XVIII from IV. The channel, which is raised above present ground level by as much as 1.10 m. is employed as a boundary wall (see figs. 3.15 and 3.16).

Approximately 59 m. to the west, the channel is cut by two paths, heading north and northwest respectively. The channel then continues in a straight line towards enclosure IV, eventually disappearing below ground level, some 35 m. to the west, as it makes its way across the eastern toe of the hill/outcrop which dominates enclosure XVII (see figs. 3.17 and 3.18). Rubble and windblown deposits have filled the channel, making it difficult to trace on the ground. In our view, the channel skirted the northern flank of the hill/outcrop at approximately 443.5 MSL. first supplying a tank (XVII/1) on the northwest toe of the hill (see below), and a raised channel running through enclosure IV (see fig. 3.19). A square well (XVIII/3), lined with masonry slabs, is located to the south of the plastered tank (XVIII/1).

A second, open channel, labelled XVIII/2, connected the plastered tank (XVIII/1) with the raised channel in enclosure IV. Its remains run north-south inside enclosure XVII, roughly parallel to the western alley (see figs. 3.19, 3.20 and 3.21). The channel skirts the west side of the hill/outcrop in enclosure XVIII. The dimensions of the channel vary quite widely (from 30 to 60 cm. in width). It is constructed from masonry slabs laid end to end in two parallel rows. The channel was originally lined with chunam, traces of which can still be seen (see fig. 3.22). After leaving the control tank, the channel passes through a crevice in a flat, sloping outcrop (see fig. 3.23). Traces of lime-plaster survive on the inside of the crevice (see fig. 3.24). It then runs along a terraced platform supported by masonry blocks. Traces of brickwork support its sides (see fig. 3.25). The remaining section of the channel is in excavation (see fig. 3.26). The conduit is last visible 20 m. east of the alley separating enclosure XVIII from IV and 50 m. south of XVIII/1 (see fig. 3.19).

The plastered tank (XVIII/1) is built, on its south side, into the edge of the outcrop through which XVIII/2 is cut. Its remaining three sides, built in excavation, are supported by rubble walls plastered on their interior face (see figs. 3.27, 3.28 and 3.29). The floor of the tank, which is also plastered, slopes down to the southwest. Although the tank has suffered severe disarticulation, two layers of lime-plaster are visible. This suggests that the tank was renovated or repaired (see fig. 3.30). Amongst the debris to the northeast of the XVIII/1 are the fragmentary remains of the conduit which connected it to XIX/25. There are also indications of terracing, and a single flight of rock-cut steps.

A single pipe outlet is set high on the west side of the tank, above a step of plastered brick (see figs. 3.31. and 3.32). In its present condition, the outlet consists of a broken, spigot-jointed pipe encased in brick and lime-mortar. Its diameter is 19 cm. The earthenware pipe is well-fired. No trace of a closing device or gate remains. In our view, the outlet was regulated by the placing of a flat stone or board in front of the pipe. Such a closure would be maintained by the pressure of water. The pipe is encased in heavy brickwork which forms a protective container (40 cm. by 50 cm. in section). Between the pipe and the brickwork, there is an additional layer or skin of chunam about 2 cm. thick. The solidity of construction suggests that the pipe was under pressure (see figs. 3.33 and 3.34). The remains of the pipe run westward through enclosure XVIII's west wall (see figs. 3.35 and 3.36). After passing across the alley separating enclosure XVIII from IV, the pipe emerges in enclosure III, running westwards towards the 'Throne Platform' (IVb/1) (see figs. 3.37 and 3.38).

Two substantial water features, the 'Queen's Bath' (XX/1) (see fig. 3.39) and the 'Octagonal Bath' (see fig. 3.40), are closely related to the open channel (XIX/25) discussed above. Although no conduits connecting the 'Queen's Bath' (XX/1) to the open channel survive on the ground, such a linkage is both logical and practicable. The orientation of the supply spout in the 'Queen's Bath' (XX/1), which is located at ground level on the eastern side of the feature, perpendicular and opposite to the open channel (see fig. 3.41), supports this hypothesis.

The union of the 'Octagonal Bath' to the open channel is more problematic, as the former is set upon higher ground than the latter, some distance to the northeast. The 'Octagonal Bath' would have been more easily fed with water, either from the lower area to the southwest by an earthenware pipeline, or by a catchment tank situated to the southeast. The disturbed and vestigial remains of a bund and associated conduits have been noted northeast of Kamalāpuram village. The bund, which joins two ranges, has been partially destroyed by the construction of a modern storage tank for water from the Tungabhadrā Right Bank Main Canal. Originally, it would have created a tank capable of supplying agricultural land inside the 'Urban Core' as well as water features to the east of the 'Royal Centre'. This tank is the most likely candidate for the supply of the 'Octagonal Bath'.

The 'Queen's Bath (XX/1) and the 'Octagonal Bath' are high-status structures. Both are constructed in an elaborate, Muslim-influenced style. They are built of rough masonry and brick covered in lime-plaster. On stylistic grounds, we propose that they were both built at the same time during the final embellishment of the 'Royal Centre' in the sixteenth century A.D. The buildings are located outside the interlocking system of walled enclosures which form the nucleus of the 'Royal Centre'. They were, however, protected by the concentric walls of the 'Urban Core'.

The 'Queen's Bath' (XX/1) (see figs. 3.42 and 3.43) is composed of an enclosed arcade around a shallow, square pool overlooked by elaborate, projecting balconies. The exterior is plain. Alexander Greenlaw's waxed-paper negatives, taken in 1856, indicate that the 'Queen's Bath' (XX/1) originally possessed a tower. Today, the internal staircase ascends to a flat roof with a fragmentary parapet of stucco. A small moat or collection channel (1.8 m. wide and 1.4 m. deep) runs around the outside of the bath (see fig. 3.44). Water passed into the bath at ground level via a monolithic conduit on its eastern side. The chunam floor of the bath was replaced with concrete in the first quarter of this century (APRSC 1908-9:16). Four stone blocks are set into the original, plastered, bottom of the bath, each provided with a recess suitable for the fixing of a post or upright timber. Archaeological reports indicate that there were originally five, forming a quincunx (ARSC 1909-10:3). This suggests that the 'Queen's Bath' (XX/1) was originally provided with a central structure or perhaps a raised awning.

The bath is provided with an outlet in the middle of its southern side (see figs. 3.45 and 3.46). Water left the tank through two separate, though connected, pipe outlets. One is located level with the bottom of the bath, whilst the other is set up on a step, 59 cm. wide, which runs around the while bath, 83 cm. above the floor. Both outlets are constructed of spigot-jointed pipework. The lower outlet is the larger and is composed of pipes 19 cm. in diameter, set into a stone housing. The upper outlet is located directly above the lower and is perpendicular to it. The pipe is 10 cm. in diameter and feeds into the larger conduit below. The upper outlet is set behind a small step.

The design of the outlet suggests that it served a more complex purpose than the mere passage of water out of the bath. In our view , the upper outlet allowed silt free water to leave the bath, whilst the lower facilitated the flushing out of dust and silt. Either outlet could be closed by the placing of a board or flat stone in front of the pipe, the closure being held by water pressure. Our inspection of the outlet suggested that the device now in situ replaced an earlier outlet. We suggest that the earlier outlet took the form of a simple notch cut through the main step of the bath. It was probably regulated by a hand-operated gate. The moat running around the outside of the bath is located at a higher level than the outlet and must have been filled by some other means.

The 'Octagonal Bath' (see fig. 3.47)) is not so completely preserved as the 'Queen's Bath' (XX/1), much of the chunam covering having disintegrated. It consists of an open colonnade surrounding a stone-lined, octagonal tank. An octagonal fountain, constructed of dressed stone masonry and bonded with lime-mortar, is located in the centre of the bath (see fig. 3.48). Water filled the pool through eight small holes drilled at regular intervals at the top of the inside face of the bath. The bath was extensively repaired by the Archaeological Department of the Southern Circle, who plastered over the central fountain (ARSC 1919-20:16). A ruined masonry building, which may have served as a changing room, is located next to the bath. Both are set apart from the high status remains of the 'Royal Centre'. However, Patil (1985b:129) has observed two unexcavated 'palaces' close to the 'Octagonal Bath'.

Two further water features are located to the east of enclosure IV. Between the 'Queen's Bath' (XX/1) and the Chandraśekhara temple lies an undecorated, masonry structure, which was heavily restored this century (ARSC 1919-20:17). It is rectangular in plan (3.75 m. by 6.75 m. and 4 m. high) and apparently solid (see fig. 3.49). Longhurst (1917:53) identified this structure as a 'water tower'. We have found no evidence on the ground to suggest that its original function was connected with water supply. A small bath of plastered brick (XIX/12) is located to the north of the 'Queen's Bath' (XX/1). A flight of steps on the east side gives access to the bath, which stands isolated on a flat, consolidated deposit of granite splinters and soil (see figs. 3.50 and 3.51).

3.2.2 Features within enclosure IV

The greatest density of evidence for domestic supply is found within enclosure IV. The enclosure has been traditionally identified with the residence of the king. More recently, it has been argued that this area witnessed royal performance or the demonstration of ritual power (Fritz *et al.* 1984). We have observed that enclosure IV is a palimpsest, representing in its surviving form the terminal structural alterations of the sixteenth century A.D.

The supply of water must be viewed as having a dual function. Firstly, on a practical level, considerable quantities of water were required for sanitary and culinary purposes. The strict rules concerning bathing and food preparation in South Indian Hindu culture made flushing, washing and rinsing an essential practicality for everyday life. Secondly, water was inexorably linked with ritual sanctity, royal and religious. Where temples could not be built near to natural water sources, such as rivers and lakes, the presence of water was symbolically indicated by architectural and iconographic motifs. It must be assumed that kings, who to a greater or lesser extent buttressed their fragile political positions by active association with powerful devotional cults, also sought to build their residences near water, or at least provide them with running water. The relatively extravagant use of water, in a semi-arid environment, by Vijayanagara monarchs may be considered indicative of an active concern to achieve and maintain ritual status, thus simultaneously providing a pleasant living environment, and demonstrating the control of great wealth.

Enclosure IV is divided in half, along an east-west axis, by a raised channel. This prominent feature runs from the midpoint of the eastern wall of enclosure IV to the axially placed well (IVa/33) in its centre. Water was conducted from the eastern enclosure wall, westwards across the edge of a flat outcrop which protrudes into it, and represents the most westerly extension of the hill/outcrop in XVIII. As the outcrop falls away, the channel's command is maintained by the use of masonry piers (see figs. 3.52, 3.53 and 3.54). The channel was restored and partially rebuilt by the Archaeological Department of the Southern Circle in the first quarter of this century (ARSC 1909-10:4). Dense undergrowth around the channel was cleared at this time (ARSC 1919-20:17, 1920-1:14).

Approximately 20 m. from the eastern wall of the enclosure, an unusual configuration of holes is drilled into the rock of the raised channel (see fig. 3.55). The holes are iron stained. We suggest that they held an iron fitting or clamp in place, over the top of the channel. Its purpose would have been the regulation of water in the channel. We have also noticed small plaster 'lumps' or obstructions inside the channel, upstream of take-off points (i.e. outlets to subsidiary channels). The creation of affluxes in the supply facilitated minor adjustments to the height of the head and the velocity of the flow.

The supply of water to enclosure IV has been discussed in section 3.2.1. Taking into account the quantity of perennial water required and in the absence of any other substantial supply features, we suggest that the large domestic supply channel (XIX/25) furnished enclosure IV with its supply.

The raised channel is generally well preserved, a short section having been removed to allow the passage of vehicular traffic through the enclosure. It is constructed from dressed granite and rests on plan, square-sectioned piers (up to 1.25 m. high). The piers are supported by stone bases. The channel is U-shaped in section and composed of regular 4 m. lengths with fitted joints. The average span is approximately 3.75 m. Joints are placed above the supporting piers (see fig. 3.56), with the exception of one crudely-carved pillar, located ten piers to the west of the modern road. The lower, outside edges of the channel are chamfered, a feature characteristic of temple pillar decoration in the Vijayanagara period.

The raised channel is provided with a remarkable series of religious icons, carved in shallow relief on the sides of its supporting piers. There are five icons depicting Śaiva and Vaiṣṇava subjects. The carvings face south, west and east and are distributed in an apparently random order on four of the seventeen piers between the modern, unmetalled road and the supply chute to the 'stepped Tank' (see fig. 3.57). From east to west the icons are:

1. Śiva and Parvatī mounted on Nandi, orientated to the west (see fig. 3.58).
2. Viṣṇu on Garuda, orientated to the east (see fig. 3.59).
3. Sugrīva standing, orientated to the south (see fig. 3.60).

4. Hanuman standing, orientated to the west (see fig. 3.61).
5. Nandi in a shrine, before a linga, orientated to the east (see fig. 3.62).

Although it is clear that the carved piers formed part of the original raised channel, the reconstruction of the conduit by the Archaeological Department of the Southern Circle (ARSC 1909-10:4) may have altered the position and orientation of the piers.

Enclosure IV slopes gently from north to south. The raised channel cuts diagonally across the enclosure and at the axial well (IVa/33), divides into two branches. One branch heads north towards the complex arrangement of baths, basins, conduits and drains between the 'Throne Platform' (IVb/1) and the 'Hundred-columned Hall' (IVa/1). The more fragmentary remains of a second branch of the channel continue westwards for about 30 m.

West of the modern road and running parallel to it, are the remains of a second supply conduit. This feeder channel transported water from the raised channel to the 'Great Tank' (IVc/1). The conduit consisted of two parallel rows of roughly-hewn, granite blocks, set in a thick, lime-mortar foundation (see fig. 3.63). In comparison with the raised channel, from which it drew its supply, the feeder channel may be considered on stylistic grounds to be a later addition. In 1986/7 the feeder channel was conjecturally restored using modern materials and original fragments.

The 'Great Tank' (IVc/1), which is by far the largest water feature in the 'Royal Centre', lies in the southeast quadrant of the enclosure, parallel to the south wall of the compound. Water entered the northeast corner of the tank from the feeder channel discussed above. This appears to have been the only inlet. The tank is rectangular (67 m. by 22 m.) with entrance steps on the west and east side (see figs. 3.64 and 3.65). It was originally constructed from a mixture of dressed masonry rubble and brickwork (see fig. 3.66). It was originally plaster-lined. The most striking feature of the 'Great Tank' (IVc/1) is its shallowness. A plastered platform, reconstructed in 1987, runs around the outside of the tank, and probably supported a covered colonnade. Between the tank outlet and the south wall of enclosure IV is a well (IVc/3). The water table was 6 m. from the surface in January 1988.

The tank is provided with an outlet in its south side, which takes the form of an open, square drain with a broad lip around the mouth of the opening. This suggests that it was originally provided with a cover or lid. The remains of a sluice are located directly above the drain and are partially built into the stone lining of the tank (see fig. 3.67). The absence of a covering for the drain below indicates that the outlet has been subject to major modification. The sluice appears to have been made redundant by the building of the simple drain at floor level. Judging from the height of the earlier outlet it would have taken off water at a level raised above the present floor of the tank. The tank is well suited for the settling out of silt. In our view, this was an important function of the tank before the construction of the covered drain. The earlier sluice would have served two major functions. Firstly, acting as an overflow valve to limit the amount of water held in the tank and, secondly, allowing silt free water to be skimmed off the supply in the tank. We suggest that, in its initial form, the tank served a dual function, providing a ritual bathing area and a means of removing dust and soil that had blown into the open domestic supply channels supplying enclosure IV. The exclusion of silt was of vital importance to the conveyance of water by pressure-fed pipes.

According to archaeological reports, the tank was filled with a great deposit of silt. When the silt was cleared in an attempt to uncover the original chunam floor, a vaulted chamber containing human remains was discovered beneath (ARASI 1904-5:30; ASMG 1904-5:36). The report states:

> ... underneath the floor at one place, where the slabs were loose, a vaulted chamber was discovered and human bones and a perfect human skull were exhumed (ASMG 1904-5:36).

The remains were those of a single male of short stature and the skull was identified as being of the dolichocephalic type. It was the opinion of the excavators that the remains represented a human sacrifice made during the construction of the tank.

About 50 m. to the west of the channel supplying the 'Great Tank' (IVc/1), a supply chute takes off to the south from the raised channel. It is composed of single, rectangular granite slab identical to the spans of the raised channel. Two deep notches are cut into the eastern side of this supply chute (see fig. 3.68). We suggest that it was placed in its present position by the Archaeological Department of the Southern Circle during the restoration of the raised channel (ARSC 1909-10:4). In our view, the supply chute originally formed part of the second branch of the raised channel, the fragmentary remains of which continue westwards from the axial well (IVa/33).

In 1984, the Archaeological Survey of India excavation team uncovered a chlorite-lined tank of unusual design abutting the supply chute. The 'Stepped Tank', as it has come to be known, is constructed from finely carved chlorite slabs and measures 22 m. by 22 m. and is 7 m. deep (see figs. 3.69 and 3.70). Each slab is engraved with a tripartite Kannada cypher (see below). A complex arrangement of steps gives access to the tank from all four sides. An entrance portico of granite is located in the middle of the west side and appears, stylistically and in terms of constructional materials, to postdate the tank. The tank is well preserved, although its south and east sides have suffered some constructional disarticulation in their uppermost portions. The interior surface of the tank shows traces of thin plasterwork or whitewash. The geometric design of the tank is reminiscent of Hoysala step-wells of the eleventh and twelfth centuries A.D., found in the Mysore area.

There is some evidence to suggest that the 'Stepped Tank' may, in fact, have been brought from another location in pieces and reassembled. The Kannada inscriptions on each block indicate with letter, number and letter, the direction, layer and the position of each piece (see fig. 3.71). Although the style of these characters corresponds to that of the fourteenth to sixteenth centuries A.D., the tank's design is more typical of the constructional style of religious monuments of the eleventh to twelfth centuries A.D. In our experience, masons' marks at Vijayanagara are solely restricted to architectural fragments from temples. The presence of assembly instructions on a secular structure strongly suggests that it was reassembled. Interestingly, the chloritic stone used in the tank's construction is not found locally. The nearest source of chloritic schist of this type is Dharwad district. In view of this evidence, we suggest that the Kannada inscriptions were not primary masons' marks intended to aid in situ fitting, fixing the constructional date of the tank within the Vijayanagara period, but secondary instructions incised upon a monument constructed in the eleventh century A.D., in order to permit the reassembly of the tank in a different location sometime between the fourteenth and sixteenth centuries A.D.

The supply chute which fed the tank drew its supply from the raised channel at a command of about 1m. above ground level. This command is substantially more than would be required to fill the 'Stepped Tank'. It has been suggested that the supply chute represents a modern addition. We suggest that the tank was originally filled with water from the well situated in its southwest corner (IVa/34). Subsoil water is near the surface here and may have also have filled the tank through its base, which consists of chlorite slabs laid over a bed of sand. This hypothesis may explain the depth of the tank which was perhaps intended, if water staining is to be taken as an accurate guide, to be only partially full.

It is a general observation of wells, tanks and bāvlīs that they were often excavated to a considerable depth to allow them to reach the water table, or to allow channelled water to flow into them. The actual depth of water in them is frequently very shallow. At Vijayanagara, the water table is today very close to the surface. However, the building of the Tungabhadrā Dam has significantly altered the hydrological environment by raising the water table. In our view, wells in the city may have dried up during the winter months prior to the construction of the dam. This suggests that the conservation of rainwater and the recycling of supply may have been important in the 'Royal Centre'.

A number of plaster-lined baths and tanks were originally connected to the raised channel by smaller conduits. These were built of plastered rubble, dressed stone and a mixture of the two (see figs. 3.72 and 3.73). The most important baths and tanks are:

IVa/14 Plaster-lined, rectangular tank of small size (Fritz *et al.* 1984:24,54).

IVa/18 Plaster-lined, stone tank with steps and of small size (Fritz *et al.* 1984:24,54) (see fig. 3.74).

IVa/23 Plaster-lined, square tank of medium size with associate brick chamber (Fritz *et al.* 1984:24,54) (see fig. 3.75).

IVa/24 Finely finished, rectangular, stone tank (Fritz *et al.* 1984:24,54) (see fig. 3.76).

IVa/31b Plaster-lined tank with a stone coping (Fritz *et al.* 1984:) (see fig. 3.77)

IVa/32 Plaster-lined, brick tank with stepped sides (Fritz *et al.* 1984:25,54) (see fig. 3.78).

IVb/2 Finely-finished, square, stone tank (Fritz *et al.* 1984:25,54) (see fig. 3.79).

IVb/3 Square, plaster-lined bath with central pavilion (Fritz *et al.* 1984:25,54) (see fig. 3.80).

IVb/4 Plaster-lined, masonry bath of considerable size abutting IVb/5 (see fig. 3.81). Uncovered in 1987.

IVb/5 Small, square, rock-cut stone basin (Fritz *et al.* 1984:25,54) (see fig. 3.82).

These were connected to the raised channel by conduits which survive in fragmentary form throughout the enclosure. Conduits of three types are visible:

1. Two parallel rows of dressed or undressed masonry with an internal lining of plaster, covered with masonry slabs (see figs. 3.83 and 3.84).
2. Conduits of dressed stone, constructed from three pieces and without a lining of plaster (see fig. 3.85).
3. Conduits of plaster, entirely in excavation (see fig. 3.86).

Although the constructional styles of these types are quite different, their dimensions are similar (24 cm. wide), which suggests that a standard was followed. Differences in construction may in certain cases be attributed to aesthetic and functional considerations. However, in our view such differences are largely the result of many phases of modification and improvement. The conduits used the principle of gravity flow to transport water from the raised supply channel to the various baths and tanks in the enclosure. Drains are constructed in a similar manner to supply conduits. Spigot-jointed pipelines are not found within enclosure IV.

Two tanks require further discussion. The design of IVb/3 (see fig. 3.87), which incorporates a square plan with a ruined, central pavilion (originally supported by four pillars), bears some resemblance to that to the larger XIV/2, known as the 'Water Pavilion'. IVb/3 has a stepped, two-level section and was filled with water by a stone conduit of covered type 1 (see above) in its southwest corner. Water entered the bath at ground level and flowed down into it through a makara-spout. A slab drain is set into the floor in the northeast corner of the bath (see fig. 3.88). The slab covers a rectangular hole in the plastered floor. It is provided with a 10 cm. wide opening. A lip, 2 cm. wide, runs around the edge of the opening and presumably held a cover or plug. Tank IVb/4 was uncovered in 1987. It abuts a small outcrop into which a basin (IVb/5) is cut (see figs. 3.89 and 3.81). The bath is effectively rectangular in shape with a constricted western end. An elaborate flight of steps is located on the west side. The vertical face of the next but lowest step is carved in relief with a procession of elephants. The remains of a supply conduit run parallel to and 5 m. away from the bath's south side. The fragments of this supply channel are of the dressed stone type 2 (see above). The bath was supplied by the raised channel via this stone conduit. A circular drain is situated in the northeast corner of the bath's floor (see fig. 3.90). The drain consists of a hole cut into a stone slab. The slab is set 10 cm. below the floor of the tank in a plaster recess. The circular opening is provided with a 2 cm. lip for a covering or plug.

An open channel (IVd/17) runs for over 100 m. parallel to the south wall of enclosure IV (see fig. 3.91). It is covered by heavy masonry slabs resting on uprights (see figs. 3.92 and 3.93). The channel is further shielded on its northern flank by a ruined wall. IVd/17 is, in part, cut into a low, granite outcrop which protrudes into the southwest corner of enclosure IV. Although now isolated, the channel appears to have carried water from the 'Great Tank' (IVc/1 westwards out of the enclosure and into the water feature XXII/3).

In addition to the supply carried into enclosure IV by channels, perennial water was also available from a number of wells, which gave access to the water table. These include IVa/10b (see fig. 3.94), IVa/34 (see fig. 3.95) and IVc/3 (see fig. 3.96). A single well (WA/1) (see fig. 3.97) is located to the west of enclosure IV in the alley surrounding it. It is likely that more wells will be located in and around enclosure IV.

3.2.3 Features to the south and southwest of enclosure IV (enclosures XXI and XXII)

In the northeast corner of enclosure XXII, there is a concentration of water features. They include the 'Octagonal Fountain' (XXII/1) and the remains of an important water control device (XXII/3) (see fig. 3.98) forming one end of a platform (XXII/2). The lie of the land and the position of the pipe inlet suggest that water flowed east-west into it. This point marked the beginning of a closed pressure system, indicated by the structural strength of the inlet and its associated pipework.

XXII/3 is the single most important domestic supply feature at the site. It is constructed from brick and rough masonry set in thick lime-mortar (see figs. 3.99 and 3.100). It represents part of a much larger device for the transportation of water from open channels under gravity flow into a sealed pipe system under pressure. The feature is raised up above ground level and rests on the eastern end of a rectangular, two-level platform, partially faced with stone (see fig. 3.101). Water entered the device through a series of pipe openings at different levels, set into a heavy 'wall' of plastered brick and masonry. During its passage through the 'wall', water was led into separate pipe conduits. Today, four earthenware pipes emerge from the west side of the feature, which is encased in brick and lime-mortar. Earthenware fragments and impressions left in chunam suggest that there were originally more.

In our view, XXII/3 formed the west side of a large tank for the storage and control of water. We suggest that XXII/3 was the main outlet for the tank. The remaining sides of the tank have not survived, though some footings are located southeast of XXII/3. There is a spread of brick and chunam rubble for some 12 m. east of the device (see fig. 3.102). Similar debris is built into the modern dry-stone wall which runs along the south side of the road, in front of the 'Octagonal Fountain' (XXII/1) (see fig. 3.103). This suggests that the tank, having fallen into disrepair, was vandalised and used as a source of building material in the post-Vijayanagara period. In 1986, a modern rubble wall, aligned east-west, hid a considerable portion of XXII/3 and its platform from view. The wall was dismantled in 1987.

We suggest that the platform (XXII/2) and its associated features were originally encased by a covering of chunam and brick, the greater part of which has been removed or simply fallen apart. Traces of the covering are visible in the form of plasterwork on the masonry of the platform and brick-rubble (see fig. 3.104).

Three pipe outlets are set into the plastered, eastern side of XXII/3 at different levels (see figs. 3.105 and 3.106). The outlets are arranged as one single and a pair. Fragmentary impression of pipework in lime-plaster suggest that an additional pair of pipe outlets was located to the north. The single outlet consists of a pipe, 16 cm. in diameter, set into the bottom of a recessed notch. The notch is cut into the vertical, plastered face of XXII/3 (see fig. 3.107). Water entered the pipe and passed into a raised tank behind. It was then diverted into a series of pipelines taking off at two different levels (see below). A 6 cm. wide projection is located on the right hand side of the notch. If one accepts that XXII/3 represents the west side of a tank, the projection must logically have held a gate to control the exit of water from the proposed tank. Closure of the gate would have allowed water to fill the tank to a level well above the low outlet. Opening the gate would have allowed water to be forced under pressure through the pipe, by the action of gravity on the surface of the water in the tank. The pressure in the pipe system when the water was static, generated by the weight of the head of water, would be proportional to the depth of the water from the atmospheric surface to the bottom of the tank. Apart from facilitating closure of the outlet, the notch may have also been used to exclude silt from the pipe system. The insertion of a board in front of the notch would have allowed water to collect behind it. When the water level reached the top of the board, it would slop over it and into the single pipe outlet behind, leaving any silt in the tank. Water would then flow slowly, under pressure generated by the head behind the board, through the heavily encased pipes, without the need for maintenance (i.e., the clearing of silt blockages from the system).

The double outlet, consisting a pair of 16 cm. diameter pipes, is located 78 cm. to the north of the single outlet (see fig. 3.108). The pipes are set at an angle of 45°. Immediately behind the double outlet, there is a horizontal surface (30 cm. by 46 cm.). At the junction of the horizontal and the angled surface into which the pipes are set, two holes are drilled, one into each of the vertical walls on either side of the outlet (see fig. 3.109). The holes are 4 cm. in diameter. The depth of the left hole is 9 cm. whilst the right is 5 cm. deep. In our view, they held a pivoted rod or shaft to which was attached a board for closing the double outlet. The board would have rested on the horizontal surface or shelf behind the outlet and would thus have had dimensions slightly smaller than the shelf. When a closure was required the board would be swung down on its pivot over the outlets.

Excavations on the final day of the 1987/8 season, revealed a raised tank (77.5 cm. by 1.45 m. and 1 m. deep) on top of XXII/3 (see figs. 3.110 and 3.111). The tank is constructed from plastered brickwork. It is rectangular in plan with a small projection in its southwest corner. Water entered the tank via the single pipe let into the notch on the east elevation of XXII/3 (see above). It was thereafter diverted into four separate pipelines. A small stone, ideally situated to hold a gate, is fixed inside the tank just to the south of the inlet (see fig. 3.112). Three of the pipe outlets are set into the floor of the raised tank (see fig. 3.113). The plastered floor slopes to the east. A low chunam platform runs along two-thirds of the tank's western side. Two pipes, arranged as a pair, are located in the northwest corner of the platform. The mouths of these pipes, which are of the same diameter (16 cm.), protrude slightly from the top of the platform and are protected by a skirting of chunam. A small recess is located in the side wall of the tank to the immediate west of the pair (see fig. 3.114). We suggest that it held a covering or gate. A single pipe is located 30 cm. south of the pair on the southeast corner of the platform. The three outlets allowed water to pass vertically through the floor of the raised tank and supplied the heavily encased, horizontal pipelines which flow west towards the 'Octagonal Fountain' (XXII/1). The impression of the fourth pipe outlet is clearly visible in the plasterwork at a point about 50 cm. above the floor of the tank in its southwest corner (see fig. 3.115).

The east side of the tank shows signs of alteration. A channel originally led through to the notch on the east face of XXII/3. This has been blocked by a plastered insert (see fig. 3.116). We suggest that at some stage, water from the large tank, the east face of which was formed by XXII/3, would have filled the raised tank located on top by means of an open conduit. Later, the open channel between the two tanks was replaced by a single length of pipe. The open channel was filled in and plastered over. The changes made to the raised tank suggest that XXII/3 was modified to increase its efficiency.

The 'Octagonal Fountain' (XXII/1) was supplied by one of the pipe conduits taking off from XXII/3. The command of XXII/3 over the fountain is approximately 1 m. The diameter of the pipes is standard at 16 cm. This diameter is very wide compared to the likely size of the nozzle in the 'Octagonal Fountain'. We may conclude that there would be little friction in the pipe, as the water would flow very slowly. The sluggishness of the flow would cause silt to settle out more readily in the pipe than anywhere else in the system. The removal of silt from sealed and heavily encased pipes would have caused serious maintenance problems. It was, therefore, of the utmost importance that silt be removed before the water was allowed to flow under pressure through the pipe conduits. The problem of silting in the pipes was not just restricted to those conduits which supplied fountains, but affected all sealed conduits, because of the large diameter of pipes chosen. The design of the outlets which fed these pipe conduits was, in our view, dictated by the need to exclude silt from the pipe supply wherever possible.

Three pipes, each 16 cm. in diameter, emerge from the control feature and run west along the top of the platform. The pipes are covered by a 2 cm. skin of plaster and are encased, as a group, in brick and lime-plaster. A fourth pipe runs in a southwesterly direction towards the 'Bhojanśāla'. The pipes leave the second and highest level of the platform by its southwest corner, dropping down to the level below (see fig. 3.117). At this point the group divides into a pair and one single conduit. The latter leads directly to the 'Octagonal Fountain' (XXII/1). The former leave the platform and continue for a short distance in a southwesterly direction.

We suggest that the platform may have been reused as a supporting structure for the water control feature (XXII/3) and connected pipe conduits. In part, this observation is based upon the extremely crude way in which the pipe conduits are attached to the platform. However, the strongest evidence for the reuse of the platform is the presence of what appears to be a masonry-lined channel on its first level (see fig. 3.117). This suggests that the platform's primary function was the support of a stone-lined channel or channels above ground level. A similar style of construction is employed in the 'Bhojanśāla' (discussed below). In our view, XXII/3 and associated pipe conduits utilised elements of an earlier water feature.

The 'Octagonal Fountain' (XXII/1) is located several metres to the west of the platform supporting the pipe inlet. To its immediate south, the modern road from Kamalāpuram to Hampi runs southeast-northwest. The fountain is constructed from a mixture of block masonry and brickwork, covered with chunam. The structure is octagonal in plan and is provided with a centrally placed, octagonal basin. This originally housed a fountain, removed before the beginning of this century (ARASI 1908-9-23:23, APRSC 1908-9:16). A rectangular, free-standing basin is located south of the central compartment, under the open arcade. Each face of the exterior facade is pierced by an arched entrance. These give access to a passage-way around the central chamber, similarly pierced by eight archways (see figs. 3.118 and 3.119). The style of the building is Muslim-influenced and is identical to that of the 'Queen's Bath' (XX/1) and 'Octagonal Bath'. The 'Octagonal Fountain' (XXII/1) was restored at the beginning of this century (APRSC 1908-9:16).

Enclosure XXI contains one isolated feature close to the water remains in XXII. South of XXII/3, on the opposite side of the modern road, are the remains of an open channel 30 m. long, known locally as the 'Bhojanśāla'. The channel is cut into a low, earthen embankment (approximately 60 cm. high, 5 m. across and 30 m. long), faced with granite slabs. The stone is finely dressed. The conduit is rectilinear in section (1m. across and 60 cm. deep) and shows no trace of plaster-work (see figs. 3.120 and 3.121). Although the construction of the modern, metalled road has destroyed any trace of the connecting supply conduit, this feature was, in our view, connected to XXII/3 by an earthenware pipeline. An early plan (Mackenzie 1800e) indicated that a collection-basin was originally abutted to the northwest end of the channel. No mention of the basin was made when the 'Bhojanśāla' was cleared of earth and reset in the first quarter of this century (ARSC 1919-20:16). No trace of the basin survives on the ground today.

The channel is partially flanked by heavy chlorite slabs carved in imitation of traditional eating plates (see figs. 3.122 and 3.123). Differences in style, size and workmanship suggest that the slabs do not form a homogeneous group. Mackenzie's plan (1800e) shows a greater number of carved slabs than are now extant. To us, it seems likely that the chlorite fragments of similar design now lying inside enclosure IV were originally part of the 'Bhojanśāla'. These must have been moved after 1800. We have noted three granite copies of the carved chlorite slabs used as a lining for the channel. This suggests that it may have been at some stage rebuilt. The 'Bhojanśāla' appears to have been a feeding place for the court brahmans. The stone plates on either side of the channel could easily and conveniently have been cleaned after use with water drawn from the conduit.

The Southwest end of the 'Bhojanśāla' terminates abruptly, close to the northern boundary of a cane field. The final destination of the supply which flowed through the channel is unknown. However, the fragmentary remains of a large, 'terraced' structure have been noted in dense undergrowth to the southwest.

3.2.4 Features within enclosures I, III, V and IX

Two water features are located in enclosure I, which forms a plaza in front of the Rāmachandra temple. A well (Ib/9), with a rotated square opening, is situated next to the east wall (see fig. 3.124). To the southwest of the well, there are the ruined remains of a stone tank (Ic/12) (see fig. 3.125). The tank is provided with steps on its north, east and south sides. It was originally filled by a conduit from the west.

The narrow, rectangular enclosure adjoining the northern compound wall of IV and labelled III, contains several water features. In the southeastern corner of the enclosure, a fragment of pipework encased in brick and chunam (IIc/5) runs east-west, parallel to the enclosure wall shared by III and IV. We suggest that it represents a continuation of the pipe system which takes off from the control tank in the northwest corner of enclosure XVIII (see section 3.2.1).

The remaining water features, located in the northwest of the enclosure, are in two groups. The first group is made up of two small recti-linear tanks (IIIa/3 and IIIa/4), a larger tank of plastered masonry (IIIa/5) (see figs. 3.126 and 3.127) and a small fragment of a conduit which passes through the west enclosure wall of III. The second group is made up of two monolithic stone troughs situated close to the north boundary wall. The smaller monolith is lying overturned. Ten metres to the east, a massive stone trough of fine quality (IIIa/7) rests at an angle of 45° on four thin stone wedges (see fig. 3.128). A small, circular drainage hole is provided in the eastern end of the trough.

Enclosure V, which is situated to the west of IV and north of XXII, has been identified as an area of royal or noble habitation (Fritz *et al.* 1984:25-26). It contains a well (V/4) (see fig. 3.129), a plaster-lined tank (V/10) (see fig. 3.130) and a large, rubble-lined tank (V/12) (see fig. 3.131).

Enclosure IX contains a single water feature, a well (IXa/3), located close to its eastern wall.

3.2.5 Features within enclosures XIV and XV

Enclosure XIV is an irregularly-shaped, walled compound, located to the north of the main complex of interlocking enclosures. It was identified by earlier writers as the 'Zenāna'. The walls which delimit XIV postdate the structures within the enclosure (Fritz *et al.* 1984:33-34). The largest water feature in the enclosure is the 'Water Pavilion' (XIV/2) (Fritz *et al.* 1984:33-34,111,53) located in its southwest corner. The structure consists of a rectangular stone basement, set inside a shallow, stone-lined tank (see fig. 3.132). The tank was originally plastered with chunam (ARSC 109-10:4). The basement has four, axial projections. A small bridge supported by pillars, now ruined, gave access to the central pavilion from the south. Flights of steps lead from the edge of the tank to its bottom on the east and south sides. An additional staircase is situated on the east side of the central structure. The pavilion does not survive above basement level. The association of water with this structure is reinforced by the presence of carvings of aquatic subjects on its base. Although no evidence for perennial supply survives on the ground in the immediate environs of this structure, the remains of a pipe system linking enclosure XIV to XV were extant at the beginning of this century (ARSC 109-10:4). No trace of the pipeline survives today. A small well (XIV/10) is located 40 m. to the east of the 'Water Pavilion' (XIV/2).

East of the 'Queen's Palace' (XIV/3) is a large, stone-lined, rainwater collection tank (XIV/9). It is cut into a flat, exposed outcrop which makes up the northeast corner of the enclosure. The tank (15 m. by 20 m.) is constructed from crudely dressed slabs forming a stepped section (see fig. 3.133). A flight of steps is let into its western side.

To the south of XIV, the large, open area of

enclosure XV contains three notable water supply features: a step-well (XVa/8) (see fig. 3.134), a partly collapsed step-tank (XVb/12) (see fig. 3.135) and an 18 m. section of spigot-jointed pipework (see fig. 3.136). The pipeline, which is encased in lime-mortar and brick, is located in the south-west corner of the enclosure (XVb), and runs southeast-northwest. The pipes are 16 cm. in diameter and are protected by an inner 2 cm. skin of chunam and an outer casing of brickwork. The presence of the pipework in enclosure XV suggests that the sealed pressure system, for which there is evidence in enclosures III, XVIII, and XXII, extended much further north.

3.2.6 Isolated domestic water features in the 'Royal Centre'

The major surviving features of the domestic water supply system have been discussed above. However, we note the existence of isolated and fragmentary remains throughout the 'Royal Centre.' These include fragments of conduits or drains (Ic/16 and 17, in the southeast corner of XXIb and in the compound of structure XXVI/1), tanks (XVa/3 and XXVI/1), wells (XXVI/1) and step-tanks (NQy/3) (see fig. 3.137). Other step-tanks of elaborate design are located inside the 'Urban Core', to the east of the 'Royal Centre' (NTb/2 and NT1/1).

3.3 Domestic water features outside the 'Royal Centre' and temple tanks

The evidence for domestic water features outside the 'Royal Centre' will be discussed together with temple tanks. The evidence has been divided up as follows:

1. Domestic and ritual tanks.
2. Wells.
3. Large temple tanks.

3.3.1 Domestic and ritual tanks

Amongst the standing buildings and architectural debris of Hemakutam hill, four rain-water collection tanks are let into the solid rock of the hill itself. The tanks are essentially natural features, but masonry and rubble barricades have been added to increase their storage capacity.

The first catchment tank is the most northerly of the four. It is formed by a natural semi-circular depression (7.5 m. wide and 20 m. long) partly cut into by a low masonry wall, running east-west (see fig. 3.138). The wall, which is a modern addition, is constructed from dressed granite blocks cemented together in regular courses. A barricade of rubble, which incorporates Vijayanagara period architectural fragments, is laid across the southwest side of the tank.

The second tank abuts a small shrine facing east. A narrow flight of steps leads down from the shrine to the tank which is ellipsoid in shape (5 m. wide and 15 m. long) (see fig. 3.139). The tank is overhung on the north and east sides by the boulder into which it is partly cut. A rubble wall cuts across the south side of the tank, partially blocking a 30 m. long narrow crevice filled with water. A crevice runs south-north, sloping gently towards the tank. A Sanskrit epigraph dated A.D. 1498 indicates that the tank was constructed by two brothers, Virūpākṣa-paṇḍita and Vināyaka-paṇḍita and was dedicated to Virūpākṣa (ARSIE 1934-35:351; Gopal 1985:154). The inscription also indicates that trees were planted on the hill.

A third tank (10 m. wide and 25 m. long) is formed by two masonry walls which block both ends of a wide crevice which runs east-west (see fig. 3.140). The dilapidated walls are constructed from undressed masonry blocks. A small shrine on a raised plinth overlooks the north side of the tank.

The fourth tank in the group is located in the southwest corner of the walled area on top of the hill. It is the smallest of the tanks, measuring 10 m. by 5 m. It abuts the west side of an early shrine and is constructed from a mixture of undressed masonry block and dressed slabs (see fig. 3.141).

In functional terms, the catchment tanks located on Hemakutam hill are inefficient. Water loss by evaporation from shallow, open basins is severe. In our view, they were not a primary source of supply for the hill, except perhaps in times of siege or natural disaster. We suggest that the tanks primarily served a ritual function. However, this does not preclude the possibility that the hill was a centre of settlement, using the water channel (described below) as the primary source of subsistence water.

We have observed a water conduit which runs along the western side of the hill at ground level, flowing south-north (see fig. 3.142). The channel is partly cut from solid rock and partly lined with masonry blocks. It is supplied by a subsidiary channel of the Hiriya canal (see section 4.4). The channel flows into a covered control tank (2.8 m. by 3 m.), below the northwest corner of the hill, at ground level. The tank is constructed from slab masonry (see fig. 3.143). Two separate channels emerge from the tank. The first flows northwest into the Vidyāraṇya maṭha and temple kitchens which abut the west end of the Virūpākṣa temple complex. The second flows eastwards under the west prākāra wall of the temple. On entering the temple precincts, the channel flows across the roof of a maṇḍapa. The site on which the temple is built, slopes down from the toe of Hemakutam hill towards the river. Thus, ground level outside the southern prākāra wall is approximately 4 m. above floor level inside the temple. After passing through the temple, the channel reappears outside (see fig. 3.144) and flows along the northern toe of the hill (see fig. 3.145). The channel, which is partially covered by masonry slabs, utilises natural features such as crevices in the rock. The channel is obscured by masonry debris and thick undergrowth on the hill. We suggest that it eventually flows back into the temple complex.

A rainwater collection tank is located on the south face of the south range of the irrigated valley, beneath a pair of prominent shrines perched on a high boulder. A small temple nearby is identified as the Parameśvara Maṭha (Nagaraja Rao 1985:39). the tank is rectilinear in plan (20 m. by 14 m.) and steeply lined with masonry blocks to a depth of 5.5. m. (see fig. 3.146). It is partly excavated from solid rock on its north and south sides, the remaining earthen parts are supported by masonry linings. Steps are provided in the southwest corner.

3.3.2 Wells

We have noted the presence of numerous wells outside the 'Royal Centre', located in valley bottoms and in the flat plain. The water table is at present 6 to 7 m. below the surface in the winter months. This situation has, without doubt, been affected by the construction of the Tungabhadrā Dam. The wells vary in form from simple square openings above unlined shafts to complex geometric step-wells of considerable depth (see fig. 3.147).

There is some epigraphical evidence for the digging of wells at the site. In a field outside Kamalāpuram village, known as Papajagaḷūra-hola, there is a maṇḍapa which contains an inscription (ARSIE 1904:18; Gopal 1985:179). The inscription is written in Kannada and refers to the building of a well and a rest-house by Ahamada Khāna, a servant of King Devaraya II. It is dated A.D. 1439.

A Kannada inscription dated A.D. 1411 records the construction of a step-well on the main path from the 'Royal Centre' to Matanga Parvatam, approximately 175 m. south of the Hiriya kāluve (Nagaraja Rao 1985:31). The epigraph mentions that the well, dedicated to Ganeśa, was built by Devarāya (son of Naraharideva of Rukuśākha, and Bhāradvāja Gotra.

The number of wells in the 'Urban Core' suggests that water-lifting was an important element in the domestic supply system.

3.3.3 Large temple tanks

The oldest temple tank at Vijayanagara is the Manmatha Gundam tank, which abuts the northern prākāra wall of the Virūpākṣa temple. The tank may be dated to the twelfth century A.D. (see section 1.4). It is set at a lower level than the surrounding structures to the east, including the modern Janata housing estate, which represent post-Vijayanagara extensions to Hampi village. The shrines on the west and northwest of the tank appear to be contemporaneous with the building of the tank's enclosing revetment.

In our view, the tank is supplied in the same manner as a bāvlī. That is, by the percolation of water into the lower portion of the tank from the water table. A substantial quantity of water is held within the tank in winter, despite evaporation. This may be explained by the building of the Tungabhadrā Dam, which has raised the water table.

The tank is rectangular (approximately 55 m. by 35 m.) and set below ground level in a hollow, the sides of which are revetted. The hollow is surrounded by a low wall (see figs. 3.148 and 3.149). At the foot of the revetting, a paved walkway runs around three sides of the tank. On the remaining south side, a steep revetment rises from the edge of the tank to the bottom of the Virūpākṣa temple's prākāra wall. The walkway is connected to ground level by three flights of steps, in the north, west and northeast. The northern staircase is let through a modern, dry-stone wall at the top of the revetment. This modern addition holds back over 1 m. of accumulated soil. The staircase on the west side of the tank is more elaborate and is provided with an ornate entrance structure, surmounted by a Drāviḍa roof finial. Two pairs of Nandis, two at the top and two at the bottom of the flight of stairs, are orientated to the east (see fig. 3.150). An outlet is located in the northeast corner of the tank at the same level as the paved walkway (see fig. 3.151). Its position suggests that excess water in the tank was discharged into the river.

A large temple tank is located to the east of the Viṭṭhala temple, halfway along its bazaar. This tank may be viewed as an integral part of the temple's original design. The Viṭṭhala tank measures 62 m. by 37 m. and is provided with a central pavilion, 7 m. high, capped by a Drāviḍa roof structure (see fig. 3.152). Its heavily revetted sides are stepped in section, and the whole structure is constructed from large rectangular blocks. Today, the tank remains dry in the winter and is not provided with a perennial supply. The large body of rainwater held during the cooler months following the wet season, is subject to heavy seepage and evaporation losses.

The Tiruveṅgaḷanātha and Bālakṛṣṇa temples are provided with tanks of similar design to that of the Viṭṭhala tank. The Tiruveṅgaḷanātha tank is dilapidated, its bazaar having been turned over to agricultural use. Both tanks are located on the west side of their temples' bazaar, looking away from the shrine (i.e. north in the case of the Tiruveṅgaḷanātha temple and east in the case of the Bālakṛṣṇa). The Bālakṛṣṇa tank is the better preserved of the two and holds water for most of the year (see fig. 3.153). Its dimensions are 53 m. by 28 m. It is fed with perennial water by an inlet channel which takes off from the Hiriya kāluve. The channel is composed of well-finished, U-sectioned, lengths of masonry. The dimensions of the conduit are variable (20-25 cm. deep and 23-38 cm. wide).

An interesting temple tank is located just to the southwest of the Vidyāraṇya maṭha and temple kitchens, which abut the west end of the Virūpākṣa temple complex. The Lokapāvana tank (14 m. by 16m.) is set inside a cloistered compound approximately 45 m. by 30 m. (see fig. 3.154). The compound is paved. North of the tank is a ruined shrine (6 m. by 3 m.) which houses a chlorite linga and Durgā image. The tank is fed by a natural spring. An overflow outlet is located in the northwest corner of the tank. Excess water is passed into a black-lined channel and flows out of the compound into a field to the west. A subsidiary channel of the Hiriya kāluve flows around the west side of the compound. The outlet conduit from the tank passes under this perennial channel at a lower level.

CHAPTER 4

Observations in the Field: 2. Agricultural Supply

4.1 The Kamalāpuram kere

To the southwest of Kamalāpuram village, a large man-made reservoir taps the run-off from the Sandur hills. The tank irrigates an area of about 182 hectares (Francis 1904:282) to its north. Water is held back by a sinuous bund of earth, revetted on its upstream face. The bund runs from east to west, across a shallow, natural drainage. The Kamalāpuram tank is also fed with perennial water by the Rāya kāluve. A considerable body of water is held in the tank throughout the year.

In addition to irrigating the agricultural area to the north, the tank also supplements the supply of the Hiriya kāluve (see section 4.5.1). The sinuous irrigation channels taking off from the tank come together to form a single conduit approximately 1 km. north of the tank. This carries water northwards through the 'Urban Core'. The channel fills a short moat on the outside of the defensive wall of the 'Urban Core' and eventually discharges into the Hiriya canal between the third and fourth outlets in the irrigated valley (see sections 4.4. and 4.5.1).

The Kamalāpuram tank appears to have supplied the 'Royal Centre' with perennial water during the Vijayanagara period. This is discussed separately in section 4.1.2.

The tank is at present known as the Kamalāpuram kere. However, a Kannada inscription found near the tank suggests that this is not its original name. The epigraph, dated A.D. 1440, is carved on a boulder located to the northwest of Kamalāpuram village. It refers to the granting of land for the services of the God Kariya Tiruveṅgaḷanātha, at Aṅjanagiri, by a Queen Tirumaladevi (ARSIE 1922: 697; Gopal 1985:180). Aṅjanagiri is described as being inside Vijayanagara-paṭṭana, close to the Chikkarāya kere. The epigraph, which is inscribed on a large boulder, appears not to have been moved. In our view, since there are no other tanks located in the immediate area, the Chikkarāya kere must be the Kamalāpuram tank. This inscription suggests that the tank was in use during A.D. 1440. We feel that it was constructed earlier. The Rāya canal, which supplies the tank, was almost certainly constructed at the same time. In our opinion both the tank and canal were built at the beginning of the fifteenth century A.D. (see sections 4.2.1 and 5.4.2).

The bund is approximately 2 km. in length (see fig. 4.1). Its upstream face, which is oblique, is protected by a stone revetment. The revetment shows signs of alteration and improvement. The downstream face, which is also oblique, is block-revetted in several areas. The modern, metalled road from Hospet to Kamalāpuram runs parallel to the bund, and for part of its journey eastwards runs along the top of it (see fig. 4.2). We suggest that the road was a major route into the 'Urban Core' during the Vijayanagara period.

The sinuous nature of the bund may be explained by the incorporation of natural features such as outcrops. West of the first outlet, a detour to the southwest allows the bund to join a low outcrop which protrudes southwards into the tank (see fig. 4.3). The earthen parts of the bund abutting the outcrop are not revetted. The height of the bund is greatest (approximately 6 m.) between the second and fourth outlets (see fig. 4.1). The extreme eastern end of the bund's upstream face is revetted with small, neatly dressed blocks. It is constructed without chunam (see fig. 4.4). The angle of revetting is steep, approaching 90 in many places. The rest of the bund is more heavily revetted, and is set at an oblique angle (see fig. 4.5). The lower two-thirds of the revetment are composed of boulders and large, unworked blocks piled on top of one another so as to form a sloping face. No cement or mortar is used. The top third is made up of small, neatly dressed granite blocks laid in regular courses, and resembles the revetting found at the eastern end of the bund.

A masonry structure is located between the third and fourth outlets. It is built against the downstream face of the bund (see fig. 4.1). The structure is three-sided and filled with packed earth, its fourth side being formed by the bund (see fig. 4.6). Its dimensions are as follows:

Height (from base in field on north side)	— 6 m.
Width (north and south sides)	— 16 m.
Breadth (west and east sides)	— 22 m.

Its constructional style shares many similarities with the defensive bastions built in and around the 'Urban Core'. Its position, which flanks the road running along the bund, suggests that its purpose was defensive, or perhaps connected with the collection of import and export duty. Amongst the eroded debris on top of the structure, there is a large, upright slab (1.60 m. high and 70 cm. wide) carved with a standing image of Hanuman (facing right). Slabs carved with religious icons are often found flanking important entrances and doorways during the Vijayanagara period.

4.1.1 The outlets on the Kamalāpuram kere

The tank is provided with four main outlets and a spillway (see fig. 4.1) These are spread, at fairly regular intervals, along the length of the bund. The medieval headworks of outlets one and four have been entirely superseded by modern replacements.

The first outlet is located in the northeast corner of the tank, close to Kamalāpuram village. The capstan-operated pipe inlet is modern (see fig. 4.7). It is housed within a tubular structure of plastered masonry (1.90 m. in diameter), which abuts the upstream face of the bund. Water is not let directly through the pipe into an irrigation channel, but led first into a high-sided tank on the downstream face of the bund. After passing through the tank, water flows into a vertical-sided channel. built of cement and regularly-sized granite blocks. The channel flows northwest and is bridged by the metalled road. It is used by the modern inhabitants of Kamalāpuram as a dhobi-ghāt. On the north side of the road, the channel divides into three branches, the centre and left hand branches (looking northwards) flowing northwest into the midst of the fields between the tank and the wall of the 'Urban Core' (see fig. 4.8). The remaining channel flows northeast towards the 'Royal Centre', running parallel to the main street to Kamalāpuram village.

Approximately 375 m. to the west, a second outlet allows water out of the tank and into the fields to the north (see fig. 4.9). The outlet is not marked on the Survey of India 1929, 1947 or 1976 maps. A pipe or tube is used to convey the water through the bund. This conduit protrudes 7 m. into the tank. The flow is regulated by a metal gate connected to a capstan. The capstan is supported by a pair of slabs, held above water level by two sturdy piers of granite. A modern concrete walkway gives access to the capstan from the bund. The masonry frame, which supports the capstan and is identical to other datable examples of the medieval period, has not been replaced. In our view, the modern capstan replaces a simple metal mounting for the original gate. On the downstream face of the bund, water pours into a open tank 6 m^2 (see figs. 4.10 and 4.11). The tank is located 5 m. from the toe of the earthwork, and 24 m. to the west of the outlet. The tank is 3 m. deep and functions as a secondary means of regulating the flow leaving the bund (see fig. 4.12). A stone slab is used to control the flow entering the outlet (a rectangu-lar hole let into the northwest side of the tank). The tank is protected from the incursion of wind blown deposits by a 60 cm. high coping of dressed granite.

The open tank is set upon an earthen platform which abuts the path at the bottom of the bund. Two sides of the platform are held by steeply-battered, surcharged walls. A paved walk- way runs around three sides of the tank. The toes of the retaining walls are approximately level with the field system below. Water leaves the tank and passes through the backing and retaining wall on the northeast side, flowing into the end of an irrigation channel (perpendicular to the northwest side of the platform) (see fig. 4.13).

Approximately 500 m. to the west, a third outlet of a similar design protrudes into the tank (see fig. 4.1). A masonry frame supports a modern capstan which regulates the pipe outlet below (see fig. 4.14). The style of the frame is identical to that of the second outlet. The capstan, which is 12 m. from the upstream face of the bund, is reached by a modern concrete walkway. The height of the bund is 7 m. at the outlet. Water passes into a open tank, 5 m.^{2}, which is situated on an earthen platform supported by retaining walls (see figs. 4.15, 4.16 and 4.17). A fragmentary paving of granite slabs 40 cm. wide surrounds the tank. The tank is constructed without chunam and is provided with three outlets. It is 60 cm. deep. The quantity of water leaving the tank is controlled by the placement of granite slabs in front of the outlets. A coping of dressed granite, 55 cm. high, runs around the edge of the tank, 50 cm. from its lip. Two functional outlets allow water to flow to the north and west respectively. The outlet on the east side appears not to function. The northern outlet supplies an earthen channel which shows indications of having once been rubble-lined. The western channel flows parallel to the bund towards the masonry structure located between the third and fourth outlets (see section 4.1). The channel is revetted with small dressed blocks covered with chunam. To the east of the tank, a staircase of rectangular slabs is let into the downstream face of the bund.

The fourth outlet is situated approximately 300 m. to the west (see fig. 4.1). Unlike the second and third outlets which have retained some original features intact (the sluice-frame and control tank), the fourth outlet has been entirely replaced. However, its design imitates the form of the medieval outlets. A modern concrete and masonry frame supports a capstan operated gate (see fig. 4.18). Water passes into a modern tank, abutted to the north side of the bund (see fig. 4.19). A concrete staircase leads down to the east side of the tank from the road. The tank has two rectangular outlets on its north and west sides. Water passes into three irrigation channels, one on the west side and the remaining two taking off to the north. The channels are provided with vertical, modern revetments of plastered masonry. The lining gives way to unlined, earthen banks after approximately 10 m.

A masonry structure is located on the top of the bund, 2.5 m. to the south of the metalled road and approximately 4 m. to the east of the fourth outlet. It is composed of two horizontal slabs resting on three partially-buried piers (see fig. 4.20). The slabs are 2 m. long, 50 cm. wide and 30 cm. thick. The grassed area around the structure shows signs of disturbance. We suggest that its function may be related to the control of water leaving the tank.

Some 250 m. to the east of the fourth outlet, the tank is provided with a large spillway or waste weir (see figs. 4.1, 4.21 and 4.22). The device allows excess water to escape safely from the western end of the tank without danger to the bund. A safety valve of this type was needed to ensure that the water level did not exceed the full supply level of the tank, which was predetermined by the height of the bund and the positions of the major outlets. The device is 73 m. wide and is flanked by wing walls which divert the escaping flow across the irregular surface of the spillway after it has passed over the weir crest. The major portion of the spillway is constructed from large blocks of undressed masonry laid without chunam. Masonry needles, used to support a temporary weir crest to hold back the supply in the tank, are located in a line across the apron, close to the weir crest (see fig. 4.23). A modern section of spillway, 55 m. in length, is abutted to the west end of the device.

The spillway has been subject to two alterations. Firstly, the old crest has been superseded by a wider and taller replacement of cemented masonry. The original crest of large slabs is visible behind (i.e., immediately downstream) the modern replacement. The purpose of this addition must have been to raise the amount of water held in the tank by increasing the height of the weir along its whole length. Secondly, the paved apron has been grouted.

Excess water from the tank flows across the apron and passes into an escape channel. A bridge carries the metalled road over the conduit. The escape channel discharges into a rock-cut conduit which flows to the northwest (see fig. 4.24). The rock-cut channel is flanked on its right bank by the fragmentary remains of a defensive wall.

The eastern wing wall of the spillway shows no signs of modification. It is constructed from undressed masonry laid in irregular courses without lime-mortar. Immediately to its east, on the downstream toe of the bund, are the fragmentary remains of a trabeated structure (see fig. 4.25). The remains lie in an area of disturbed ground.

A minor outlet of modern design and without a controlling gate is located approximately 50 m. to the west of the spillway. The Rāya kāluve flows into the Kamalāpuram tank 500 m. southwest of the minor outlet The source of supply for the Rāya kāluve is discussed in detail in sections 4.2 and 5.4.

4.1.2 The supply of water from the Kamalāpuram kere to the 'Royal Centre' (domestic supply)

The first outlet on the Kamalāpuram kere supplies three channels. Two flow north and northwest respectively. The remaining channel flows northeast, parallel to the main street of Kamalāpuram village, between the fields and the back of the buildings on the west side of the village (see fig. 4.1). The channel is earthenlined with occasional sections of rubble revetting. After passing behind the Archaeological Office, the channel turns sharply to the southeast and flows for 30 m. into the higher ground between the Office and the Inspection Bungalow. To compensate for the change in level, the channel is cut into the bedrock, which is close to the surface (see fig. 4.26).

After passing between the Archaeological Office and the Inspection Bungalow, the conduit makes a sharp turn to the northeast. The channel, which is 3 m. wide and has steeply oblique sides, remains rock-cut in deep excavation (2.4 m). Running parallel to the main street of the village, it flows in front of the Inspection Bungalow of approximately 70 m., crossing under the Kamalāpuram–Hampi road as it turns off to the northwest (see fig. 4.27). The channel bifurcates on the north side of the road (see fig. 4.28). One branch continues to flow northeast, parallel to the main street, whilst the other turns to the northwest and flows along the side of the Kamalāpuram–Hampi road, towards the 'Royal Centre' (see fig. 4.29).

The ground between the main street of the village and the enclosure wall of the 'Urban Core' slopes down markedly. Thus, as the northwest branch of the channel flows towards the 'Royal Centre', its command is maintained by a gradual shallowing of the excavated trench through which it flows. The rock-cut portion of the channel hits the surface 44 m. from the bifurcation, and from there on flows northwest for approximately 75 m. at ground level, retained by small earthen banks. The channel is provided with vertical, modern revetting on its left bank for some of its length. As the ground falls away further, the channel flows into a raised conduit, whose supporting piers increase in height to maintain the command of the channel as the ground slopes away (see fig. 4.30).

The raised channel flows northwest for approximately 125 m., running parallel to the Kamalāpuram–Hampi road. As the road slopes upwards in front of the wall of the 'Urban Core', the height of the raised channel decreases proportionately. The channel flows through the gateway in the wall at ground level and turns northeast away from the road and into a cane field (see fig. 4.31) the raised portion of the conduit is constructed from a mixture of masonry rubble and brickwork covered by lime-plaster. It is 21 cm. wide and 46 cm. deep. The supporting piers are composed of neatly dressed blocks and raise the channel up to a maximum height of 1.7 m. above ground level as the land falls away beneath.

Although the constructional style of the raised channel indicates that it is modern, the rock-cut conduit that supplies it appears to be more ancient. It is cut through a buried deposit of granite and has steeply oblique sides, without any lining. We suggest that it was constructed during the Vijayanagara period, and that the raised channel seen today replaces an earlier, dilapidated structure of a similar type. This hypothesis is supported by the presence of rubble footings next to the modern raised channel, as it passes through the gateway into the 'Urban Core'.

Today, the channel irrigates the fields between the enclosure walls of the 'Urban Core' and the 'Royal Centre'. The channel is the only perennial supply feature with sufficient command to have provided the large domestic supply channel XIX/25 (see section 3.2.1) with water during the Vijayanagara period. We therefore suggest that the Kamalāpuram tank supplied enclosure IV via the channel. Water features to the east of the 'Royal Centre' appear to have been supplied with water for domestic use by the ruined tank northeast of Kamalāpuram village, the bund of which joined two prominent ranges.

4.2 The Rāya and Basavanna canals

Before embarking on an account of the Rāya and Basavanna canals in their present condition, it is necessary to discuss the important changes which have taken place since their construction in the medieval period. Without such a discussion, a thorough understanding of their design and chronology would be impracticable.

4.2.1 Postconstructional changes to the Rāya and Basavanna canals

The Rāya and Basavanna canals irrigate the agricultural zone between Hospet and Kamalāpuram. The canals, which were both constructed in the medieval period, were subject to a number of major alterations during the nineteenth and twentieth centuries. Today, the canals are supplied by the modern Tungabhadrā Dam at Mālāpuram (see fig. 4.32). The dam was designed to feed a High and Low Level Canal System on both banks of the river, supplying water to Raichur and Bellary districts, as well as Āndhra Pradesh. The Low Level System incorporated the old Rāya canal, using it for the purpose for which it was originally designed, the supply of water to the land between Hospet and Kamalāpuram. The Basavanna canal had originally irrigated a large tract of agricultural land between Vallabhāpuram and Amarāvati (see fig. 4.33). The building of the Tungabhadrā Dam drowned practically all of the old canal, which was extended in two stages, to irrigate the land between Hospet and the Kamalāpuram tank.

The Rāya and Basavanna canals are now abutted to a single supply channel called the 'R and B' canal, which takes off from the dam between the Tungabhadrā Right Bank Main Canal and the Tungabhadrā High Level Canal. The flow supplied to them is regulated. The new dam stores the surface run-off from an area of 28,167 km², making it possible to compensate for seasonal fluctuations in the river flow.

Before the construction of the Tungabhadrā Dam, the canals were fed with water by four diversion weirs, situated upstream from Mālāpuram (see figs. 4.33 and 4.34). The weirs, which were originally built in the medieval period, were modified in the nineteenth and twentieth centuries to increase their efficiency. They were subsequently destroyed by the construction of the new dam between 1940 and 1956. Information relating to the medieval diversion weirs is contained in pre-Tungabhadrā Dam literature and on the 1928/129 and 1947 Survey of India maps.

The old anicuts were located at Koragallu, Vallabhāpuram, Rāmanagadda and Kūruvagadda respectively (see fig. 4.33). They can be divided into two groups on the basis of their function and design. The Koragallu and Vallabhāpuram devices acted as a pair in series, connecting a rocky island in the centre of the river to either bank. The whole width of the river was spanned. The Rāmanagadda and Kūruvagadda anicuts worked in parallel, taking advantage of the river's bifurcation at Ānaveri. The Rāmanagadda weir was located at the point where the river flow divided around the island of Kūruvagadda. It forced water between the island and the right bank of the river. The Kūruvagadda anicut, located downstream of the Rāmanagadda device, completely dammed the right branch of the river, near the north end of the island.

The Basavanna canal was supplied by the weir at Vallabhāpuram (see fig. 4.33). The canal took off from the right bank of the river 500 m. upstream of the anicut. It flowed northeast towards Mālāpuram, roughly parallel to the river

(Kelsall 1872:132). The anicut, which was 300 m. long and straight in plan (Francis 1904:91), was constructed of loose stone. The Vallabhāpuram anicut was known as a landing stage for teak floated down the Tungabhadrā from Mysore during the southwest monsoon (Pharoah 1855:83; Kelsall 1872:100; Francis 1904:7). The Koragallu weir fed a series of canals which irrigated the agricultural land north and northeast of Koragallu village on the left bank. The anicut was also constructed of loose stone. It was 366 m. long and irregular in plan (Kelsall 1872:231). An inscription recording the construction of the anicuts during the reign of Kṛṣṇadeva Rāya was found at Vallabhāpuram (Kelsall 1872:231; Francis 1904:38; Venkayya 1906:201). The epigraph, which is now lost, was dated A.D. 1521.

The Vallabhāpuram anicut was replaced (by the Madras Irrigation Company) with a new weir, 15 m. downstream, between 1846 and 1848 (Kelsall 18672:231). Between 1927 and 1928, the new anicut was raised (ARMI 1927-8:18). The old Koragallu anicut, which at the time fell under the jurisdiction of the Nizam of Hyderabad, was not replaced, although it may have received minor repairs.

The Rāmanagadda and Kūruvagadda anicuts supplied the Rāya canal, which took off from the river just upstream of the latter (see fig. 4.33). The Rāmanagadda anicut took the form of a low, irregular deflecting wall, and was constructed of loose masonry held together by stone clamps and pegs (Francis 1904:91). The weir took advantage of the uneven nature of the river bed, integrating rocks and boulders which lay in its stream. The Kūruvagadda anicut, 4.8 km. further downstream, joined the right bank of the river to Kūruvagadda island near the village of Hosakote. It was 229 m. long and concave towards the flow in plan, and was constructed from rough masonry without mortar (Francis 1904:22). The Rāmanagadda and Kūruvagadda anicuts were repaired between 1891 and 1892 (ARMI 1891-2:63).

The supply of water to the Rāya canal by the Kūruvagadda anicut needs further discussion as both the 1928/1929 and 1947 Survey of India maps (57 A/7 and 57 A/8) omit to show a connection between the two features. The connection is verified by two separate sources based on personal inspection and published in 1872 (Kelsall) and 1904 (Francis). However, the Survey of India maps show the Kūruvagadda anicut supplying two canals which flow northeast towards the river, flowing into a single channel shortly before rejoining it. On the same maps, the Rāya canal is supplied by a large seasonal cross-drainage, called the Gauripuram Vanka, which flows northwest towards the river. The Gauripuram Vanka collects the surface run-off from the southwest side of the Sandur hills and discharges it into the river. It is inconceivable that the Rāya canal, a perennial device, could have been supplied by the Gauripuram Vanka, as the former requires a supply more regular and substantial than that of a seasonal drainage. There is no literary or archival evidence to suggest that the connection between the anicut and the canal, confirmed by Kelsall and Francis, was severed before the construction of the Tungabhadrā Dam. We can, therefore, concluded that the omission of a connection between the Kūruvagadda anicut and the Rāya canal on the survey of India maps is the result of a draughtsman's error.

Two drainages run down to the river from the Sandur hills. The Gauripuram Vanka, described above, flowed into the river, west of Mālāpuram (see fig. 4.33). In so doing, it cut across both the Rāya and Basavanna canals. Both made a detour to the southeast to cross the drainage without losing their command. The Gauripuram Vanka discharged part of its supply into the canals. The drainage water contained silt brought down from the Sandur hills, which entered the canals with detrimental effect (Francis 1904:92). A second drainage flows down the valley of Sandur (see fig. 4.33). Water originally passed down the valley and directly into the river between Mālāpuram and Amarāvati. After the construction of the Rāya canal, its supply passed into the canal, north of Amarāvati. With the extension of the Basavanna canal in the nineteenth century (see below), the drainage was connected to the Basavanna and Rāya canals.

Apart from its connection to the new dam at Mālāpuram, the Rāya canal has remained essentially true to its original form. The canal has three functions. Firstly, it irrigates a large proportion of the agricultural land between Hospet and Kamalāpuram village. Secondly, the canal supplies the Kamalāpuram tank with a perennial supply, supplementing the tank's supply of surface run-off (Pharoah 1855:83; Kelsall 1872:16; Francis 1904:88,282). Thirdly, it supplies the Kālaghatta canal with waste-water (Francis 1904:90). The canal's efficiency has been improved by the addition of modern revetting and capstan-operated outlets. The 1929 and 1947 Survey of India maps (57 A/7) indicate that the Rāya canal originally supplied the Hosūru canal with an additional supply (see section 4.3). This connection is now severed.

The Basavanna canal has been subject to a number of changes. The original canal, constructed in A.D. 1521, flowed eastwards from the Vallabhāpuram anicut and ended 750 m. to the northeast of Malapuram (Kelsall 1872:231). The canal was cleared from 1856 to 1858, improving its discharge (SRMG XXVII:60, XXXVIII:174-175, LIV:342-343, LIX:11). From 1891 to 1895, a number of minor improvements were made, including the fitting of new sluices (ARMI 1891-1892:63, 1893-1894:74, 1894-1895:69). Engineers lengthened the canal by some 2 km. before the beginning of this century (Francis 1904:92), taking the canal east of Amarāvati village (see fig. 4.33). Between 1925 and 1937, the canal was extended further to a point south of the Kamalāpuram kere (ARMI 1925-6:16, 1926-7:17, 1927-8:18, 1928-9:44, 1929-30:47, 1930:1:58, 1933-4;15, 1934-5:17, 1935-6:14, 1936-7:13).

On a chronological note, the name 'Rāya kāluve' suggests that the canal was built during the reign of King Kṛṣṇadeva Rāya (A.D. 1509-1530). Although almost any substantial remains are attributed by local tradition to Kṛṣṇadeva Rāya and given the affix 'Rāya', the association of the large bund southwest of Hospet with this King has proved accurate (see section 4.7). However, in our view the Rāya canal was constructed much earlier, probably during the reign of Bukka II (A.D. 1399-1406) or Devarāya I (A.D. 1406-1422). This hypothesis is based on Nuniz' account of the building of a large anicut to supply the city with water (see section 5.4).

The old anicuts had been built in an area of great irrigational potential. During the nineteenth century, British engineers toyed with various proposals for modification and replacement of the old works. The Madras Irrigation Company, on the suggestion of Sir Arthur Cotton, began work in 1859 on the High Level and Low Level Lines of the Upper Bellary Project (Francis 1904:93-94). Two new canals were to be built. The high Level Line was to be fed by a dam at Vallabhāpuram, replacing the modified, Medieval anicut. The canal was to flow to Bellary and Hagari. The Low Level Project consisted of a new dam at Hosūru and a canal to take water south of Bellary to Hagari. The Madras Irrigation Company abandoned work on both schemes in 1866 due to administrative problems and the old Tungabhadrā canals, as they were then known, fell under the jurisdiction of the Public Works Department of the Madras Presidency.

The possibility of improved irrigation from Tungabhadrā was subsequently re-examined on several occasions, by the Irrigation Commission which superseded the Madras Irrigation Company (Francis 1904:94). Engineers in the nineteenth century were not slow to appreciate the potential for building a large dam at Mālāpuram. However, little progress was made (RIIC 1901-1903). Administrative and financial problems prevented any more than modification and extensions of existing works until 1940, when the construction of the new dam at Mālāpuram and the Low level Canal Project began in earnest, under British supervision (ARMI 1946-7:34-36). The Low Level Project was completed in 1952. The High Level System was inaugurated in 1956 by the governor of Āndhra Pradesh and represented the culmination of the Tungabhadrā Project.

4.2.2 The Rāya and Basavanna canals in their present condition

It has been demonstrated that the Rāya and Basavanna canals were modified by British engineers between 1856 and 1973 to increase their efficiency. The present condition of the canals indicates that they were again modified during the construction of the Tungabhadrā Dam.

In its present condition, the Basavanna canal is a modern construction (see fig. 4.32). After separating from the 'R and B' canal, it flows along the north side of the Tungabhadrā Colony. From this point, it flows east to the village of Amarāya (see figs. 4.35 and 4.36) and on through the centre of Hospet, crossing under its main street (Station Road). On the east side of the town, the canal turns to the northeast and flows under the Hospet–Kamalāpuram road (see figs. 4.35 and 4.37).

After passing through Hospet, the canal flows north to the village of Bhattarahalli, where it is crossed by the main line of the Southern Central Railway line from Bellary to Hubli (see figs. 4.35 and 4.38). The canal then turns east and flows parallel to the Tungabhadrā Right Bank Main Canal for some 3 km. At the village of Mallappannagudi Tānda, the canal turns southeast, flowing under the Hospet–Kamalāpuram road in deep excavation and between the villages of Mallappannagudi and Hosa Mallappannagudi (see figs. 4.35 and 4.39).

The canal continues to flow southeast for approximately 4.5 km., terminating south of the Kamalāpuram tank. Apart from the two sections excavated through buried rock, one near Mallappannagudi Tānda and one between Mallappannagudi and Hosa Mallappannagudi, the canal is built entirely in shallow excavation. It is at present just over 16 km. in length (excluding the 'R and B' canal, which is 750 m. in length) and is provided with 15 modern, capstan-operated outlets.

Two types of linings are visible in the Basavanna canal:

1. A revetting of small, dressed blocks, trapezoidal in section and bonded with chunam (see fig. 4.40). The revetments have been regrouped and are partially covered with lime-plaster. Many parts have become disarticulated.
2. Vertical revetting composed of small, dressed blocks laid in regular courses and capped with a thick layer of cement (see fig. 4.41).

The first type is, judging by its condition and the number of renovations made to it, the older of the two types and may be dated to the second lengthening of the canal between 1925 and 1937. The second type, which has been used to replace the oldest and most dilapidated sections of the canal, appears more modern and is most likely contemporaneous to the building of the Tungabhadrā Dam. This type of lining is predominant in the section of the canal from the 'R and B' canal to the west side of Hospet; i.e., that which represents the first extension of the medieval canal from Mālāpuram to Amarāvati as reported by Francis (1904:92). We suggest that the existence of either type in the other canals permits us to date those sections to the nineteenth and twentieth centuries. Further, since the modern linings found in the Basavanna canal span the entire period of post-medieval alteration and improvement to the Tungabhadrā canals, the revettings of the other canals which do not correspond to these modern types may be dated to the medieval period.

The Rāya canal was cleared from 1856 to 1858 (SRMG XXVII:60, XXXVIII:174-175, LIV:342-343, LIX:11), modified from 1891 to 1895 (ARMI 1891-2:63, 1893-4:74, 1894-5:69) and again from 1926 to 1930 (ARMI 1926-7:17, 1928-9:44, 1929-30:47). After separating from the 'R and B' canal, it flows northeast along a sinuous route to the outskirts of Chittavādigi village, now a suburb of Hospet (see figs. 4.32 and 4.32). On the west side of Chittavādigi, the drainage from the valley of Sandur pours into the canal. The canal then flows around the north side of Hospet, running almost parallel to the Tungabhadrā Right Bank Main Canal.

Approximately 1 km. to the east, a large subsidiary channel takes off from the Rāya canal and flows northwards towards the village of Karakallu (see section 4.3). There it divides into three separate channels, two of which pass into the river north of the village, whilst the third turns eastwards and flows into the Kālaghatta canal. A further five subsidiary channels take off from the Rāya canal between Chittavādigi and Nāgenahalli (see fig. 4.35). These are of smaller size. All the subsidiary channels irrigate the agricultural land between Hospet and the river, which slopes gently to the north.

Close to the village of Nāgenahalli, the canal's route becomes more sinuous (see fig. 4.35). It turns north past the village and then east into a range of hills to the north and east of Benakāpura (see figs. 4.42 and 4.43). The canal flows through a valley in the range. The outcrops on either side of the valley rise to a maximum height of 568 MSL. Some sections of the canal are supported by earthen embankments which rest against the sides of the range. The command of the canal is maintained at just below 460 MSL for the majority of its length, but 2.25 km. east of Kadirāmpuram, it passes down through a 250 m. rock-cut section to a lower command. The canal hugs the foot of the range, flowing southeast towards the Kamalāpuram tank. Approximately 1.25 km. northeast of Nāgenahalli Tanda village, the canal enters a second and final 250 m. rock-cut section, allowing it to pass across the saddle of a low, sheet rock. The saddle represents the easternmost extension of the range. The canal continues to flow around the outside, lower edges of outcrops and exposures of rock (see fig. 4.44). The canal is held in embankment against outcrops wherever necessary, but is otherwise in excavation. After passing under the Hospet–Kamalāpuram road (see figs. 4.45 and 4.35), the canal divides into two separate channels. Both flow into the northwest corner of the Kamalāpuram tank. The canal is, at present, some, 24.5 km. in length (excluding the 'R and B' canal) and is provided with 27 modern, capstan-operated outlets.

The Rāya canal is furnished with several types of lining, but for the majority of its length its banks are dilapidated and overgrown. In some areas, the banks have been prevented from collapse by the construction of crude revetments of rubble (see fig. 4.46). These appear to be modern. Long sections of the canal appear to have no lining at all, whilst in other sections the fragmentary remains of an old vertical lining of irregular granite blocks has survived (see fig. 4.47). We date this to the Vijayanagara period. There are indications that the canal lining was originally plastered with chunam (Francis 1904:92). However, we have been unable to locate any trace of the chunam covering. Vertical revettings of a similar type are also found in the Hiriya kāluve (see section 4.5.1). The two types of post-medieval revetting extant in the Basavanna canal (see above), are also found in certain areas. The old revetting was replaced between 1891 and 1895 (ARMI 1891-2:63, 1893-4:74, 1894-5:69), 1926 and 1930 (ARMI 1926-7:17, 1928-9:44, 1929-30:47) and during the construction of the Tungabhadrā Dam.

4.3. The Hosūru and Kālaghatta canals

An anicut is located 500 m. southwest of the village of Hosūru (see fig. 4.48). It blocks the passage of water flowing between a small island and the right bank of the river (see fig. 4.49). In older literature, the Hosūru weir is known by the name Bella (Kelsall 1872:232, Francis 1904:92). The old anicut has been replaced by a taller, modern device built immediately downstream of it. The new weir is constructed from small, dressed blocks of granite bonded with cement and capped with a crest of concrete. A three-gated sluice is provided in the centre of the device and allows excess water to be safely passed downstream. The old weir is difficult to observe as it is almost totally submerged, but appears to be constructed of dressed masonry and covered with a layer of chunam. Its crest is fitted with masonry needles to support a temporary weir crest.

The Hosūru canal takes off next to the anicut and flows northeast, irrigating the field between the villages of Hosūru and Bande. The canal was cleared between 1856 and 1858 (SRMG XXVII:60, XXXVIII:174-175, LIV:342-343, LIX:11). It was improved from 1891 to 1894 (ARMI 1891-2:63, 1893-4:74), and again during the construction of the Tungabhadrā Dam. The 1929 and 1947 Survey of India maps (57 A/7) indicate that the supply in the Hosūru canal was boosted by the Rāya canal via a channel some 2.5 km. in length, which joined the Hosūru canal just southeast of the village (see section 4.3). This connection no longer exists (see fig. 4.48). The Hosūru canal originally divided into two channels close to the village of Yippetavaru. The two channels ran parallel for some 2 km., the western channel terminating at Bande, whilst the eastern channel flowed northeast and discharged into the river. At present, only the western channel of the Hosūru canal is extant. The eastern channel has been abandoned, the remaining 1.75 km. of it having been utilised as a second escape channel for the head of the Kālaghatta canal (see fig. 4.48). We suggest that this alteration may have become necessary to compensate for the greater quantities of water being allowed into the Rāya canal after the construction of the Tungabhadrā Dam to which it was abutted. The second escape channel would have allowed greater quantities of waste water to pass into the river if so required.

The Kālaghatta canal is interesting in that it is not directly supplied by an anicut catchment drainage tank. It appears that the canal was specifically designed to conduct the excess water from the Rāya canal into the Hiriya canal and at the same time irrigate the thin strip of agricultural land adjacent to the right bank of the river, which could not be reached by the Rāya canal itself (see fig. 4.50). The canal, which begins south of the village of Bande, is approximately 8 km. in length and is supplied by subsidiary channels taking off from the Rāya canal, which flow northwards the river (Francis 1904:90).

At present, two escape channels at the head of the Kālaghatta canal allow excess water to pass into the river north and northeast of Karakallu (see fig. 4.50). The canal is provided with five, capstan-operated outlets. The canal divides into two channels approximately 500 m. southeast of the deserted village of Kallaghatti. The channels came together to form a single canal south of the Hiriya anicuts and flow eastwards for 250 m. before discharging into the Hiriya kāluve (see section 4.4).

4.4 The Hiriya anicuts and canal

The Hiriya canal irrigates an extensive area of agricultural land from a point 2.5 km. west of Hampi to the village of Bukkasāgara, near Kampli (see fig. 4.51). In particular, the canal supplies all the agricultural land in the irrigated valley, whilst maintaining a high command over the valley floor. The canal is fed by an arrangement of six weirs in series, known as the Hiriya or Turtha anicuts, which block the flow of the Tungabhadrā between a low, fertile island and the river's right bank. In this section, the Hiriya anicuts and canal are discussed in detail. The complex section of the canal which runs through the irrigated valley is covered separately in section 4.5.

The supply of the Hiriya canal is boosted at two points (see fig. 4.52). Firstly, the Kālaghatta canal passes water from the Rāya kāluve into the Hiriya canal 750 m. southeast of the anicut (see section 4.3). Secondly, the outlets and waste weir of the Kamalāpuram tank discharge water into a single canal which connects with the Hiriya canal in the irrigated valley (see sections 4.1 and 4.4.1).

The Hiriya anicuts, which were originally constructed in the medieval period, have all been rebuilt in modern times. Five small and one large diversion weirs divert water from the river into a single channel at a point 2.5 km. west of Hampi (see fig. 4.53). Kelsall provides a description of the anicuts before they were modified for the first time between 1891 and 1895 (ARMI 1891-2:63, 1893-4:47, 1894-5:69, 1895-6:73). He states:

> This is taken across the river about one mile west of the old city of Hampi, and is formed of a number of bits of masonry connecting islands and rock. The channel taken from it winds about very much, especially where it passes through Hampi and varies a great deal in width (1872:232).

The whole width of the river channel is blocked by the weirs, which as Kelsall indicates, join together outcrops and shoals. Outside of the monsoon season, the entire flow of the river channel is diverted into the Hiriya canal, leaving the river channel almost entirely dry downstream of the weirs.

The weirs may be divided into two groups, the series of five to the north and the larger, single weir to the south. The five weirs to the north are all relatively small and are irregular in plan, connecting with convenient obstruction protruding from the bed of the river. The sixth and southernmost weir is much larger and regular in plan.

The five smaller anicuts take the form of low walls, constructed from neatly dressed blocks of granite, laid in regular courses and bonded with cement (see fig. 4.54). Predominantly, the weirs are provided with vertical faces on the upstream and downstream side. However, isolated sections have a vertical upstream face and a gently curved downstream face. The fragmentary remains of the original weirs are visible just down-stream and adjacent to the new devices, at a slightly lower level. These consist of rectangular slabs laid in regular courses, with their short faces against the flow. Small holes are drilled at irregular intervals on the top surface of each slab (see fig. 4.55). The majority of the holes are empty, but from an examination of those which survive in their original condition, it is apparent that all were originally used to retain iron clamps set in lead. Rather than being bonded with mortar, the slabs were held in place by short iron rods bent to an angle of 45° at both ends, each of which was inserted into a hole drilled in a slab. The joint was made secure by filling the hole with molten lead. The heavy bonding of the slabs suggests that the visible surface is the outer casing of the old weir and was intended to withstand the full force of the water. The inner sections of the old weirs appear to be constructed of dressed slabs. Francis reports that the old anicut at Siruguppa, 52 km. north of Bellary, was bonded in a similar fashion (1904:91).

The modern weirs were constructed on top, or immediately upstream, of the old anicuts. The modern weirs have crests approximately 50 cm. higher than those of the old devices, to compensate for the deposition of silt upstream.

The sixth and southernmost anicut of the Hiriya series is separated from the group of five smaller devices by a rocky island, which it joins to the right bank of the river (see fig. 4.53). The anicut seen today is a modern construction and takes the form of a low wall built across the channel. It is constructed from small, dressed blocks of granite, bonded with cement, and capped with a coping of concrete (see fig. 4.56). The anicut is built on the solid rock of the channel bed (see fig. 4.57). The fragmentary remains of the old weir are visible below the surface of the water, upstream of the anicut. Dressed slabs, some with drilled holes identical to those discussed above and evidently part of the old device, are visible amongst the rubble immediately downstream of the weir (see fig. 4.58). In our view, these fragments result from the partial demolition of the medieval anicut during the construction of the new weir.

An undated inscription is located on the north side of the anicut. It can be reached only by coracle as it overhangs the upstream side of the weir (see fig. 4.53). The inscription, which is badly worn, is incised on the vertical face of a large boulder (see fig. 4.59). It is written in a southern script characteristic of the fourteenth to sixteenth centuries A.D. Some words are misspelt and the diacritical marks are particularly difficult to read. The language is confusing, but some words appear to be written in Sanskrit. It appears to read:

1st line. *cinthāyaka devaṃ nāma*
2nd line. *rajaśikhara śitau Virupākṣaha*
3rd line. *Śivāyahada jayahada*

Three religious names are mentioned, Rajaśekhara, Virūpākṣa and Śiva, which suggests that the inscription may have been a prayer. We suggest that it was written during the construction of the Hiriya anicuts and was intended to confer the blessing of Śiva upon the work.

We have noted a configuration of square-sectioned holes, each 13 cm. by 14 cm. and 14 cm. deep, on the north side of the southernmost anicut, amongst the boulders downstream of the device (see fig. 4.60). The holes are arranged in a line running away from the anicut (see fig. 4.53). We suggest that they were intended to house the vertical supports of a temporary weir crest.

The canal takes off just upstream of the southernmost of the six anicuts (see fig. 4.53). The canal was modified between 1891 and 1895 (ARMI 1891-2:63, 1893-4:74, 1894-5:69, 1895-6:73), and again during the construction of the Tungabhadrā Dam. It is not provided with a head work for controlling the amount of water entering it. Instead, close to the inlet, the canal is provided with an escape weir, which allows excess water to flow harmlessly over the crest of the weir and back to the river, downstream of the anicuts (see fig. 4.61). The escape weir is constructed from rough slabs without mortar, and has been modified by the addition of a modern crest of concrete. Rather than running parallel to the river, the canal, in excavation, flows southeast in order to maintain a high command (see fig. 4.52). After 750 m. the canal makes effectively a 90° turn to the east and flows parallel to the river, towards the village of Kṛṣṇāpuram. From this point onwards, a number of modern outlets allow water to pass out of the canal to the agricultural land in the north.

The first outlet allows water to flow down into an original water feature (see fig. 4.52). It is located approximately 250 m. to the north of the outlet and has survived without modification. The function of the device is to allow water to pass down to a lower command without increasing velocity. The channel flows through the device, which consists of four courses of large slabs laid without lime-mortar lengthways along the direction of the flow (see fig. 4.62). Gaps are left between the slabs, allowing the water to pass through the device which acts like a spillway, breaking the force of the water. The channel is guided into the device by a short length of revetted conduit (see fig. 4.63).

As the canal flows east, it is held in embankment against the north side of a low range by an earthwork (see fig. 4.64). The right bank of the canal is, for some distance, carved from a solid outcrop of sheet rock. A path or track runs along the top of the embankment.

The canal skirts the sheet rock and passes eastwards into level ground. It then flows along the north side of the village of Kṛṣṇāpuram, approximately 750 m. south of Hampi (see figs. 4.65 and 4.52). The village is located on flat ground. As the canal flows towards the village, it is no longer held in embankment, but constructed in excavation. At Kṛṣṇapuram, a single outlet takes off from the right bank of the canal, passing water into a channel flowing southeast. The main canal passes under the modern bridge, which carries the Kadirampuram–Hampi road just to the west side of Kṛṣṇāpuram village. From this point, it flows eastwards in deep excavation towards the irrigated valley (see fig. 4.66).

Approximately 150 m. east of Kṛṣṇāpuram village, the canal flows through a dilapidated Vaiṣṇava temple (see fig. 4.52). The temple is constructed on both sides of the canal, passage across the conduit being facilitated by a simple slab bridge (see fig. 4.67). A three-sided slab is located in front of the ruined temple, on the left bank of the canal. An inscription is incised on each face of the slab. The first side is dated A.D. 1524 and refers to offerings made to the God Raghunātha, 'installed' on the Hiriya kāluve (Nagaraja Rao 1985:31-34). The inscription on the third side is undated. It also mentions the God Raghunātha of the Hiriya canal (Nagaraja Rao 1985:34), and records the donation of irrigated land to the brahmans engaged in the services of the God. These inscriptions are important for three reasons. Firstly, they provide us with the original name of the canal used in the Vijayanagara period. Secondly, they indicate that the canal was built no later than A.D. 1524. Thirdly, they suggest that water supply features were, in the Vijayanagara period, associated with deities.

After passing through the temple, the canal flows eastwards in excavation for a further 75 m. before bifurcating in front of a large, exposed boulder. The channels flow around either side of the boulder, and in so doing drop down to a lower level, resulting in considerable turbulence (see figs. 4.68 and 4.69). The channels, which are partially lined with blocks, laid without chunam, converge downstream of the boulder. The canal then flows in excavation towards the Kamalāpuram–Hampi road, 75 m. to the east (see figs. 4.70 and 4.71) and into the irrigated valley (see fig. 4.51). The canal from the Kamalāpuram–Hampi bridge to the foot of Matanga Parvatam is discussed in section 4.5.1.

As it enters the irrigated valley, the character of the Hiriya canal changes. It is no longer in excavation, but held in embankment against the south side of the valley by a substantial earthwork. The canal follows a contour on the southern range, keeping a high command above the agricultural land below. After passing through the valley, the canal flows eastwards, eventually discharging into the Tungabhadrā, 2 km. northeast of the village of Bukkasāgara, crossing over the Kampli canal in the process. The southern Chika channel, which takes off from the Hiriya kāluve to the west of Matanga Parvatam (see section 4.5.1), discharges into the river at Talarigattu (see fig. 4.72). The water level in the channel at the outlet has been raised by the construction of a temporary weir of boulders and vegetal debris (see fig. 4.73). This allows water to be channelled off to irrigate terraced fields adjoining the right bank of the river.

The canal, which has been fitted with modern outlets, has had dilapidated sections of lining replaced. Two types are visible and correspond with those found in the Basavanna and Rāya canals. We date them to the 1920s–1930s and 1940s–1950s respectively. The original lining, which consists of rough blocks and small boulders without bonding, has been observed in isolated sections. It is vertical in section. Some parts of the canal are without a lining of any sort.

4.5 Agricultural water features in the 'Irrigated Valley'

The Irrigated Valley may be defined as the area contained by the two parallel ranges north of the flat plain, stretching from the Kamalāpuram–Hampi road bridge in the west to Talarigattu in the east. The valley is approximately 3.5 km. in length. Our investigation concentrated on the part of the valley between the Kamalāpuram–Hampi road bridge and a point immediately south of the Tiruveṅgaḷanātha temple, at the foot of Matanga Parvatam (see fig. 4.74). The complex nature of the hydrological remains in this area demands a detailed examination of the surviving features.

4.5.1 The Hiriya kāluve (Turtha canal) from the Kamalāpuram–Hampi road bridge to the base of Matanga Parvatam

The Hiriya kāluve (modern name Turtha or 'swift' canal) cuts the Kamalāpuram–Hampi road running west-east, 250 m. north of the Uddhāna Virabhādra and Chaṇḍeśvara temples. The bridge over the canal is modern. We suggest that it replaced an earlier device. On the west side of the bridge, the canal is entirely in excavation. However, on its east side and throughout the irrigated valley, the canal is held in embankment against the south range of the valley by an earthwork. The canal hugs a contour, maintaining a command of 1 to 7 m. above the floor of the valley, which is used for intensive agricultural production (see figs. 4.7 and 4.75). As the floor of the valley falls away, the height of the embankment on which the canal rests increases proportionately. The effectiveness of the canal has made the irrigated valley an area of great agricultural importance. The cultivation of factory crops, predominantly bananas for the Bombay and Goa markets, has replaced the production of traditional staple crops.

The maintenance of the canal revetments in the irrigated valley is particularly important for two reasons. Firstly, to prevent the collapse of the linings, a problem which is aggravated by the swift flow. Secondly, to obviate the possibility of damage to the supporting embankment by seepage from the canal. Some sections, particularly those where there is some degree of turbulence in the canal, have been protected in modern times by a thick layer of lime-plaster. We suggest that part of the canal were also lined with chunam in the medieval period.

The canal linings show a stylistic development which can be compared with datable linings in the Basavanna and Rāya canals. The old, vertical revetting survives in isolated sections and consists of heavy block, laid in regular courses without lime-mortar. Two types of replacement revetting are visible. The first is oblique in section and consists of small, neatly dressed blocks laid in regular courses and bonded with chunam. It has, in places, been covered with cement. The second is vertical and made up of small, neatly dressed blocks laid in regular courses, bonded with cement and capped with a coping of concrete. We suggest that these linings correspond with the modification of the 1920s–1930s and 1940s–1950s respectively (see sections 4.2.1, 4.2.2 and 4.4). We have noted that the canal banks are considerably higher than the water surface of the conduit, which suggests that its builders allowed for a reasonable margin of error during the construction of the canal. Throughout the valley, the width of the canal remains fairly consistent at 6 to 7 m.

An unpaved path follows the north bank of the canal from the modern road bridge along the whole length of the 'Irrigated Valley'. The track is located on top of the earthen embankment into which the canal is cut on its north side. The provision of a route for pedestrian and vehicular traffic along the embankment appears to be part of the original design feature of the canal. This, together with the dimension of the canal itself, explains the massive appearance of the embankment.

The design of the canal uses the terrain. Natural features, in particular outcrops and boulders, act as elements of the design, reducing the constructional cost. The canal is not so much imposed on the landscape as determined by it, using geographical peculiarities to their best advantage. The canal winds gently around and between natural boulders, using them to impede flow or cause turbulence when a conversion of potential into kinetic energy is required. The design of the canal may be considered organic and functional, utilising natural resources with a minimum of superfluous labour and expense.

There are no outlets on the south side of the canal, east of the Kamalāpuram–Hampi bridge. Water is taken off through the north bank of the canal and taken down to the irrigated tracts of land at the bottom of the valley. Land to the south of the canal is watered by the canals supplied by the Kamalāpuram tank (see sections 4.1 and 4.1.1).

The first outlet after the road bridge, is located immediately to the east of a ruined temple (see fig. 4.74). The outlet is modern and consists of a capstan-operated gate. It supplies water to the land between the canal and the Bālakṛṣṇa bazaar. It takes off from the north bank of the canal just upstream of a small rapid (see fig. 4.76). The rapid takes the water level in the canal to a lower command with no increase in velocity.

The second and third outlets are located close together on the north bank of the canal (see fig. 4.74) and are of similar design to the first outlet. The second outlet feeds a small channel which flows northeast towards the Bālakṛṣṇa temple bazaar. It is unlined. The channel which takes off from the third outlet, heads northeast across the valley floor towards Matanga Parvatam. It is known locally as the Chika kāluva and is earthen-lined (1.6 m. wide and 50 cm. deep). It bifurcates at the eastern end of the Bālakṛṣṇa bazaar into a larger and a smaller channel. The channels flow eastwards, parallel to one another, but at different commands. The smaller or northern channel keeps a high command. It flows around the base of the Matanga hill and is, for some of its journey, rock-cut. The southern channel divides again at the base of the hill.

Downstream of the third outlet, a channel supplied by the Kamalāpuram tank flows into the Hiriya canal, supplementing its supply (see section 4.1 and 4.1.1). Moving eastwards along the canal, the valley bottom falls away quite sharply, leaving the canal raised up in high embankment on its north side. The earthwork is steep and without a protective facing.

Approximately 50 m. downstream of the Kamalāpuram inlet channel, a fourth outlet takes off from the canal (see fig. 4.77). It supplies an irrigation channel known locally as the Kiriyagava kāluve. The outlet, which is modern, consists of two sets of gates separated by a dividing wall. The first set allows water to pass directly out of the Hiriya canal and down to the Kiriyagava channel heading northeast, whilst the second diverts the flow into the same channel via a cascade and stilling pool. The channel flows northeast towards Matanga Parvatam.

A further 50 m. to the east, a fifth outlet allows water to pass down from the canal into the productive agricultural area below. The outlet is provided with a modern, capstan-operated gate. The canal maintains a command of 6 to 7 m. over the fields below.

At the base of Matanga Parvatam, the valley is divided by a large earthwork, the top of which is approximately level with the banks of the canal (see fig. 4.78). About 40 m. west of the earthwork, a sixth outlet of modern type takes off from the canal. Between the outlet and the earthwork, a paved area or road is visible on the south bank of the canal (see fig. 4.79). The disarticulated surface of the road is made up of large blocks of granite. It stretches from the vertical revetting of the canal in the north, right up to the base of the outcrop which dominates the southern side of the valley. In our view, the road was part of the route northwards across the irrigated valley, which allowed pedestrian and vehicular traffic to move from the 'Urban Core' to the 'Sacred Centre'

A portion of the road and the right bank of the canal adjacent to it have broken away to reveal a second, earlier phase of construction (see fig. 4.80). Behind the missing section of stonework, a crudely built wall or vertical revetment is visible (see fig. 4.81). This feature would have been totally obscured by the construction of the right (i.e., south) bank of the canal. The function of the earlier, exposed structure is at present unclear.

4.5.2 The Matanga Parvatam revetted earthwork: The Bhūpati bund

Perpendicular to the road (discussed above), a large earthwork divides the Irrigated Valley below Matanga Parvatam (see figs. 4.82 and 4.83). Today, this structure serves no other purpose than to allow passage across the valley without descending into the deepest part near the Hiriya kāluve.

The earthwork shows signs of alteration. Although the southern end which abuts the embankment of the Hiriya kāluve is relatively well preserved, the northern end has been dismantled, giving the feature a ramp-like appearance in its north-south section. Some of the damage to the north end may be attributed to the erosion of the unprotected, earthen core (see fig. 4.48). A large rain gully (up to 5 m. deep), running through the centre of the earthwork, has formed at its southern end (see fig. 4.85).

The earthwork is provided with revetting on both faces. On the upstream face (i.e., the west) it is protected by oblique step-revetting laid upon a foundation of boulders (see fig. 4.86). The top three courses of the revetment are vertical. Their vestigial remains stand isolated, like a wall. At its highest point the revetting is approximately 15 m. above the valley floor. On the downstream face, the remains of a vertical revetting, 6 to 10 m. high, are visible (see fig. 4.87).

The width of the bund is greatest where it abuts the Hiriya kāluve. A revetted platform, which supports a dilapidated Śiva temple, protrudes from the downstream face at this point. A maṇḍapa, which protects rock-cut carvings of Śaiva subjects, is located just to the northeast of the temple in the adjacent field. An inscription is located on a boulder protected by the maṇḍapa. It is undated, but contains a reference in its fourth line to a tank called 'Bhūpatikere' and a donation to the God Mallikārjuna by Kola Rāmayya (Nagaraja Rao 1985:34). The tank referred to in the inscription, in our view, is that which was formed by the bund across the valley. The Bhūpati tank is also mentioned in an inscription dated A.D. 1534 in the north gopuram of the Tiruveṅgaḷanātha temple (ARSIE 1904:16; Devakunjari 1983:60; Gopal 1985:161-162; Rajaśekhara 1985:112, 119). The epigraph also records the building of the Tiruveṅgaḷanātha temple and Achyutāpura by mahāmaṇḍaleśvara Hiriya Tirumalarāja Voḍeya. The tank is given as the southern boundary of Achyutāpura.

Inscriptional evidence suggests that the Hiriya canal was built before A.D. 1524 (see section 4.3). Since the earliest references to the tank are contemporaneous with the construction of Achyutāpura, we suggest that the canal was built before the tank. To us, it seems likely that the tank was built as a storage receptacle for the water supply to Achyutāpura. Thus the reservoir was formed on land previously irrigated by the canal, water from the canal being used to supplement the supply of surface run-off to the tank. In our view, the northern end of the bund was broken down after the abandonment of Achyutāpura in A.D. 1565, to facilitate the cultivation of the valuable agricultural land occupied by the tank. The quantity of the core missing from the bund suggests that earth from it may have been used to reconstruct the left embankment of the Hiriya kāluve in the 1920s–1930s and 1940s–1950s.

4.5.3 The bridges below Matanga Parvatam

Four bridges are located beneath Matanga Parvatam which give access across the valley canals. The most southerly and largest of the bridges spans the Hiriya canal, linking the road with the bund to its north (see figs. 4.88, 4.89 and 4.90). The bridge has three spans supported by two piers. No lime mortar is used in its relatively simple construction, which utilises undressed, rectangular blocks. Its design may be compared to other Medieval constructions in the Karnataka region (Deloche 1984:26,88). It is noteworthy that the canal, which is quite shallow (50 to 80 cm. deep), drops noticeably on the eastern side of the bridge as it flows over a low, sloping boulder.

A Śiva temple (10m. by 5 m.) is located immediately in front of the bridge on a revetted platform which protrudes from the downstream face of the bund (see figs. 4.85 and 4.91). The temple has partly collapsed into a rainwater gully. Between the path which runs along the left bank of the canal and the temple, a paved ramp descends to the valley floor (see fig. 4.92). At the bottom of the ramp, an upright slab (1.54 m. high and 76 cm. wide) bearing a short in-scription, referring to Achyutāpura, leans against the east face of the revetted platform supporting the temple (see fig. 4.93). The epigraph marks the outer boundary of Achyutāpura, constructed in the first half of the sixteenth century A.D.

The second bridge below Matanga Parvatam is situated north of the Hiriya canal bridge, between the broken down northern end of the bund and an outcrop to which the bund must have been originally connected. The bridge spans the Kiriyagava channel, which takes off from the Hiriya canal (see section 4.1.1) (see figs. 4.94 and 4.95). The unlined canal is 20 to 30 cm. deep and varies in width from 1 to 4 m. The area around the bridge contains a quantity of disarticulated masonry, presumably originating from the bund, and is thickly overgrown. The passage of the Kiriyagava kāluve through this part of the valley was facilitated by the removal of a large section of the revetted bund. the bridge is modern.

The north end of the Kiriyagava bridge abuts a flat-topped outcrop which protrudes from the north side of the valley. The top of the outcrop is approximately level with the crest of the bund. The route northwards across the valley at this point divides in two, one track running northwards over the outcrop, whilst the other follows a course around its west side. The tracks come together on the saddle of the outcrop just to the south of the third bridge.

The third bridge, spanning the southern Chika channel, is constructed across the low saddle of the outcrop (discussed above). The channel flows over the outcrop at a high command. The complex design of the bridge will be discussed separately in section 4.5.4.

The fourth bridge consists of a single slab of granite laid over the northern Chika channel, which runs eastwards along the north side of the valley towards the Tiruveṅgaḷanātha temple. A small shrine of trabeated construction is located in front of the bridge. It contains a Śaivite rock carving depicting a linga, Vīrabhādra, Ganeśa and two female devotees.

4.5.4 The third bridge below Matanga Parvatam: The old spillway or waste weir

The third bridge below Matanga Parvatam is a far larger and more complex device than would be required to merely provide a passage across the northern Chika kāluve. At present, the channel, which is 1 m. wide and 40 to 65 cm. deep, flows under the most northerly of three spans supported on two piers (see figs. 4.96 and 4.97). We suggest that it was originally designed as a waste weir for the Bhūpati tank. The construction is made up of rectangular slabs.

The waste weir has undergone some modification. On the downstream face (i.e., the east), wing walls flank a sloping arrangement of steps formed by masonry slabs, laid in regular, though angled courses on an inclined foundation. A clear passage for the southern Chika canal has been opened through the left side of the device by the removal of some slabs (see figs. 4.98 and 4.99). This passage allows the canal to flow through, rather than over, the spillway.

The modification made to the spillway or waste weir suggests that it ceased to function in its original capacity after the demolition of the north end of the Bhūpati bund. The spillway, which would have acted as a safety valve for the tank, would have allowed excess water to escape harmlessly. When the water level in the tank reached the height of the spillway's sill, any excess water would flow over it.

After passing through the modified waste weir, the southern Chika channel is diverted, by slabs, over a large, flat-topped boulder. It then flows eastwards in embankment against the north side of the valley. It is supported by a dilapidated dry-stone wall. The northern Chika channel, which runs parallel to its southern branch at a higher command, flows northeast towards the Tiruveṅgaḷanātha temple. It is also held in embankment, its right bank forming the major pedestrian route from the base of Matanga Parvatam to the Tiruveṅgaḷanātha temple.

4.5.5 The old sluice on the Hiriya kāluve

An important outlet is located on the left bank of the Hiriya kāluve, approximately 30 m. downstream of the old bridge. A rectilinear slot in the left bank of the canal allows water to pass out of the canal (see fig. 4.100). Water is taken off at a point of considerable turbulence. It is carried from the channel into a complex and partially ruined structure partly built into the canal embankment. The outlet has a command of some 4 m. over the fields below.

The structure consists of three major elements and is flanked on its western side by a large, natural boulder (see figs. 4.101, 4.102 and 4.103). It shows signs of considerable alteration and reuse. The device is heavily overgrown. It was cleared of vegetation at the end of the 1987/8 season, revealing an extraordinary complex of features (see figs. 4.104 and 4.105). The south side of the structure abuts and partly cuts into the embankment of the Hiriya kāluve. An examination of the feature reveals severe structural dislocation. Today, water issues into the west and central bay of a three-bayed structure, at a point some 1.3 m. above the level of the valley floor. No trace of a closure or gate survives and water flows out of the structure without regulation. In its present condition, the sluice serves no other function than to break the force of the water leaving the canal at a command of some 4 m. above the valley floor. An investigation of the conduit by which water is conveyed into the masonry structure was prevented by the accumulation of fallen debris.

Fritz (1988: personal communication) has indicated:

> This is a complex of several massive features. The most important is a large, three-bayed building with the remains of an east-west channel on its roof (a). Immediately to the west are the remains of a ramp-like structure (b). This sits on a base of now tumbled blocks. To the north, and stretching between the northwest corner of (a) to a large boulder to the west is a fortification wall (c). Another wall or terrace to the southeast stretches from the north side of the boulder towards the probable course of the ramp. Finally stretching east from the northeast corner of (a) are traces of another fortification (?) wall (d).

The structure is built on top of, and partly cut into, an exposure of friable bedrock which slopes down to the north. The three-bayed structure (a) is composed of four heavily built, north-south piers. It shows signs of having originally extended further to the east. The eastern side of (a) has been extensively robbed for stone. Its south face is finished but partially hidden by the canal embankment, which suggests that the canal was constructed after the three-bayed structure.

At present, the three-bayed structure is 16 m. by 14 m. Its piers are constructed of heavy blocks laid in regular courses without lime-mortar. The partially collapsed piers are filled with a core of earth and rubble. The bays between the piers (approximately 2.25 m. by 13 m.) contain soil and fallen masonry. The bedrock, into which the piers are cut, rises between them, partially blocking the three bays formed by the structure. The floors of the bays have been gullied by the action of the water passing through the structure from the canal. Heavy cross-beams were laid on top of the piers. Long ceiling-beams were placed above and form a solid platform (see fig. 4.106). The gaps between the ceiling-beams are filled with lime mortar. There is also some evidence to suggest that parts of the upper surface of the platform were covered by a skin of chunam.

Two interesting features are located on top of the platform of (a). Firstly, a central channel formed by parallel, plastered blocks runs west-east across the platform. The disarticulated channel is approximately 4.27 m. wide and its bed is lined with thin slabs. The sides were at least two slabs high. Secondly, a parapet of long slabs, approximately three courses high, runs along the north side of the platform. There are frag-mentary indications, in the form of four wedge-shaped blocks, that an additional parapet may have been located on the west side of the platform.

Abutted to the west side of the three-bayed

structure (a) are the remains of a ramp formed of two component parts (b). The first component consists of about seventeen slabs side by side in two courses without chunam. The slabs lie north-northwest–south-southeast and rest on others parallel to the long axis of the structure. The central channel appears to have been led on to (a) by the ramp. The second component is a shallow and disarticulated stack of blocks which joins the canal embankment to the south side of the first component of the ramp. A fortification wall of wedge-shaped blocks (c) joins the northwest corner of (a) to the northeast corner of the natural boulder to the west. The wall, which has a stepped section, curves slightly to the south and is approximately 15 m. in length. The wall (c) is abutted to the northwest corner of (a) and may have originally extended along the west side of (a). The fragmentary remains of a second fortification wall (d) are indicated by the presence of five wedge-shaped blocks to the north of (a). The 3 m. wall appears to have joined the northeast corner of (a) to a low exposure of bedrock.

The present condition of the complex of structures (a) to (d) suggests that an open channel ran west-east over it, from a point somewhere near the canal (see fig. 4.107). Today, water flows through the notch outlet and into the bays of (a). In our view, it originally flowed into the open channel on top of (a) via the ramp (b). The three-bayed structure (a) was, therefore, used at one stage as an aqueduct, the specially-constructed ramp (b) allowing the channel to pass onto (a). It is unclear for what purpose the high command of the open channel was intended. The severe structural disarticulation of the east end of (a) does not allow us to draw any firm conclusion as to the subsequent route of the channel. At some point, ramp (b) collapsed or was removed and the canal outlet was connected instead to a pipe which directed water into the bays of (a). This hypothesis does not explain the complexity or size of (a), which predates the building of the Hiriya canal and was, therefore, reused as an aqueduct.

We have suggested that (a) predates the building of the Hiriya canal. The three-bayed structure (a) must then also be postdated by the Bhupati bund, which was constructed after the Hiriya canal and is associated, in our view, with the supply of water to Achyutāpura. We suggest, therefore, that the three-bayed structure (a) and defensive walls (c) and (d) formed part of an earlier structure, the purpose of which is, at present, unclear.

In conclusion, we note that the constructional style of (a) is reminiscent of the Setu aqueduct (see section 4.6). The Setu aqueduct is much narrower and taller than the device under examination here, but both are massively built and support a surface of slabs raised up on heavy piers. The piers of the Setu aqueduct were constructed heavily in order to resist the force of the Tungabhadrā in flood. The upper surface of the aqueduct was provided with a layer of chunam to render it waterproof. The massive design of the three-bayed structure and presence of a chunam covering on its upper surface suggest that it may have formed part of a water supply system in the irrigated valley which predated both the Hiriya canal and Bhūpati tank.

4.5.6 An inscription on the right bank of the Hiriya kāluve, close to the Matanga Parvatam bridges

Approximately 1 km. to the east of the Matanga bridges, an inscription is incised onto one side of a crevice in a large rock through which the Hiriya canal flows (see fig. 4.108). The inscription is undated, but the style of the characters corresponds to that of the fifteenth to sixteenth centuries A.D. (see fig. 4.109). The language used is Kannada. The epigraph appears to record the breaking of a large rock during the construction of the Hiriya kāluve.

It reads:

1st line. *Kārakunayya embuvanu koḍuva rabuyeniladu*

2nd line. *ee nandi banda voḍabekide*

Translation: 'The person, Kārakunayya, donated this. This Nandi rock had to break.'

A small carving of a seated Nandi bull is located beneath the left side of the inscription. The inscription refers to the removal of a segment of the outcrop by an individual called

Karakunayya. We suggest that he was an engineer engaged in the building of the canal.

4.6 The Anegundi anicut and canal

The Anegundi anicut is located near the village of Sānāpur, 1.25 km. to the north of the Hiriya anicuts. The Anegundi weir supplies a long and sinuous canal which flows north of Anegundi (see fig. 4.51) to a point 4.5 km. south of Gangāwati. The canal is approximately 19 km. in length and represents a spectacular feat of technology in maintaining a command over such a great distance. Instead of discharging into the river, the Anegundi kāluve flows into another canal which takes off from the river at Singanagundu.

The anicut, which traps the flow of water between the left bank of the river and the north side of an island, is located at the mouth of a steep-sided gorge (see fig. 4.110). The main weir blocks the passage of water into the gorge, creating a large backwater pool upstream. A smaller, subsidiary anicut is located to the southwest. The anicut, which is constructed entirely of rectangular, dressed slabs packed together to from a solid mass, takes the form of a gentle curve in plan (approximately 75 m. across). There is no evidence for the use of iron clamps to bond the structure together, the courses of stone being held with stone wedges and pegs. A similar construction method is reported by Francis for the Rāmanagadda weir (1904:91). Water is taken off into the canal at a high command on the north bank.

This anicut is the least modified of those examined. Only a concrete crest has been added to the original structure (see fig. 4.111). The weir is composed of slabs, with their flat sides in a position perpendicular to the river flow, sandwiched between further courses facing the opposite direction, in front and behind. Gaps are packed with small, stone wedges. The weir crest is 1.04 m. wide. The downstream side is provided with a 9 m. apron (see fig. 4.112) which is horizontal and ends in a vertical face (see fig. 4.113). It is composed of courses of slabs laid in alternate directions (see fig. 4.114). There are indications that the weir was originally grouted with chunam.

The water held upstream of the anicut flows into the canal immediately to the north of the weir. Water entering the canal is not regulated in any way; however, excess water is allowed to pass over the right bank of the canal and flow back into the river just downstream of the anicut. The canal is held in high embankment against the range of outcrops which flank the river channel below (see fig. 4.115). The weight of the canal is supported by courses of large, granite slabs.

The Anegundi canal follows a contour of the granite range which runs more or less parallel to the river (see fig. 4.51). It is provided with a variety of different linings, indicative of several periods of rebuilding. The old lining of rough blocks without lime-mortar has been extensively replaced by vertical revetments of small dressed blocks bonded with cement. In some areas, the earthen banks of the canal are completely unprotected. Sections of lining are plastered to protect them, particularly where there is turbulence. The canal is held against the south face of the range by a high embankment of earth. It is predominantly not revetted. The canal outlets are modern.

The canal today provides water for the agricultural land north of the river between Sānāpur and Singanagundu. Originally, it also supplied the island of Virūpāpuragadda. The fragmentary remains of a large aqueduct stand in the shallow, north channel of the river, at a place known locally as Setu (see figs. 4.51 and 4.116). The aqueduct permitted water to pass out of the canal and flow across the north channel of the river to the island without a loss of command. The structure has been disconnected from the Anegundi canal by the dismantling of its northern end. The south end of the device has also been dismantled. The severance of the aqueduct from the canal has increased the amount of water in it passing beyond Setu and suggests a desire to increase the supply to agricultural lands north of the river. Today, the island is supplied with water by an electric pump connected to the river on the south side.

The surviving section of the aqueduct is approximately 100 m. in length, with fifteen complete piers, rising to a height of between 7 and 12 m. (Deloche 1984:26). In its original form, the structure must have been some 200 m. in length. The piers are built of large, dressed blocks laid in regular courses without chunam. Long slabs, supported by corbels, are laid between the piers. The superstructure was surmounted by a channel some 2 m. in width. The sides of the channel were constructed of small dressed blocks, the inner surfaces of which were originally plastered with chunam. The aqueduct is built directly onto the rock of the channel bed.

The great structural strength of the device cannot at present be explained by the character of the river which is at this point almost dry outside the monsoon season. However, the construction of the Tungabhadrā Dam has altered the river flow. We suggest that prior to its building, the river channel at Setu may have been considerably deeper and swifter flowing.

The island of Virūpāpuragadda is dominated by a granite range, on which stands a ruined Vaiṣṇava temple. Channels appear to have transported water from the aqueduct in the northwest corner of the island to the low-lying areas extensively cultivated on its flat, southern side. Some are rock-cut. Other evidence for water supply on the island includes a number of stone-lined wells, which give access to sweet subsoil water. In addition, the remains of a low bund of earth, lined on its upstream face with boulders, are located in the central part of the island. The tank is now dry. We suggest that it was intended to trap the surface run-off from the central range. It may also have been used to store water from the Setu aqueduct.

4.7 The Rāya bund, southwest of Hospet

Southwest of Hospet, a huge bund is constructed across the valley of Sandur (see fig. 4.33). It was intended to trap the run-off from the valley. A smaller, breached bund is located at Rājāpuram, 3.5 km. further up the valley, to the southeast. We suggest that the smaller bund was broken down in order to allow the surface run-off to reach the larger bund constructed after it. The larger bund is known locally as the Rāya kere. Its building during the reign of Kṛṣṇadeva Rāya was witnessed by Paes (Sewell 1900:244-245). This is corroborated by Nuniz (Sewell 1900:364-365). Francis (1904:90-91) records a local tradition that an officer of King Kṛṣṇadeva Rāya, called Mudda, is buried in the bund.

The bund of the Rāya kere is just over a kilometre in length and joins one side of the valley to the other. It is constructed of earth and rock fragments packed together to form an earthwork of massive bulk (approximately 15 m. in height and 30 m. in width). Theoretically, the site of the bund is a perfect spot to build a catchment drainage tank as large drainage flows down the centre of the valley to the river. Its builders evidently thought so, as a great deal of labour was expended in constructing the device. However, despite the fact that the bund was completed and heavily revetted on its upstream face, the tank is completely dry. Instead of collecting upstream of the bund, the run-off percolates under it, rendering the device totally useless. According to the local inhabitants, during the monsoon the tank fills to a depth of about 1 m. but the water drains away rapidly.

We have observed that the rock on either side of the valley, at the point where the bund is built, is stratified and of a much finer grained composition than the homogeneous granite which typifies the locality. In addition, the soil at the bottom of the valley is loamy and porous. Both these factors suggest that the site of the bund is not a good one, a conclusion missed by the builders of the bund. Today, the bed of the tank is used for dry cultivation. The bund carries the road from Hospet to the southwest. A considerable quantity of fertile silt has accumulated upstream of the bund. The outlets are partially buried in this deposit. The bed of the tank is completely flat. We suggest that this is the result of deliberate landscaping and intensive ploughing.

The upstream face is step-revetted with granite blocks to a point approximately halfway up the bund (see fig. 4.117). The remaining portion of the upstream face is steeper in section and composed of packed dry-stone. The dressed masonry of the step-revetting has been partially cannabalised for building material, exposing the earthen core.

The remains of two sluice outlets, one complete and one fragmentary, survive in front of the upstream face. No trace of the outlets on the downstream face is visible. The first outlet is situated 12 m. from the foot of the bund at its eastern end (see fig. 4.118). The outlet, which is 6 m. high, takes the form of a masonry frame. Two piers (88 cm. by 86 cm. and 80 cm. by 91 cm. in section respectively) are set 2 m. apart and support a wide cornice (see fig. 4.119). The weight of the cornice, which is in three pieces, is braced by two beams lying in a position perpendicular to it, one on each side, between the top of the piers and the cornice. A projection carved in imitation of an unopened, banana flowerbud faces inwards on each beam. This decorative device is commonly found on pillar capitals of the later Vijayanagara period. The cornice is surmounted by the remains of a ruined, brickwork superstructure. The piers are joined at 1.05 m. above ground level by a granite crossbar, 30 cm. thick. A hole, 10 cm. in diameter, is let through the centre of the cornice and crossbar for the fitting of a control rod.

The masonry frame of the first outlet is extensively carved in shallow relief. The carving falls into two categories, decorative designs and religious icons. The latter are restructured to the vertical faces of the piers, whilst the former are concentrated on the cornice and supporting beams. The icons are:

1. A standing Viṣṇu in a makara-maṇḍala (northwest face of the western pier).
2. A seated Ganeśa in a makara-maṇḍala (northeast face of the western pier).
3. A young, crawling Kṛṣṇa (northeast face of the eastern pier).
4. A standing Nandi (southeast face of the eastern pier).
5. A standing Hanuman, facing right (southwest face of the eastern pier).

The decorative designs are predominantly floral roundels, rosettes and medallions.

The remains of a second outlet are located at the west end of the bund (see fig. 4.120). A single, monolithic pier stands 17 m. from the toe of the bund. The pier is 90 cm. by 89 cm. in section and approximately 10 m. in height. No trace of the second pier or cornice is visible. This suggests that either the sluice frame was not completed or sections of it were removed for reuse after its abandonment. The absence of a notch for the crossbar of the sluice frame on the pier suggests that the accumulated silt in the bed of the tank has risen to a considerable height, burying it from view.

Close to the remains of the second outlet, a rectangular, overgrown structure extends out from the upstream face of the bund by 17 m. The structure appears to be a three-sided box abutted to the bund. It is earth-filled, having an outer skin of dressed masonry and rises to a height of approximately 7 m. The purpose of this device, which is not structural, is unclear. In our view, it may have formed part of the second outlet, perhaps forming a walkaway between the masonry frame and the bund.

The contemporary accounts of Paes and Nuniz provide valuable evidence relating to the original arrangement of the tank's outlets. Paes states:

> The tank has three large pillars handsomely carved with figures; these connect above with certain pipes by which they get water.... (Sewell 1900:244).

This passage accurately describes the decoration of the masonry frame which is located on the upstream side of the bund. However, there is some confusion in the text as to the general design of the frames. Three pillars are mentioned in such a way as to give the impression that each facilitated the passage of water out of the tank. The evidence on the ground indicates that the sluice frame of each outlet was composed of two upright piers supporting a stone cornice and brick superstructure above. We suggest, therefore, that Paes used the term 'pillar' to describe the whole frame, which, from a distance, probably would have had the appearance of a single upright structure. If this hypothesis is correct. Paes is describing three outlets. Nuniz also provides a description of the tank and its outlets. He relates:

> He (the king) made a bank across the middle of the valley so lofty and wide that it was a crossbow-shot in breadth and length, and had large openings; and below it he put pipes by which the water escaped, and when they wish to do so, they close these (Sewell 1900:365).

The Portuguese word translated as 'openings' is *espacos*. Sewell suggests that another meaning of the word is sluice or weir (1900:365). The accounts of both Paes and Nuniz mention the existence of pipes. We suggest that these were the conduits that were originally let through the bund and were controlled by the gates attached to the masonry frames.

An interesting feature of the Rāya kere is the considerable depth of the outlets below the crest of the bund. For the outlets to have functioned at all, the water in the tank could not have reached a level above the top of the masonry frames, otherwise the apparatus for the raising and lowering of the gate would have been drowned. We have suggested that the bed of the tank has been raised by the washing down of silt from the valley. This suggests that the depth of the tank was originally greater. However, a full supply level in the tank, approximating the top of the outlets, cannot have necessitated a bund of such massive proportions. There are two possibilities. Either the outlets were built after it was realised that the bund would not hold a supply level anywhere approaching the full height of the earthwork and were an attempt to capitalise on a lower projected full supply level in the tank, or the outlets were part of the original design of the tank as initially envisaged and the bund was made larger in an attempt to prevent water draining through it. We feel that the second possibility is the most plausible on the basis of the contemporary accounts of the tank's construction (Sewell 1900:244-245, 364-365). To us, it seems likely that, to prevent water escaping from the tank, the bund was made higher and thicker. This may have resulted from the belief that water was percolating through, rather than under the bund. The appearance of the bund today suggests that its builders still held some hope of the structure performing a useful function in spite of what would appear to have been clear evidence to the contrary.

One possible explanation for the stubborn continuance of the project might be that, during the construction of the Rāya bund, water was held back by the functional bund at Rājāpuram, higher up the valley of Sandur. The Rāya bund could thus be constructed over a period of several years without interference from the drainage running down the valley Every monsoon. This first phase of building would be represented by the lower half of the upstream face of the bund, protected against the action of the water by step-revetting, and the two outlets. When the new bund was completed, the Rājāpuram bund was breached in the hope that the waters it had contained would be added to the run-off from the valley between the two dams. Unfortunately, all the water passing down the valley percolated, as it does today, under the Rāya bund rather than being held upstream of it. The builders would have only realised their failure after the breaking down of the Rājāpuram bund, which allowed water to flow down the valley unhindered. Sensing failure, but without understanding the reasons for it, the builders would have then increased the dimensions of the new bund in the hope of salvaging the work. This second phase of construction is represented by the upper half of the upstream face of the bund, which is crudely revetted with drystone. The second phase of building was therefore not intended to be in contact with the water held in the tank, but merely to add bulk to the existing bund.

Another reason for the continuation of building in the valley of Sandur could be that the project, and perhaps the site, was the particular brainchild of Kṛṣṇadeva Rāya and that he may have ignored the advice of his engineers. It is possible that the King's enthusiasm about this particular project may have arisen from the selection of the site by astrologers or brahmans. We feel that it is unlikely that anyone would have dared to interfere with the King's plans.

Nuniz's account, which was compiled more than ten years after the construction of the Rāya bund, surprisingly mentions that Kṛṣṇadeva Rāya sent to Goa for Joao della Ponte, a 'great worker in stone' (Sewell 1900:364-365). The man is described as having acted as a design consultant for the building project, recommending the extensive use of chunam in the construction. Nuniz states:

> ... the king told how he wanted the tank built. Though it seemed to this man impossible to be made, nevertheless he told the king he would do it and asked to have lime prepared,

> at which the king laughed much, for in his country when they build a house they do not understand how to use lime (Sewell 1900:365).

The account does not mention whether the consultant's advice was considered. However, the evidence on the ground indicates that chunam was not used. This suggests that, either the consultant's suggestions were ignored by the King, or the reference to Joao della Ponte is erroneous. There is some evidence to support the latter. Firstly, the account was written after the completion of the building project Secondly, the reference to Joao della Ponte is not corroborated by Paes who witnessed the con-struction and would most likely have known if a fellow European was working on the building project he described. Thirdly, the suggestion made by Joao della Ponte to use chunam is peculiar, as its use was well known to the Vijayanagara builders. It was used extensively for elaborate buildings and water supply features in the city. These inconsistencies suggest that the reference to Joao della Ponte is probably spurious.

The purpose of the tank was, according to Sewell (1900:162-163), two-fold: the supply of water to Nāgalāpura, the suburban extension to the city built by Kṛṣṇadeva Rāya, and the agricultural land to the northeast of the tank, between Hospet and the 'Urban Core'. The contemporary accounts on which Sewell based his assertion require discussion. According to Paes, the inhabitants used the water:

> ... to irrigate their gardens and rice-fields (Sewell 1900:244).

This must refer to the agricultural land between Hospet and the 'Urban Core'. Nuniz indicates that some of the water from the tank was also intended to be used for domestic supply. He states:

> By means of this water they made many improvements in the city, and many channels by which they irrigated rice-fields and gardens ... (Sewell 1900:365).

There is no specific reference in either account to the tank satisfying the water demands of Nāgalāpura. We suggest that the improvements made to the 'city' mentioned by Nuniz, refer to the 'Urban Core'.

The supply of water from the tank to the 'Urban Core' would have been perfectly feasible had the tank held water. It would not have necessitated the construction of a new canal to connect the two elements of the system. The Rāya canal, which we have suggested was built at the beginning of the fifteenth century A.D., transported water along the most practical route from Mālāpuram to the Kamalāpuram tank at the southwest corner of the 'Urban Core'. The canal also supplied the 'Royal Centre' indirectly, via the Kamalāpuram tank (see section 4.1.2). Water passing out of the Rāya tank must have flowed roughly along the old course of Sandur Valley drainage down to Amarāvati and Chittavādigi. From this point, the most logical route eastwards is the one occupied by the Rāya canal. In our view, the Rāya bund may have been built in the hope that its supply would supplement the discharge of the Rāya canal. The failure of the Rāya bund to supplement the Rāya canal's supply may explain the construction of the Rāmanagadda anicut, which boosted the supply passing on to the Kūruvagadda weir.

It is interesting to note that the account of Paes and Nuniz describe the Rāya tank as functional. The tank does not hold water and has not been subject to significant alterations, which may have impaired its water retaining abilities. In basic terms, the site is simply not a good one. We suggest, therefore, that the contemporary accounts are mistaken on this point.

4.8 The Mallappannagudi bund

To the southeast of Mallappannagudi village, a revetted bund runs southeast–northwest for some 800 m. (see fig. 4.35). The earthwork is 5 m. high and irregular, having a sinuous design in plan. The bund originally trapped surface run-off from the Sandur hills as it flowed past Kariganurū and Sankalapura. It is now dry, but the 1929 and 1947 Survey of India maps (57 A/7) indicate that it held water before the construction of the Tungabhadrā Right Bank High Level Canal. We suggest that the canal prevents the run-off from reaching the tank and has thus rendered it useless. A modern track runs north–south from Mallappannagudi to Kariganūru across the dry bed of the tank.

The tank has been examined by Morrison as part of a regional survey of the Vijayanagara metropolitan region (1988:8-9). Its design correlates well with the constructional details of the Kamalāpuram tank (see sections 4.1, 4.1.1 and 4.1.2). The upstream (i.e., southwest) face of the bund is revetted with large, dressed blocks, the lower courses resting on a heavy foundation of boulders. The downstream (i.e. northeast) face of the bund is without revetting. Morrison states:

> On the west side there are two granite sluice gates, one with carved donor portraits. Opposite this sluice gate is a staircase of long dressed slabs set into the tank embankment. A plastered brick and cement tunnel near the south end of the embankment also served to channel water to the east. Downstream are various channels designed to distribute water from the tank to the field below it. These include a long enclosed channel of plaster and brick with a square open tank at its outlet, and a square brick basin circa 5 by 5 metres, of unknown depth. A clay pipe, now blocked, once connected this feature to the tank (1988:8-9).

This tank is of some importance as it allows an inspection of the pipe outlets used during the Medieval period. As in the case of the Kamalāpuram tank, water was let through the bund into square control tanks, before being distributed to the fields. A pipeline and/or plastered-brick conduit ran through the bund at ground level, allowing water to pass into an open tank. The amount of water leaving the tank was regulated from a masonry frame on the upstream side of the bund.

A Kannada inscription, dated A.D. 1412, has been located on a slab in front of a temple in Mallappannagudi (ARSIE 1904:25; Gopal 1985:186-187). It records the construction of a watershed or aṛavatige at the junction of the roads leading to Bisilahaḷḷi kaṇuve and Baḍavaliya kaṇuve. Specific provisions were made for its maintenance. Heggaḍe Sovaṇṇa-aṇṇa, a devotee of the God Mailāraśankara, was responsible for the work. We suggest that the watershed referred to in the inscription is the Mallappannagudi tank. In our view, it was built in A.D. 1412 to supply the agricultural zone around the villages of Mallappannagudi, Hosa Mallappannagudi and Gālammagudi.

CHAPTER 5

Interpretation and Analysis

5.1 Analytical methods employed

The social, political and physical environment in which the water supply system of Vijayanagara developed has been examined in the first two chapters of this investigation. The physical and epigraphical evidence for water supply on the ground has been documented in chapters 3 and 4. In this, the final chapter, the evidence for water management at the site is examined as a whole and put into the wider perspective of the history and development of Vijayanagara as an urban settlement and capital.

A multidisciplinary approach has been utilised, integrating the field observations with epigraphical, literary and archival data. This methodological approach has been used because it provides the most effective way of tackling a system of domestic and agricultural supply as complex as that which survives at the site. To find out how the system worked and was modified and to ascertain what changes have taken place since A.D. 1565, it has proved necessary to examine a range of data from a variety of sources. The data base constructed from the documentation of the remains on the ground has, where possible, been tested against alternative data. Those sections of the contemporary account which refer to the management of water, some of which have been subjected to critical analysis in the field observation sections, are examined for evidence relating to the development and chronology of the system.

Some methodological constraint has been placed on the author in this investigation by the nature of the data collection method employed in the field. Although buried features could not be identified by surface survey in the examination of the 'Royal Centre', the range of material on the ground may be taken as sufficiently representative to allow a reconstruction of the domestic supply system in this part of the site. Most of the major water features, the skeleton of the domestic supply, have been preserved in at least fragmentary form. Excavations in the 'Royal Centre' will undoubtedly reveal further traces of the domestic supply system. It is unlikely, however, that any new remains will radically differ in type from those documented in this investigation. The collection of data pertaining to agricultural water supply has been complicated by, firstly, the continued use of many features after the abandonment of the city up to the present day and secondly, the building of the Tungabhadrā Dam at Mālāpuram. The most severe restriction has been the impossibility of investigation of the four anicuts of Mālāpuram except from pre-Tungabhadrā Dam literature, archives and maps.

Comparative evidence from other sites will be examined in order to establish whether water management at Vijayanagara represented an extension of existing ideas, or a new approach to the problem of irrigation and water supply in the context of the urban settlement. The establishment of a water system may be considered a pre-requisite for urban settlement in the semi-arid environment in which Vijayanagara was built. The importance of the water supply systems of Vijayanagara as an urban settlement is also examined.

The physical and epigraphical evidence for pre-Vijayanagara water supply in Bellary and Raichur districts is discussed in section 5.2. This section examines the antecedents of water supply in the immediate areas of Vijayanagara. By investigating the development of water management in Bellary and Raichur districts, it is possible to determine that much of the technology of water supply used in the Vijayanagara period had its origins in a local, indigenous tradition.

The division between domestic and agricultural water supply employed in the discussion of the field observations, has been maintained for the purpose of analysis and interpretation. Domestic and agricultural supply are discussed in sections 5.3 and 5.4 respectively. Each section begins with a brief review of the literature, which demonstrates how ideas concerning water supply at the site developed prior to our investigation. The literature review is followed by a detailed analysis of the evidence on the ground, in conjunction with epigraphical, literary and archival material. The analyses have, for reasons of clarity, been broken up according to the four specific research questions raised in section 3.1.

5.2 Evidence for pre-Vijayanagara water management in Bellary and Raichur districts

In his survey of Raichur district, Allchin (1954: 510-511) suggests that irrigation tanks of stone, directly associated with towns and villages, developed during the early historic period. He also notes the existence in this period of small catchment bunds of earth, built at the foot of hills and designed to secure a supply of water for agricultural purposes. These developments appear to have grown out of a tradition of terracing and field-levelling in the Neolithic and post-Neolithic periods. He pointed to a number of small bunds associated with megalithic graves (1954:130), indicating that surface drainage tanks, some with stone revetting, were already constructed during the first millennium B.C. Further evidence for water supply, in the form of terracotta ring wells and a single stone well, is reported by Allchin (1954:85,508) in an early historic context a Māski.

During the medieval period, there is physical and epigraphical evidence for a number of important development in irrigational technology. Large numbers of catchment drainage tanks were constructed to provide a reliable source of agricultural and domestic supply for an increasing population. Epigraphical evidence indicates that perennial river-fed canals were first built during the eleventh century A.D. These devices facilitated the extension and intensification of profitable wet cultivation. The evidence for the use of perennial canals in the pre-Vijayanagara period is restricted to the low lands on the north bank of the Tungabhadrā. During the Vijayanagara period, the construction of anicuts allowed canals to be built on the uneven lands of the right bank, where there was much greater potential for agricultural expansion. The building of canals in the Vijayanagara period must be linked with a greater investment of wealth in agricultural improvement projects (Kuppuswamy 1981:78).

The small earthen catchment bunds of the early historic period appear to have been superseded by composite bunds of earth and stone, trapping natural drainages. Tanks of this sort were built from at least the middle of the third century A.D. the construction of a tank at Chandravalli can be dated to A.D. 258 on the basis of a short inscription of the Kadamaba King, Mayūraśarman (ARASM 1929:50-58; Allchin 1954:95). Other tanks of this sort are reported by Allchin at Rayadrug (1954:101), Koṇḍapur (1954:104), Bilebhāvi (1954:129-130), Śivapur (1954:136), Rajulabanda (1954:161), Māski (1954:167), Lingsugūr (1954:183), Koppal (1954:189), Mānvi (1954:196197), Piklihāl (1954:239), Rāmaswami kā pahār (1954:532) and in the Benkal forest (1954:66, 121,124). Similar evidence comes from Devnimori in Gujarat, where a number of bunds were built in the fifth and sixth centuries A.D. (Mehta 1963:359-365).

Under the later Cālukyan kings, the agricultural hinterlands of some settlements were provided with perennial supply features. At Rodalkoṇḍa, Allchin (1954:201 noted the existence of a rain gully of dressed stone and a small, stone-lined tank. A slab inscription set up by the side of the tank and dated A.D. 1053 refers to its building by Trilokyamāla Cālukya. The huge tank at Danāyakanakere (see section 1.4) appears to have its origins in the later Cālukya period and may have been the site of Jayasiṁha II's capital, called Pottalakere (Francis 1904:29). An epigraph of the tenth century A.D. at Anegundi refers to the granting of land near Hampi by the Cālukya king, Āhavanlladeva

(Shama Sastry 1917:285-291). Inscriptional evidence indicates that another Cālukya capital, established by King Viṣṇuvardhana Vijayāditya, was located at Kampli (see section 1.4).

Perennial, river-fed canals were constructed in low-lying areas on the north bank of the Tungabhadrā in the eleventh century A.D. The Munirabad inscription of Vikramaditya Cālukya VI records the building of the Premogal canal and subsidiary channels taking off at different levels (HAS 1922:10). Water is allowed to pass into the canal through a notch cut through the left bank of the river, which operates without the need for an anicut. This design feature only permits the canal to irrigate the low ground and must be considered to predate the use of the anicut to create affluxes by blocking the river flow. The canal is without headworks. The remains of an early anicut, possibly dating to the later Cālukya period, are reported by Allchin (1954:178) on the Māski nullah, near the Mudgal–Lingsugūr road.

There is some evidence to suggest that perennial canals had been built in other parts of India from the early historic period onwards. The earliest evidence for perennial irrigation is provided by the Khāravela inscription at Hāthigumpha in Orissa (*c.* first century B.C.). It refers to the reconstruction of a canal in the first century B.C., originally built by the Nanda kings of Pāṭaliputra in the fourth century B.C. (Sircar 1965:213-221).

The Greek ambassador Magasthenes, who visited North India around 300 B.C., noted that multi-cropping was practised (McCrindle 1926:52-53) and that great officers of the state were responsible for the fair distribution of irrigation water from canals (McCrindle 1926:86). This suggests that perennial canals were in use at this time. Further evidence for perennial irrigation in the Mauryan period is provided by the Junagadh inscription of the Kṣatrapa Rudradāman dated to the second century A.D. It indicates that a bund, originally built by Chandragupta Maurya at Girnar in Gujarat, was repaired (EI VIII:36-49).

In North India, two canals assigned to the early historic period have been reported. At Rajghat, an unlined channel, 7.66 m. wide at the top and 12.35 m. deep, was found (IAR 1961-2:59)). The channel, which was dated by the excavators to between the fifth and second centuries B.C., was probably connected to the rivers Varunā and Gangā. During the excavations at the Kham Baba site, Besnagar, a brick-lined canal, 2.13 m. wide, 1.67 m. deep and 56.49 m. long, was located (ARASI 1914-5:67-71, pls. XLVIII, XLIXa). The canal was provided with vertical revetting. It was associated with levels older than the pillar of Heliodorus, set up in 165 B.C. The excavators suggest that the canal was probably part of the irrigation system of the city of Vidiśā and was connected with the Bes river.

In South India, Karikāla, an early Coḷa king (*c.* first to second century A.D.) is credited with having constructed a perennial irrigation system on the river Kāverī at his capital, Kāverīpaṭṭanam (Kanakasabhai Pillai 1890:331, 341; EI VII:119-128; Vankayya 1906:202-203). The hydraulic works constructed by Karikāla are amongst the earliest perennial hydraulic works reported in South India. At Kāverīpaṭṭanam, excavations in the 1960s revealed the vestigial remains of a perennial channel abutted to a tank or reservoir (Maloney 1985:12). The report states:

> A brick structure almost semicircular in plan, with nearly 2 m. high walls and an internal diameter of 8 m. laid bare at Vanagiri, appears to be a small water reservoir fed by an 83 cm. wide brick-built inlet channel from the river Kāverī. Since the structure is badly damaged by brick robbing, its full plan could not be traced. The earthen bund of the lake was also partially exposed. It is interesting to note that the brick structure show moulding produced by the chamfered and rounded edges. The channel seemed to have a corbelled roof. The occurrence of the megalithic Black-and-red and the Rouletted Wares in layers contemporary with the structure help to determine its date, which may be fixed around the first to second century A.D. (IAR 1963-4:20, pl. XIV)

The dating of the channel corresponds with the literary and epigraphical sources which mention Karikāla. The remains at Kāverīpaṭṭanam are extremely important as they represent the earliest evidence on the ground for a river-fed, perennial canal in South India.

Sewell (1932:14-15) has suggested that labourers from Sri Lanka may have been carried off by the Coḷa King, Karikāla, in the second century A.D. His hypothesis is based on evidence taken from the Rājāvaḷiya, a Sinhalese text. Complex irrigation works of imposing scale were constructed in Sri Lanka during the closing centuries of the pre-Christian era (Fernando 1980:1-24). It is possible that the labourers abducted by Karikāla may have been brought back specifically for their knowledge of irrigation and water supply.

At Nagārjunakoṇḍa, in Āndhra Pradesh, a canal 9.14 m. wide, 1.83 m. deep and 304.80 m. long was uncovered during the excavations. It was earthen and was not provided with a lining. The canal, which was found close to the bank of the river Kṛṣṇa near the north slope of the Nagārjunakoṇḍa hill, is datable to the Ikshvāku period (third to sixth century A.D.).

The physical and epigraphical evidence for pre-Vijayanagara water supply in Bellary and Raichur districts, indicates that much of the technology of water supply used in the Vijayanagara period had its origins in a local and indigenous tradition. The earliest evidence for water supply, in the form of wells and small catchment bunds at the foot of hills, appears in the context of permanent human settlement in the early historic period. Permanent settlements of limited size constructed wells to ensure a pure and reliable domestic supply, whilst modest agricultural water features were built to artificially irrigate terraced fields. The use of small catchment tanks may be viewed as an attempt to increase agricultural production by reducing dependence on scarce and uncertain rainfall.

By the first millenium A.D., the scale and enterprise of the irrigational technology had increased markedly, with large numbers of catchment drainage tanks of increasing size being constructed throughout the area under investigation. The wide spread of the irrigation tanks is indicative of the growth of new settlements and growing population in Bellary and Raichur districts. The construction of catchment tanks stemmed from a need to extend and intensify agricultural production through the use of perennial watering. Domestic supply appears to have depended on wells, tanks and water taken directly from rivers and streams.

The earliest evidence for the building of river-fed canal systems appears in the eleventh century A.D. at the Cālukyan settlement of Munirabad. River-fed canals were used to irrigate low-lying lands adjacent to the river, whereas tanks storing surface run-off and trapping small drainages provided a supply in areas away from perennial water-courses, particularly in uneven and irregular terrain. It would appear that at Danāyakanakere, a huge catchment tank was constructed in this period. Canal systems and large tanks required considerable investment of economic resources, both labour and capital, indicative of increasing population levels and socio-political organisation. The available evidence suggests that the use of the permanent anicut or diversion weir to supply perennial canal systems probably developed just prior to the Vijayanagara period. The construction of temporary anicuts of wood, earth and rubble prior to the fourteenth century A.D. can be inferred, although no physical or epigraphical evidence for them survives. Contemporary evidence for domestic water supply is not available, but it seems likely that wells and tanks continued to be the major sources of supply.

In our view, the essential components of the agricultural water supply system of Vijayanagara, catchment drainage tanks and river-fed canal systems, were developed prior to the fourteenth century A.D. in Bellary and Raichur districts. During the fourteenth to sixteenth centuries A.D., the refinement of contour-hugging canals allowed perennial water, supplied by permanent anicuts, to be used to irrigate areas away from the flat tracts adjoining the river. Contour canals appear to have developed out of an indigenous and ancient tradition of field terracing. Similar developments appear to have taken place in other parts of South India independently (Ludden 1985:20-23, 52-55).

The domestic water supply system of Vijayanagara employed the well, rainwater storage pond and catchment tank for the supply of the greater part of the urban settlement. This technology was developed locally. There are no antecedents in Bellary and Raichur districts for the complex domestic supply systems of the type witnessed in the 'Royal Centre' at Vijayanagara. However, there is some evidence for the indigenous development of open supply channels and pipes in South India. This evidence is discussed in section 5.3.3.

5.3 Domestic water supply

In this section, the extant evidence for domestic water supply in the 'Royal Centre', catalogued in chapter 3, will be examined in detail and tested against alternative data in order to shed new light on the design, development and chronology of the system as a whole. Before this analysis can begin, it is necessary to review briefly the literature concerning domestic supply at the site.

Domestic water features on the ground were first recorded by Mackenzie in his account of the 'Present State of the Anagoondy Country' (Mackenzie 1800a). He reported on a number of domestic water features including the tanks at Anegundi (1800a:127), the 'Queen's Bath' (XX/1) (1800a:129-130) and the 'Bhojanśāla' (1800a:130). He also briefly mentioned the ruined open channel (XIX/25) running to the east of the 'Queen's Bath' (XX/1) (1800A:130). A water-colour plan of the 'Bhojanśāla' was made at this time (1800a).

Three survey maps of Vijayanagara were produced by Mackenzie. Two water colour maps are held in the India Office Library and Records (Mackenzie 1800c, 1800d). A third finished ink map of the urban settlement and the surrounding area (1800b) is held in the National Archives, New Delhi, and has been reproduced by Michell and Filliozat (1981:2-3).

In the key to this map, Mackenzie noted the existence of four water features:

1. The 'Hannee Banna' (reference no. 4) or 'Bhojanśāla'.
2. The 'Oakula Goomdum' (reference no. 5) or 'Queen's Bath' (XX/1).
3. The 'Carinja Goondum' (reference no. 6) or 'Octagonal Fountain' (XXII/1).
4. The aqueduct or 'Carinja' (reference no. 13) by which water was taken from the Kamalāpuram tank to the 'Royal Centre' (XIX/25).

The remains were clearly visible above ground level at the time and were considered of sufficient interest by Mackenzie to merit inclusion in his survey.

There is a significant gap of 72 years between Mackenzie's report and the next references to the domestic supply features at the site. Kelsall briefly reported the raised conduit in enclosure IV and the 'Queen's Bath' (XX/1) (1872:291). He also provided the dimension of the monolithic stone trough (IIIa/7) in the 'Royal Centre' (1872:291) and remarked on the Hemakutam hill conduit flowing into the Virūpākṣa temple kitchens (1872:292). The same material was reproduced by Kelsall in a short article for the *Indian Antiquary* on the archaeological remains of Bellary district (1873:177,178).

In 1900, Sewell published *A Forgotten Empire* and in it brought together, for the first time, most of the available historical data concerning Vijayanagara, including some accounts of contemporary visitors in translation. Amongst these contemporary accounts were references to domestic supply features, includ-ing baths and conduits (Sewell 1900:91, 256). Although in individual cases Sewell attempted to identify features using the descriptions in the accounts, no attempt was made to catalogue the evidence for domestic water supply on the ground. The remaining contemporary accounts were published in translation separately by other scholars and these also contained references to domestic water supply (Dames 1918:201-202, 208).

The extant domestic water features were described briefly by Longhurst (1917), who included extracts from the contemporary accounts extensively in his detailed guide to the ruins. He describes the 'Queens Bath' (XX/1) (1917:54), 'Octagonal Fountain' (XXII/1) (1917:54) and 'Bhojanśāla' (1917:54-56). Four features in the 'Royal Centre', the 'Great Tank' (IVc/1) (1917:56), the raised channel in enclosure IV (1917:51), the 'Octagonal Bath' (1917:56) and the 'Water Pavilion' (XIV/2) (1917:82) were noted for the first time.

The remains of earthenware water-pipes are described close to the 'Octagonal Fountain' (1917:54) and elsewhere in the 'Royal Centre'

(1917:51). Longhurst erroneously concluded that the water features of the 'Royal Centre' were supplied with water drawn by bucket from a well just to the south of the enclosure wall of the 'Urban Core' (1917:51-53. The first reference to a temple tank, that of the Tiruveṅgaḷanātha complex (1917:109-110), was provided by Longhurst.

From the beginning of the twentieth century, Vijayanagara was cleared of vegetation and roads were built to give access to the Royal Centre from Kamalāpuram village. A number of monuments were restored and consolidated. In the 'Royal Centre', the 'Queen's Bath' (XX/1) (ARASI 1908-9:24; APRSC 1908-9:16; ARSC 1909-10:3), 'Octagonal Fountain' (XXII/1) (ARASI 1904-5:30, 1908-9:23; APRSC 1908-9:16), 'Bhojanśāla' (ARSC 1919-20:16), open channel (XIX/25) (ARSC 1912-3:49, 1919-20:16), the 'Water Pavilion' (XIV/2) (ARSC 1909-10:4), and 'Octagonal Bath' (ARSC 1919-20:16) were reported and consolidated. Inside enclosure IV, the 'Great Tank' (IVc/1) (ARASI 1904-5:30; ASMG 1904-5:36; ARASI 1908-9:24) and raised channel (ARASI 1908-9:24; ARSC 1909-10:4, 1919-20:16, 1920-1:14) were cleared and described. The records of the excavations and clearance at the site briefly mention that steps were taken to open up portions of the domestic supply system (ARASI 1919-20:14-15). No further details are available. In our view, the modern raised channel which runs alongside the Kamalāpuram–Hampi road (discussed in section 4.1.2) may represent part of this scheme.

A second and more detailed guide to the site was published in 1970 by Devakunjari. Inside the 'Royal Centre', the 'Queen's Bath' (XX/1) (1970:22-23), 'Octagonal Bath' (1970:41), 'Bhojanśāla' (1970:44) and 'Octagonal Fountain' (XXII/1) (1970:43-44) were described. The presence of earthenware pipes near XXII/1 (1970:43:44) was again commented on, but no details were provided. In enclosure XIV, the 'Water Pavilion' (XIV/2) (1970:35) and the rainwater collection tank (XIV/9) (1970:36) were noted. Devakunjari benefited from the results of preliminary investigations (clearing and limited excavation) conducted by the Archaeological Survey of India in the area of enclosure IV. In addition to the 'Great Tank' (IVc/1) (1970:23) and the raised channel (1970:26) in enclosure IV, Devakunjari remarked on a number of tanks, cisterns and water channels (1970:31). He proposed that the raised channel was used to feed the various water features inside enclosure IV (1970:31). However, the water features are not described individually and no attempt was made to construct a chronology of water features. With regard to the supply of water to the 'Royal Centre', he suggested that water was perhaps raised to the height of the conduits from the well or the river (1970:26). The problem of the supply of water to the 'Royal Centre' was, therefore, avoided. Outside the 'Royal Centre', Devakunjari noted the Manmatha tank (1970:57) and the Mallappannagudi stepwell (1970:70-71). On Hemakutam hill, the perennial supply conduit (1970:54) and one rainwater collection tank (1970:53) were recorded.

Michell and Filliozat made reference to the open channels and earthenware pipes of the 'Royal Centre' (1981:50,65). Brief descriptions of the 'Great Tank' (IVc/1), 'Queen's Bath' (XX/1), 'Octagonal Bath', 'Bhojanśāla', 'Water Pavilion' (XIV/2), a rainwater collection tank (XIV/9) and the 'Octagonal Fountain' (XXII/1) were included (1981:64-65). Outside the 'Royal Centre', the Hemakutam hill supply conduit (1981:50) and the step-well at Mallappannagudi were also reported (1981:65).

To date, the best catalogue of domestic water features has been provided in the preliminary survey of the 'Royal Centre' published by Fritz *et al.* in 1984. Water features on the ground were recorded in detailed descriptions of enclosures 1 to XXXI (1984:18-37). The architecture of the Muslim influenced water features, in particular the 'Queen's Bath' (XX/1) (1984:124-126), 'Octagonal Bath' (1984:144-145) and 'Octagonal Fountain' (XXII/1) (1984:126-127) is discussed. A short section was dedicated to the hydraulic features of the 'Royal Centre', particularly those in enclosure IV (1984:52-55). This surface study generated an invaluable corpus of data and provided a standard system of reference codes for surviving features. It has also provided the information base upon which the investigation devoted to domestic supply has built. The study further enabled the present writer to better understand the relationship between the fragmentary remains of domestic water supply and the complex area of the urban centre in which they are located.

5.3.1 How did the domestic supply system operate?

Domestic supply may be defined as the provision of water supply and drainage to centres of permanent settlement, such as villages, towns and cities, in order to satisfy the demands of a whole range of human activities. In general terms, domestic water is essential for drinking, hygiene, sanitation and food preparation. At Vijayanagara, as throughout India, it was also used for religious and ritual purposes, in addition to the filling of ornamental and functional pools, baths and tanks.

At Vijayanagara, a distinction must be made between the facilities constructed for the supply of water to the 'Royal Centre' and those constructed to supply the rest of the 'Urban Core'. Apart from temple complexes, there is no evidence for water supply to municipal water features which could be used by the general population of the city. The great majority of domestic water features are actually inside the walled area of the 'Royal Centre'. We may, with some certainty, conclude that they would have been reserved for the use of the King and his entourage. It was erroneously concluded in the archaeological reports that the 'Octagonal Fountain' (XXII/1) was a public water feature (ARASI 1908-9:23). Longhurst (1917:54) elaborated this hypothesis by suggesting that the monolithic trough close to the fountain basin, was used to store milk distributed to the general population during religious festivals. There is no evidence to support these hypotheses. The position of the 'Octagonal Fountain' (XXII/1), indicates that it was inside the 'Royal Centre' and therefore, was used solely by the King and his entourage.

Temple tanks appear to come nearest to representing a class of municipal water features at Vijayanagara. Most rainwater collection tanks and temple tanks dried up. However, two temple tanks were fed by a perennial source. First, the Manmatha Gundam tank acted like a bāvlī, taking its supply directly from the water table. Second, the Bālakṛṣṇa tank was supplied by a conduit connected to the Hiriya Kāluve. An open channel was used to transport perennial water from the Hiriya kāluve into the Vidyāraṇya maṭha and Virūpākṣa temple kitchens. Temple facilities were available to the general population and would have served ritual and practical needs. However the population of the city would have relied, for the bulk of their water needs, on communal rainwater collection tanks and wells. The river was undoubtedly utilised for domestic water supply in those parts of the city adjoining it. Wells and small tanks would have also enabled small garden plots to be maintained inside the 'Urban Core'. It is possible that domestic water was taken from the irrigation canals and channels which flowed through the irrigated valley and the greater metropolitan area. However, the primary purpose of the canals was clearly agricultural water supply and any use of this supply for domestic purposes would have been minor and secondary. We have found no evidence to suggest that canal water is used for domestic purposes by the modern inhabitants of the area.

In the sixteenth century A.D., elaborate steps appear to have been taken to provide Achyutāpura with perennial water taken from the Hiriya canal. Water from the canal was diverted into a bunded tank called the Bhūpati kere, at the base of Matanga Parvatam. It is unclear whether water from the tank was intended for domestic or agricultural use. Some water must have passed into the Tiruveṅgaḷanātha temple tank, which today is dry and overgrown.

In our view, the system for the supply of perennial water to the 'Royal Centre' was in essence quite simple. The domestic water system in the 'Royal Centre' at Vijayanagara utilised the principles of gravity flow and the siphon to convey modest quantities of water by pipeline and open channel. The system was primarily supplied by the Kamalāpuram tank, with a back-up system of wells and rainwater collection tanks. Our investigations suggest that a single open channel (XIX/25) brought water from the Kamalāpuram tank into the 'Royal Centre'. Open channels and pipelines were used to transport water within the royal enclosures and dispose of waste water. The destination of the supply taken into the 'Royal Centre' was a complex series of baths, tanks and fountains. The greatest concentration of these features was in and around enclosure IV, an area associated with the performance and demonstration of ritual power.

The tanks and basins inside enclosure IV form an eclectic group which must have been constructed throughout the occupation of the 'Royal Centre'. The 'Stepped Tank' shows strong affinities with the geometric temple tanks of the Hoysala period found in the Mysore area. We have suggested, on the basis of its material and the presence of 'assembly instructions' in cipher form, that this tank may have been brought from elsewhere and reassembled in enclosure IV. Rather than being filled by perennial conduits, this feature seems to have been filled like a well, the height of the water table dictating the amount of water held. It is noteworthy that, in many cases, tanks in the 'Royal Centre' have stairs on their western sides.

Although water would have been used for sanitary and culinary purposes, a large supply was also needed for ritual purification of the body and food. This may be viewed as one of the primary reasons for the construction of the supply system to the 'Royal Centre'. The conspicuous consumption of water, which was a rare commodity in the semi-arid environment, must have been seen as the appropriate course of action for a powerful king. The provision of water to the palaces provided a pleasant environment, but was also a sign of great wealth and power.

However, the supply system to the 'Royal Centre' appears only to have operated for ceremonial purposes or when important visitors came to the court. This hypothesis is supported by the design of the distribution system. The control tanks, which allowed water to be collected and the silt removed, had their outlets arranged so that the tank had to be filled and emptied. They could not work continuously. In our view, day to day needs would have been more easily and efficiently satiated by the rainwater collection tanks and wells. It is important to note that the construction of the Tungabhadrā Dam has totally altered the hydrological environment of the Vijayanagara metropolitan region. The conservation of rainfall and recycling of domestic water was undoubtedly important in the city.

Contemporary accounts indicate that the King's drinking water was brought from a special spring reserved for the purpose, in a sealed container. Rizvi and Flynn (1975:35) indicate that a similar procedure was followed at Fatehpur Sīkrī, where Akbar's drinking water was collected at Sorūn (Soron) and transported to the capital in sealed containers. The jars of drinking water are reported to have been cooled with saltpetre (potassium nitrate).

In addition to the network of open channels, certain areas of the 'Royal Centre' were provided with sealed pipelines. The earthenware pipes are composed of spigot-jointed sections, fitted together and sealed with a skin of chunam. Forbes has suggested that, properly sealed earthenware pipes can withstand a pressure of up to 50 atmospheres (1954:664). A heavy casing of brick, bonded with lime-mortar, encased the pipes. This served two purposes. Firstly, it protected the pipework from damage. Secondly, it increased the pipes capacity to work under pressure. The pipes were made in three diameters, 10 cm., 16 cm. and 19 cm., suggesting that a series of standard sizes were used. It also suggests that the pipes were mass-produced to a standard quality by specialist craftsmen. It is interesting to note that no pipework is found inside enclosure IV where the greatest density of hydraulic remains are concentrated. Archaeological reports indicate that enclosure XIV was supplied with earthenware pipes (ARSC 1909-10:4). These pipes, which must have connected the hydraulic features of enclosure IV to the pipe system in enclosure XV to the south, are not visible today.

Pipelines were used to transport limited quantities of water under pressure, from control tanks fed by open channels, to a variety of domestic features including fountains. The use of the siphon enabled water to be fed by atmospheric pressure into fountains and basins at different levels. The best example of this is the pipe inlet XXII/2, which formed the west side of a large control tank and at least six separate pipelines. To prevent the pressure-lines becoming blocked, it was necessary to exclude silt from the supply entering the pipe. The exclusion of silt from the pipelines was doubly important because the diameter of the pipes was fairly wide. The sluggishness of the flow inside the pipes greatly increased the probability of any silt in the system collecting in the pipes themselves. The thick casings in which the pipes are housed, indicates that silt could not have been easily removed from the system once it had entered the pipe. Control tanks were used to settle out silt and prevent wind-blown deposits from the supply entering the tank from open channels. They were also used as a means of regulating the supply entering the various pipes.

The basic methods of excluding silt appear quite simple. In our opinion, water was allowed to enter the control tanks, which had their outlets closed by the placement of a board or gate in front of the pipe. The supply to the tank was shut off when the water had risen to a predetermined full supply level. Silt suspended in the stationary water would fall to the bottom of the tank, allowing a silt-free supply to pass into the pipe outlet. The silt collected at the bottom of the tank and was prevented from entering the outlet, which was set up above the floor level on a step. This technique was only able to supply as much water to the pipe outlet as the tank could hold at any one time and, therefore, supports the assumption that the hydraulic features of the 'Royal Centre' were not continuously used.

Two examples of control tanks feeding pipelines survive, XXIII/1 and XXII/2. Control tank XXVII/1 is a simple device supplying a single pipeline, silt being removed from the system by the method described above. Pipe outlet XXII/2 is far more complicated and was provided with a number of outlets at different levels. The tank appears to have been used to settle out silt in the manner described above. Water was taken off at a level above the floor of the tank by at least one double outlet. However, some attempt was, in our view, made to supply a continuous silt free supply to the single outlet in the recessed notch. The placement of a board in front of the notch would have allowed water to collect behind it in the tank. When the level reached the top of the board, water would slop over the obstruction and flow into the outlet. Silt would, therefore, settle out in the tank. This method would allow a constant supply to be taken from the tank. Water passed from the recessed outlet into a raised tank to the immediate west of XXII/2; and was fed into three pipelines through outlets in the floor of the tank. The floor of the tank is angled away from the outlets. A fourth outlet was located at a level above the floor of the tank. This is the only example of a tank with outlets let into the floor. It could have operated under a variable head, providing a constant supply.

The 'Great Tank' (IVc/1) high outlet, which was at some point replaced by a slab drain, allowed water to be skimmed off the supply in the tank at a level close to the full supply level. Although the high sluice must have worked as an overflow valve, it was also perfectly suited to excluding silt from the water leaving the tank. The large surface area of the tank would have facilitated the efficient settling out of silt. Water leaving the 'Great Tank'(IVc/1) passed, not into a pipeline, but an open channel (IVd/17). However, IVd/17 is the only conduit which can have supplied the large control tank, the west side of which was formed by XXII/2. It appears, therefore, that attempts were made to exclude silt from the supply entering the pipelines (starting from XXII/2) at two separate points. We suggest that, in certain cases, the baths and tanks of the 'Royal Centre', in addition to facilitating ritual and pleasurable pursuits, also served a secondary and practical function of removing silt from water entering the pipelines.

At the beginning of this century, copper pipes were reported in the 'council room', the old name for the 'Lotus Mahal' (XIV/1):

> A curious feature was discovered in the interior. That is, that in some of the angles of the piers, copper tubes project from the walls. They are about the same diameter as modern speaking tubes, and may have served the same purpose, or to enable those in the upper storey to hear what is being said below (ARASI 1904-5:29).

The pipes projected from the plasterwork of the four, central piers of the lower storey. Shortly after this report was published, the plasterwork was chipped away and the pipes stolen (APRSC 1907-8:12; ARSC 1909-10:4). Although the original function of pipes in this instance, appears to have been to facilitate verbal communication inside the 'Lotus Mahal' (XIV/1), it may be inferred that copper piping was also used for water supply in the 'Royal Centre'. Copper piping visible on the surface is likely to have been pillaged from the site early on as a result of its value. We note that the original fountains in the Tāj Mahāl at Agra were connected by copper pipes (*The Times of India*, 25th June, 1973).

The conservation and reuse of rainwater was important throughout the 'Royal Centre'. Although water was brought into this exclusive area by a perennial conduit, the 'Royal Centre' must have also depended for its water needs on the collection of rainwater and the use of wells giving access to subsoil water. Small, plastered conduits, which run around the bases of some of the structures in enclosure IV, appear to have conserved rainwater, and collection tanks are scattered throughout the 'Royal Centre'. It seems likely that copper piping was used as a guttering and channel system to collect and store rainwater. Although expensive, metal piping would have aided the efficient collection of rainfall and it seems quite possible, therefore, that it was used for this purpose.

There is some evidence to suggest that standard dimensions were used in the construction of the water system. The slab outlets of tank IVb/3 and IVb/4 are similar in design and size, measuring 10 cm. and 11 cm. wide respectively. Although three, different constructional styles of secondary, open conduit survive in enclosure IV, they have a standard width of 24 cm. It has already been mentioned that three pipe diameters, 10 cm., 16cm., and 19 cm., were employed in the 'Royal Centre'. The choice of pipe width appears to have been decided largely by the amount of water required by a given feature. However, in at least one case, the width of pipe would appear to have been determined by the specific purpose of the conduit. The outlet of the 'Queen's Bath' (XX/1) consists of two pipes at different levels. The lower pipe, which is located at floor level, has a diameter of 19 cm, whilst the remaining pipe has a diameter of 10 cm. and is located 83 cm. above the floor of the bath. The difference in the widths can be explained by the different function of each pipe. The lower and larger of the two pipes would have allowed any silt which had collected in the bath to be flushed out. However, the higher outlet, which may be presumed to have acted as an overflow valve, skimmed of silt free water from the supply in the bath. The lower of the two gates could be regulated by the placement of a board or gate in front of the pipe. It is likely that the remaining outlet would be kept permanently open in order to prevent the supply in the bath rising above the full supply level.

No epigraphs mentioning the engineers who built the water features of the 'Royal Centre' survive at Vijayanagara. However, Rizvi and Flynn (1975:21) indicate that Faṭhu'lāh Shīrāzī, an important philosopher, scientist, mathematician and engineer, who had been resident at the court of Ibrahim 'Ādil Shāh (A.D. 1579-1616) at Bījāpur moved to Fatehpur Sīkrī in A.D. 1582. This scholar was famed for his mechanical contrivances and, therefore, it seems quite likely that he was responsible for at least some of the water features at both royal capitals. The movement of Faṭhu'lāh Shīrāzī from Bījāpur to Fatehpur Sīkrī indicates that skilled engineers moved regularly between wealthy centres of royal patronage. It may be possible to infer that engineers specialising in hydraulic systems also travelled to Vijayanagara.

5.3.2 When and by whom was the domestic supply built?

The 'Royal Centre', in which the greatest density of domestic supply features are concentrated, witnessed religious activity in the pre-Vijayanagra period. However, royal performance and residence, with which the water features are intrinsically linked, cannot in our opinion be dated to the 'Royal Centre' earlier than the first quarter of the fifteenth century A.D. (see section 2.3). There is no evidence for abandonment between the first quarter of the fifteenth century A.D. and A.D. 1565, the 'Royal Centre' being continuously rebuilt during its occupation. The domestic supply features were intended to serve the high status dwellings and public buildings associated with the king and his entourage and form an integral part of the 'Royal Centre'. Therefore, it may be established that the domestic supply features are contemporaneous with the buildings of the 'Royal Centre' and were constructed between the first quarter of the fifteenth century A.D. and the middle of the sixteenth century A.D.

The dating of individual domestic supply features in the 'Royal Centre' has proved difficult for three reasons. Firstly, their construction was not recorded in epigraphs. Secondly, the 'Royal Centre' is stratigraphically complex and has suffered severe post-depositional changes. Thirdly, the buildings and associated domestic supply features in the 'Royal Centre' were modified and extended many times, the fabric of earlier structures in some cases being reused. The 'Stepped Tank', which appears to have been reassembled at the site, is perhaps the best example of this phenomenon.

However, from an examination of abutted and modified features, it is possible in limited areas, to establish a sequence of building. The supply conduit to the 'Great Tank' (IVc/1), is constructed in a different manner to the raised channel to which it is crudely abutted in an aesthetically unpleasing way. The raised channel, which runs east-west through enclosure IV, originally supplied a complicated arrangement of baths, basins and tanks via a series of branch channels (see section 3.2.2). It is clear that the supply channel to IVc/1 was constructed after the raised channel. Since IVc/1 has no means of supply other than the channel connecting it to the raised channel, it was, without doubt, built after the raised channel. The position of the 'Great Tank' (IVc/1) in the southeast corner of the enclosure demanded the construction of a new, feeder channel abutted to the raised channel, which remained the main source of supply. With regard to the dating of the raised channel in enclosure IV, the Persian traveller Razzak, writing in about A.D. 1442, refers to polished conduits of stone in the area of the King's Palace (Sewell 1900:91). If the stone conduits described by Razzak are those in enclosure IV, the raised channel must have been built between about A.D. 1425 and 1442.

The constructional style of the 'Great Tank' (IVc/1), which employs a mixture of dressed stone and plastered rubble, is identical to that of the 'Queen's Bath' (XX/1), 'Octagonal Bath' and 'Octagonal Fountain '(XXII/1). This strongly suggests that they were built at the same time, as part of an extensive building programme inside the 'Royal Centre'. It has been established that the 'Great Tank' (IV/1) was constructed after the raised channel. If the 'Queen's Bath' (XX/1), 'Octagonal Bath' and 'Octagonal Fountain' (XXII/1) can be considered contemporaneous with the 'Great Tank' (IVc/1) on the basis of a similarity of constructional style, it must be concluded that all four were built after the raised channel.

Fritz *et al.* (1984:122-123) have indicated that the architectural style of the 'Queen's Bath' (XX/1), 'Octagonal Bath' and 'Octagonal Fountain' (XXII/1) represents a syncretic creation influenced by the Muslim style of the Deccani Kingdoms. It has been established that the first Muslim-influenced monuments were built at Vijayanagara from the second quarter of the fifteenth century A.D. (Fritz *et al.* 1984:122). However, the Muslim-influenced structures associated with water supply represent a more complex and ornate style than that of these early monuments. We therefore propose that the 'Queen's Bath' (XX/1), 'Octagonal Bath' and 'Octagonal Fountain' (XXII/1) date to the sixteenth century A.D., at which time extensive improvements were made to the city by Kṛṣṇadeva Rāya and Achyuta Rāya.

There is considerable evidence for alterations made to some water features after their construction. Both the 'Queens Bath' (XX/1) and 'Great Tank' (IVc/1) are provided with replacement outlets. These were clearly modifications made after the structure had been completed and suggest that either the outlets originally fitted did not function correctly or there was a change in the use of waste water let out of the tanks which necessitated a different sort of outlet. The replastering of water features, for which there is evidence in the control tank XVIII/1, indicates that routine maintenance was carried out to ensure the efficient functioning of the domestic supply system of the 'Royal Centre'. We have also noticed 'lumps' of plaster inside the raised channel of enclosure

IV. These additions appear to have been designed to create small affluxes in the flow, close to the take-off points of subsidiary channels. Minor adjustments to the height of the head and the velocity of the flow in the raised channel may have been necessary, when channels supplying new tanks were abutted to it. There is evidence for the rebuilding of the pipe outlet XXII/2 which served as one side of a control tank supplying a series of sealed pipe conduits. The supply to a small collection tank on top of the outlet was transferred from the main control tank to a different sources. It is clear that XXII/2 was modified in response to new demands on the domestic supply system.

It is tempting to use differences in the constructional style of domestic supply features in the 'Royal Centre' for the purpose of dating. Provisionally, it is acceptable to suggest that the use of dressed stone was superseded by a degenerate constructional style dependent upon the use of plastered rubble masonry. The use of plastered rubble masonry seems to indicate a desire to build quickly and for effect. However, factors other than a switch from one building style to another could explain the dichotomy. For example, a desire to build quickly and cheaply in different periods, may have determined the used of a certain constructional style at a given time. Political instability and/or financial considerations may have dictated such a proviso. In addition to this, many of the supply features, such as pipes and control tanks, were clearly not intended to be seen, either because they were outside royal enclosures or had been skilfully covered to hide them from view. It is therefore important to appreciate that, in certain cases, the dichotomy in the constructional styles of water features may be the result of two contemporaneous methods of building, rather than a change from one building style to another.

In our view, the extension and elaboration of the domestic water features of the 'Royal Centre' may correspond with periods in which war was successfully fought. Portable wealth, taken during warfare, provided the Vijayanagara kings with an opportunity to express their power and prestige in physical terms in the 'Royal Centre' at the capital. The extensive programme of building carried out by Kṛṣṇadeva Rāya at Vijayanagara, occurred after the conclusion of his successful military campaigns in the Raichur Doab and the east. To us, it seems likely that the 'Queen's Bath' (XX/1), 'Octagonal Bath', 'Octagonal Fountain' (XXII/1) and 'Great Tank' (IVc/1) date to this period. If this hypothesis is correct, the pipe system, which was connected to all of them, was also built at the same time.

The dating of domestic water supply features outside the 'Royal Centre', such as domestic catchment tanks and wells, is reliant on the existence of dedicatory and donatory inscriptions. Individual wells can be dated in this way (see section 3.3.2). It is possible in most cases to date the large temple tanks, which in three cases were constructed in the bazaars of religious complexes as an essential part of their design. These tanks can be dated to the construction of the basic features of the temples. Thus, the Bālakṛṣṇa, Tiruveṅgaḷanātha and Viṭṭhala temple tanks, which are all of similar design, can be dated to the first half of the sixteenth century A.D. The Manmatha tank, which serves the Virūpākṣa temple complex, is a notable exception and was modified in the Vijayanagara period. Epigraphical data indicates that the tank was originally constructed in the twelfth century A.D.

A dichotomy exists between the domestic water features of the 'Royal Centre', which were constructed in order to serve the king and his entourage, and the tanks and wells provided by individuals and temples for the general use of the community. It has been suggested above, that the kings at Vijayanagara did not provide municipal water features for domestic uses. This was left to individuals and religious establishments. It appears, therefore, that the king's duty to provide and maintain facilities for the population of the capital did not extend beyond agricultural water supply and the financing of large scale irrigation projects. This goes against the general Indian tradition.

5.3.3 Does the technology used represent a development of new ideas or merely a continuation of an existing tradition?

Water supply at Vijayanagara was undoubtedly influenced by earlier technology and contemporary developments. We shall briefly examine the evidence for the development of domestic supply in South Asia, before discussing contemporary influences. The history of domestic water supply is, of course, a massive subject and we shall arbitrarily select a few examples to illustrate our discussion.

In the urban settlements of the Indus civilisation, there is extensive evidence for domestic water supply and drainage from the middle of the third millennium B.C. (Bromehead 1942: 186; Clark 1944;12). Communal, brick-lined wells in paved plazas, gave access to the water table and provided the primary source of domestic water. Pottery jars were used to store the water once lifted. A complex network of open and covered conduits of brick, controlled by gates, served to carry off standing rainwater and channel domestic wastewater into soakage pits. Private houses were often provided with small bathrooms. Waste-water and sewage were carried out of the building and passed, either into soakage jars outside, or into the municipal system of drainage channels. Evidence for this phenomenon comes from the two, largest settlements at Mohenjo-dāro (Marshall 1931:i,278-282) and Harappā (Vats 1940:13-14). A number of smaller urban sites, including Kalibangan (IAR 1960-1:132, pl. XLIIa,b; 1961-2:43-44, pls. LXIIIa,b, LXIVb, LXVa,b; LXVIIb; 1962-3:27,30, pls. LIVa, LVb, LVIII; 1967-8:45, pls. XIII, XIV, XVb, XVII, XXI), Lothal (IAR 1965-7:15, pl, LVa,b; 1957-8:12, pls. X, XI, XIIa, b; 1957-58:13-14, pls. XIIIa,b, XIVa,b; 1959-60:17, pl. XIIa; 1961-2:10, pl. XVIIIa), Surkotada (IAR 1971-2:16, pl. XXIV) and Rangpur (Rao 1963:32-33, 36, 41-44, pl. IX) have also yielded similar evidence. At Kalibangan, one wooden drain was recovered (IAR 19612:43-44, pls. LXIIIa,b, LXIVb, LXVa,b, LXVIIb), suggesting that organic materials may have been used in the design of the domestic drainage systems of the Harappan culture.

There are indications that Harappan drainage systems were well maintained. Mackay (1938: 229) states:

> That the drains of Mohenjo-dāro were cleaned out periodically is attested by the little heaps of greenish-gray sand that we frequently find alongside them. The more finely levigated clays would be readily carried off by the rush of water whereas the heavier particles of sand were deposited.

Marshall also thought that the soak pits were cleaned out at the end of each season (1931:i. 278-282).

There is evidence at Harappan sites for the use of earthenware pipes. Marshall (1931:i.280-181) and Mackay (1938:93,98,170,426,651) noted their presence at Mohenjo-dāro. At Chanhu-dāro, drainpipes of three types ware found during the excavations (Brown 1943:215-216, pls. Xb, XIb, XVIe, LXXXVIII). Fragmentary remains of pipes carved from brick and coarse earthenware were recovered at Harappā (Vats 1940:377, pl. LXXII:48 and 49). The pipes were used to convey waste water under buildings and through structural walls. In some cases, pipes were vertically fixed to carry rainwater from guttering into street drains. At Kalibangan, there is also evidence for the use of earthenware half-pipes (IAR 1960-1:32, pl. XLIIIa,b). Neither the horizontal nor vertical pipelines were pressure sealed. Both appear to have been used solely for the purposes of drainage.

There is a considerable body of evidence for domestic water supply during the early historic period (*c.* 500-200 B.C.) In the context of urban settlement, wells lined with terracotta rings or brickwork provided the primary source of supply. Water was, in some cases, stored after lifting in brick-lined cisterns and tanks. The larger settlements and capitals were provided with brick and occasionally stone-built conduits, both open and covered, which carried waste water into soakages pits and moats. A number of sites in the northern plain have produced such evidence. They include Kauśāmbī (IAR 1956-7:47, pls. LXIIIa,b, LXIV; 1957-8:47, pl. LIVb; 1961-2:51, pl. LXXXIXa; Sharma 1969:32-35), Mathura (IAR 1954-5:16, pl. XXVIII), Rājghāt (IAR 1960-1:37, pl. LXIb; 1961-2:58, pl. XCIa,b), Sohagaura (IAR 1961-2:56, pl. LXXXIXb), Prahalāpur (IAR 1962-3:41, pl. LXXIIIb), Draupadi-ghāt (IAR 1960-1:33, pl. LIIIa), Peddabankur (LAR 1968-9:2, pl.Ia; 1970-1:2, pl. IIIa), Ayodhya (LAR 1976-7:52-53, pl.

La) and Hastināpura (Lal 1955;15-16,26, pls. IV, VIII, IX, X, XIa,b, XIb) in Uttar Pradesh, and Buddhiyana hill (LAR 1960-1:4, pl. Ia) in Bihar. Investigations at Pāṭaliputra in Bihar, the Mauryan capital (fifth to second century A.D.) have revealed not only ring-wells (Sinha and Narain 1970:15-19), but also the remains of a large timber-lined drain let through the defensive palisade (ARASI 1926-7:138, pl. XXXI; Wheeler 1948; pl. XXXVI, fig. 4.) The drain, which was over forty feet in length and jointed with iron strips, appears to have been part of a drainage system flushing into a defensive moat. Magasthenes, who visited Pāṭaliputra in around 300 B.C., reported that the city was provided with a moat into which sewage could be flushed (McCrindle 1926:65-67). Several writers have suggested that the moat was connected to a perennial, river-fed channel (Dutt 1925:91-95; Mate 1970:63).

The use of technology of this type was widespread. Similar evidence has also been found at Ujjain (IAR 1955-6:19, pl. XXVIIb; 1956-7:27, pls. XXVIIIb, XXXIa; 1956-7:27, pl. XXXb) and Besnagar (IAR 1964-5:19, pl. XV; 1975-6:30. pl. XLIIa) in Madhya Pradesh, Rupar (LAR 1953-4:7, pl. IIb; 1954-5:9, pl. IX) in the Punjab, Ganwaria (IAR 1974-5:40; pl. XXXXIb) in Gujarat, Tamiluk (IAR 1954-5:20, pl. XXXVIa, b) in Bengal, Śiśupālgarh (Lal 1949:68, pl. XLIa) in Orissa, Prakash (IAR 1954-5:13, pl. XXIa) in Maharashtra, Purana Quila, Delhi (IAR 1970-1:8, pl. XIXb), Tilaura-Kot (IAR 1961-2:74, pl. CXXIVb) in Nepal and the Bhir mound at Taxila (ARASI 1920-1:19; 1924-5:47, pl. VIIIb).

There is limited evidence for a more complex development of water supply in the early historic period. At Ujjain, a limestone basin with attached earthenware pipes was recovered (IAR 1964-5:18, pl. XIVb). Earthenware pipes were also found during the excavation at Kauśāmbī (Sharma 1969:38,41, pls. XIXb, XXb) and Chandraketugarh (IAR 1957-8:51, pl. LXXI). These discoveries are important for two reasons. Firstly they demonstrate a continued use of pipelines in domestic supply after the demise of the Harappan culture. Secondly, the evidence from Ujjain suggests that pipes may have been used in this period for water supply rather than drainage.

Terracotta ring-wells and brick-built drains and tanks continued to be built from the second century B.C. to the fourth century A.D. in the context of urban settlement. Evidence for this comes from a number of sites, which include Masaon (IAR 1964-5:43, pl. XXXV), Peddabankur (IAR 1968-9:2), Bhīṭa (ARASI 1911-2:38,39), Piprahwa (IAR 1974-5:39, pl. XXXIIIb) and Mathura (IAR 1974-5:50, pl. XLIV) in Uttar Pradesh, Chirand (IAR 1964-5:7, pl. VII) in Bihar, Ganwaria (IAR 1976-7:50-51, pls. XLIIb, XLIV, XLVIa) and Karvan (IAR 1974-5:16, pl. XVIIb; 1975-6:15, pl. XIVb) in Gujarat and Sanghol (IAR 1977-8:43) in the Punjab. There is one instance of earthenware pipes used in Sirkap. Taxila, in this period (ARASI 1927-8:61, pl. XVIIIb). In addition, a network of stone built drains was found near the palace, at the same site (Ghosh 1948:46-47, pl. VIII).

At Śringaverāpur in Uttar Pradesh, a bath complex has been dated to the first century A.D. (IAR 1978-9:57-59; 1979-80:73-74). Two interconnected baths of brick were filled with water by an unlined channel connected with the river Gangā.

The evidence of domestic supply in the middle historic period suggests that temples began to be provided with channels for the transport of ritual libations out of the vimāna. At Karvan in Gujarat, the linga of a ruined temple was found to be connected to a pranāli and brick-built drain (IAR 1975-6:15, pl. XVIIb). Objects recovered during the excavations were dated to between the second and eighth centuries A.D. The excavation of ruined temples, which in certain cases has facilitated an examination of foundations, suggests that the use of channels in temples continued into the medieval period. At Mahāsthān in Bengal, a late Gupta pillar was reused as an open channel to conduct libations out of the shrine (ARASI 1928-9:90, pls. XXXVIIa, XXXVIIIa,b. XLb). The site is dated by its excavators to between the eighth and thirteenth centuries A.D.

Religious complexes, such as monasteries, were also sometimes provided with drainage systems. The Buddhist temple and monastery at Paharpur, Bengal, dated to between the tenth and twelfth centuries A.D., was served by a series of open drainage conduits, including a terracotta open drain made up of short sections (ARASI 1930-4: pt. 1, 122,127; 1930-4: pt. 2, pls. LVIIb, LXId, LXIIe). The development of drainage conduits in temples and religious complexes appears to be contemporary with the use of domestic supply technology in urban centres.

Although the evidence for pre-Vijayanagara domestic water supply in Bellary and Raichur districts is restricted to catchment tanks, temple tanks and wells, some sites in South India have yielded more complex domestic supply features. At Nagārjunakoṇḍa in Āndhra Pradesh, the remains of a brick-built, drainage system and ornate tanks of plastered brick were discovered (IAR 1956-7:37, pl. LVa,b; 1957-8:8-9). These were dated to the Ikshvāku period (*c.* third to sixth century A.D.). An interesting series of channels and drains, cut into laterite blocks associated with Rouletted Ware, were recovered at Dharanikota in Āndhra Pradesh (IAR 1964-5:2, pl. 1a,b). Closer to Vijayanagara at Vadagaon Madavāpur in Karnataka, a brick-built drain dated to the Śatavāhana period (*c.* first to fourth century A.D.) was discovered (IAR 1972-3;26, pl. XXXIb). This evidence suggests that much of the technology employed in the domestic supply system of the 'Royal Centre' at Vijayanagara was developed in an earlier period at a limited number of sites in South India.

Several Deccani capitals contemporary with Vijayanagara, were provided with complex supply and drainage system. At Bidar fort, constructed in A.D. 1430 by Ahmad Bahmanī I of Gulbarga (A.D. 1422-1435), the fragmentary remains of an open channel system, supplying gardens and ornate pools, survives in the royal enclosure (ARHAD 1928-1929; 7-8). A channel connects a well from which water was drawn and a reservoir on the top of the Sola Khamb mosque constructed in the Bahmanī period (ARHAD 1928-1929:8). Water stored in the reservoir was apparently distributed to other cisterns. Evidence for a water-lift and sealed pipe system supplying a series of fountains at different levels also survives (ARHAD 1928-1929:8,11). Excavations in the Medieval palace complex at Golkoṇḍa in Āndhra Pradesh, unearthed pipelines and conduits leading to basins, cisterns and toilets (IAR 1970-1:1, pls. I,IIa). Water was drawn from rock-cut wells and stored in cisterns. A network of covered channels, pipes and drains connected the various elements of the water system (IAR 1970-1:94, pl. XCa,b; 1971-2:1-2, pl. IIa; 1972-3:1; 1973-4; 7, pl. VIIa).

The fortified city of Bījāpur was provided, from the sixteenth to seventeenth centuries A.D., with an extremely elaborate domestic supply system. Perennial water was supplied from three separate sources. Firstly, a series of large catchment tanks outside the walls of the citadel. The most important of these was the Begam Talāo located to the south, which filled an earthenware pipeline supplying the city (Cousens 1916:121). The pipes were spigot-jointed and sealed with chunam. Control towers for maintenance and the exclusion of silt were provided at regular intervals along the length of the pipeline. Although the tank may have been constructed earlier, epigraphical evidence suggests that the pipeline was constructed between A.D. 1651 and 1652 (Cousens 1916:121-122; Rozter 1984:137). Secondly, a qanāt leads from Surang Bāurī, near Torweh, into the city. It was built by 'Ali 'Ādil Shāh I (A.D. 1557-1580) and supplied a number of important water features (Rozter 1984:174-193). Thirdly, a series of aquifer-fed, step tanks inside the fortified walls served the municipal needs of particular quarters of the city (Rotzer 1984:138-151). Water, lifted from the tanks by cattle-powered drag-lifts, was used to supply pipelines in the environs of the tank.

At Bījāpur, no distinction was made, in terms of water supply, between the residential areas of the city and the royal palaces. Although water was supplied directly by pipe and channel into the palaces and houses of the nobility, extensive municipal features were provided for the everyday needs of the population. Although it is not possible here to discuss in detail, we note that the Muslims derived many of their ideas relating to water supply from the Romans. Unlike Vijayanagara, a perennial system supplied the whole city, rather than just a 'Royal Centre'. In the sixteenth and seventeenth centuries A.D., a system of pipes and open channels inside the walls of Bījāpur was used to distribute water in precise quantities, at carefully calculated heads, to private and municipal features. Municipal water features are conspicuously absent at Vijayanagara. The hydraulic features of Bījāpur reached a high level of ornamental sophistication and even open channels were in some cases decorated with ridged patterns, to make the water flow in attractive patterns (Cousens 1916:120).

A qanāt (or infiltration gallery) is a long adit or network of adits excavated in compact, alluvial soil or permeable rock. Its purpose is to collect and transport groundwater from a source of supply to a point of demand. Shafts, sunk vertically at regular intervals along the length of the qanāt, allow spoil to be removed during construction, and may also serve as wells when it is finished. At Bījāpur, the qanāt was excavated in permeable basalt, chunam being used to seal its lower portion. Its presence provides for hydraulic technology brought by the Pardesis, the foreigners at the court of the Muslim Kingdom of the Deccan, from the Middle East. The qanāt is thought to have originated in the eighth century B.C. near Urartu, in the highlands of Armenia and Turkestan (Garbrecht 1980:306-311; Costa 1983:274) and is found throughout the Middle East. Qanāts do not suffer from evaporation. Their supply is also protected from wind-blown deposits. However, their construction demands accurate levelling, to ensure the creation of the correct gradient necessary to bring the water from the point of supply to demand. The unusual hydrological environment at Bījāpur, which is characterised by water-bearing, fissured basalt, made possible the construction of a qanāt of some size, stretching from the village to Torweh into the City. The construction of qanāts at Vijayanagara is impossible due to the nature of the granitic bedrock which is far more suitable for the building of diversion weirs on perennial rivers and the use of embanked canals.

Decccani forts were in some cases provided with water supply systems. The old Hindu fort of Koilkoṇḍa, 23 km. to the southwest of Mahbūbnagar, contains extensive evidence for domestic supply. The fort, perched on a high hill, was modified and enlarged under the Qutb Shāhs of Golkoṇḍa. An inscription indicates that the garrison of the fort successfully helped Ibrahim Qutb Shāh ascend the throne of Golkoṇḍa in A.D. 1550. The water features, which appear to date to the reign of Ibrahīm Qutb Shāh (A.D. 1550-1581), consist of three reservoirs located at different levels on the side of the hill (ARHAD 1928:2-4). The three reservoirs were connected by conduits which allowed excess water to pass from the highest in the series down into the middle tank, any excess flowing down to the lowest receptacle in the series. Water from the highest and largest tank was raised by a water-lift and passed into a massive aqueduct feeding the quarters of the garrison commander. At the top of the hill on which the fort is built, a natural depression has been modified for the collection and storage of rainwater.

In Western India, there is extensive evidence for domestic supply at the medieval city of Champaner. Excavations in the palace area revealed a pleasure-garden, with water channels and decorated tanks of plaster (IAR 1972-3:12, pl. XIXb; 1974:14-15, pls. XIVa,b, XVa,b).

In North India, domestic water systems were built at some Muslim capitals of the fourteenth to sixteenth centuries A.D. At Tughlakabad, Ghiyās-ud-dīn (A.D. 1320-1325) constructed a series of rainwater storage tanks for domestic supply inside the citadel walls (Hearn 1906:108). The citadel of 'Ādilābād constructed by Muhammad-bin Tughlak (A.D. 1325-1351), is joined to Tughlakabad by a fortified bund. 'Ādilābād is provided with a number of drainage channels (Waddington 1946: pl. XIII). Wastewater appears to have been passed through the defensive wall. Close to Tughlakabad at Lal-Kot, south of the Qutb Minar Complex, a lime-plastered, street gutter with interconnected house drains was found in the medieval layers (IAR 1957-8:25, pl. XXVIb).

Fatehpur Sīkrī, the fortified capital of Akbar from A.D. 1571 to 1591, was provided with an elaborate domestic supply system. The city is located 37 km. from Agra, in Uttar Pradesh. The interlocking arrangement of courtyards, which made up the citadel zone are located on top of a lofty, sandstone ridge. In this area were located the palaces or Daulat Khana, the Zenāna, treasury, offices, royal workshops or Kārkhānas, ornamental gardens, and houses of the king's officers. The houses and gardens of the nobles and merchants clustered around the side of the ridge. Water was raised up some 45 m. to the citadel by a complex series of windlasses at different levels (Hussain 1970:46-47, 96-97; Rizvi and Flynn 1975:111-113, 120-121). The supply was secured solely from two large step-wells, the North Bā'oli and Shā Qulī's Bā'oli, at the foot of the ridge and was lifted, in up to five stages, by pairs of cattle. The wells are located on the north and south side of the ridge respectively. An extensive open channel and sealed pipe system spread throughout the royal palaces. Glazed and unglazed pipes were produced to a standard diameter of 11.4 cm. (Rizvi and Flynn 1975:121-122). Overflow from the channels and pipes was stored in a special tank, called the Sukh Tāl, and reused.

Channelled and piped water on the ridge fed fountains, basins, tanks, gardens, toilets and kitchens in the citadel (Husain 1970:53-54,69,76,78; Rizvi and Flynn 1975:20, 30-31,57). Several public and private baths (ḥammāms), provided with hot and cold water, were located on the ridge (Hussain 1970:42-46,76; Rizvi and Flynn 1975:26,65,91,93; IAR 1976:48-49, pls. XXXIXa,b, XLa,b; 1979-80:71-72). In at least one case, a residence was provided with a piped, underfloor, heating system (Rizvi and Flynn 1975:21). The perennial supply on the ridge was supplemented by the ingenious conservation of rainfall. In the Jāmi'Masjid, the great mosque to the southwest of the palaces, consummate skill was employed to conserve rainwater by a complex arrangement of gutters, hollow pillars and underground channels at different level (Vats 1950:99, pls. XXXVI, XXXVII). It was then diverted into a huge, underground cistern known as the Birkha (Hussain 1970:104,116-117; Rizvi and Flynn 1975:72-73,77). Rainwater on the west side of the mosque was channelled down to the Jhālrā reservoir, located some 25 m. below, from which it could be lifted to other tanks (Husain 1970:116; Rizvi and Flynn 1975: 90-91).

The evidence cited above, suggests that domestic supply systems employing open channels and pipes became widely used in India from the early fifteenth century A.D. onwards in the context of urban settlement. At Vijayanagara, apart from the supply of perennial water by pipe and open channel to the 'Royal Centre', domestic water supply relied upon wells and the collection of rainwater in tanks. In the Muslim cities, hydraulic technology was used to supply high status dwellings and municipal features with perennial water. The presence of a qanāt at Bījāpur indicates that the hydraulic technology of Muslim cities in the Deccan was influenced by that of the Middle East. Hydraulic technology developed in the dry environment of the Middle East could have been used in the Deccan, which possessed a similarly arid climate, with little modification. Contact between Vijayanagara and the Muslim kingdoms of the Deccan is attested by political history and the borrowing of artistic and architectural styles. The movement of artisans and engineers must also be presupposed. It is, therefore, possible that the supply system of the 'Royal Centre' at Vijayanagara was influenced indirectly by foreign technology.

However, archaeological evidence attests to an indigenous and ancient tradition of drainage channels and pipe conduits in South Asia. Extensive evidence for this tradition also comes from Nepal and Sri Lanka. In particular, at Sigiriya (*c.* sixth century A.D.) in Sri Lanka, the royal water gardens include underground stone ducts, and fountains (verbal information from Dr. F.R. Allchin). In our opinion, the supply system for the 'Royal Centre' at Vijayanagara represents the culmination of this tradition. At Vijayanagara, where the hydrological environment was characterised by a shortage of water, great emphasis was placed upon the supply and conservation of water, rather than drainage. In the 'Royal Centre', the well-established tradition of constructing open channels and pipes was used to procure a perennial supply.

5.3.4 To what extent did the success of Vijayanagara as an urban centre depend upon the operation of its domestic water supply system?

On a practical level, Vijayanagara was provided with domestic water from wells and rainwater collection tanks built by individuals. Temple tanks also provided a valuable source of water for the religious and day to day needs of the inhabitants. The bulk of the urban population would thus have been supplied with water for drinking, cooking, washing, sanitation and ritual activities by fairly basic means. There is no evidence on the ground for a municipal drainage system. The technology involved in the supply of water to the bulk of the population was no different from that found in villages and towns.

The perennial supply system to the 'Royal Centre' functioned solely to provide water to a restricted, walled areas associated with the king and his entourage. Although the water brought into the 'Royal Centre' was in part used for practical purposes, such as food preparation and sanitation, large quantities were required for ritual activities. Much was also expended in filling baths, pools and fountains and for watering the royal gardens. The raison d'être for the construction of the supply system to the 'Royal Centre' was not dictated by practical considerations any more than the prestige gardens of Roman dignitaries, or the water gardens of Sigiriya in Sri Lanka.

The Vijayanagara kings appear to have expended wealth and scarce resources as a means of acquiring symbolic prestige. Conspicuous consumption of water in a semi-arid environment by an elite, may be interpreted as a method by which the symbolic prestige of the urban settlement in which they dwelt could be increased. The portrayal of the urban settlement as successful was of great importance in attracting scarce population and trade (see section 1.5). It may be concluded, therefore, that although the general population of Vijayanagara did not directly benefit, in a practical sense, from the existence of the perennial supply system to the 'Royal Centre', it was essential for the generation of royal and locality charisma, upon which the success of the urban settlement in a politically unstable environment depended.

5.4 Agricultural water supply

A summary examination of the field data relating to agricultural water supply at Vijayanagara makes it clear that previous writers seriously underestimated the complexity of the system and its history. There has been a marked tendency in recent literature, when discussing the agricultural supply system, to ignore the changes which have occurred to it since the medieval period. In addition to this, rather than examining the extensive evidence on the ground, earlier writers based their hypotheses concerning agricultural water supply on information contained in the translated accounts of contemporary visitors, particularly Paes and Nuniz. The wealth of information relating to the Tungabhadrā canals, contained in Public Works Reports and archival material, has not been examined previously.

The agricultural water features of the city were first noted by Mackenzie (1800a). He mentions the Kamalāpuram tank and suggested that it originally watered the royal gardens (1800a:133-134). He stated:

> From the peculiar situation of this city in narrow valleys whence the river is pent up by rocks and hills, the bottoms between which are cultivated with plentiful crops of paddy watered by many conduits carried from the Tommbudra (1800a:134-135).

Mackenzie also recorded an important local tradition, which stated that several canals and the tank at Danāyakanakere were built by a minister of Kṛṣṇadeva Rāya called 'Moondada Maick' (1800s:133). The same tradition was also noted by Francis (1904:90-91).

Although they are not labelled or referred to in the key, a number of agricultural water features are included on Mackenzie's 'Plan of the Ancient City of Beejnagar' (1800b) (Michell and Filliozat 1981:2-3). The Kamalāpuram tank and a number of canals, including the Hiriya kāluve, are represented. The canals are not accurately drawn. Although the Hiriya canal is correctly shown to be abutted with the channels taking off from the Kamalāpuram tank, no connection with the river is indicated.

The next reference to the agricultural supply features of Vijayanagara is provided by Newbold (1845:518), who reported that some of the larger, flat-bottomed valleys were irrigated by canals. Ten years later, Pharaoh's Gazetteer referred to the anicut-fed canals on the Tungabhadrā (1855:83,85). The Koragallu, Vallabhāpuram, Kūruvagadda, Hosūru, Hiriya and Rāmanagadda anicuts were best catalogued and described by Kelsall (1872:27, 231-232) and Francis (1904:78, 90-92). With regard to dating, Kelsall concluded that all the anicuts, with the exception of the modern replacement at Vallabhāpuram, were built during the medieval period (1872:232). A brief reference to the Tungabhadrā anicuts was included in the imperial Gazetteer of India (IGI 1908:169, 437; 1909:34, 99).

The Koragallu inscription, which refers to the construction of the weirs at Koragallu and Vallabhāpuram and the Rāya canal by Kṛṣṇadeva Rāya in A.D. 1521 was first recorded by Kelsall (1872:231). Sewell (1900:162) included details of the inscription taken from Kelsall. Buckley concluded that the majority of the old weirs on the Tungabhadrā were constructed by Kṛṣṇadeva Rāya, but did not directly refer to the inscription (1905:13). The inscription received no attention after Sewell, apart from brief references in Francis (1904:38), Venkayya (1906:210) and the Imperial Gazetteer of India (IGI 1908:437).

References to the agricultural supply system are contained in the Administration Reports of the Madras Public Works Department, the Records of the Madras Government and occasionally Parliamentary Returns. From the mid-nineteenth century, the medieval catchment drainage tanks and anicut-fed canal systems, upon which the success of local agriculture was based, began to require maintenance. They had operated efficiently since the medieval period, but by this time were in need of renovation and repair. Details of the work carried out by British engineers of the Madras Presidency were recorded in great detail. However, no particular interest was paid to the builders or the chronology of these works.

The Kamalāpuram tank was mentioned by Pharaoh (1855:83), Kelsall (1872:16) Francis (1904:88,282), Longhurst (1917:6) and Devakunjari (1970:72). All asserted that the tank's supply was boosted by the Rāya canal. A Government of Madras Archaeological Survey Report (1902:15) indicates that one of the tank's outlets supplied water to Hallupatnam through channels and pipes. Hallupatnam is the old local name for the 'Royal Centre', which is no longer used. Menon reported the existence of two sluice outlets dating to the Vijayanagara period (1987:78). It was suggested in the Annual Reports of the Archaeological Survey that the tank provided perennial water for the 'Royal Centre' (ARASI 1904-5:29; 1908-9:23). Morrison reported the Mallappannagudi bund (1988: 8-9).

Sewell (1900) provided translations of several contemporary accounts of Vijayanagara. They contained two important descriptions of irrigational building projects. The first was the construction of a large catchment drainage tank during the reign of Kṛṣṇadeva Rāya, witnessed by Paes in A.D. 1521 and also recorded by Nuniz some 14 years later (Sewell 1900:244-245, 364-365). Kelsall (1872:16-17) noted the large, dry tank southwest of Hospet. Sewell correctly identified this tank as the Rāya ker described in the contemporary accounts (1900:162-163). This identification was followed by Longhurst (1917:48-50), Michell and Filliozat (1981:50) and Menon (1987:76-78).

The second description records the building of an anicut and canal (Sewell 1900:301-302). It is contained in the dynastic history of Nuniz and not corroborated by Paes. The account is important for three reasons. Firstly, it enables us to date the construction of one canal on the Tungabhadrā to the beginning of the fifteenth century A.D. Secondly, it refers to the sponsorship of an irrigational scheme by the king. Thirdly, it provides the only contemporary description of anicut building.

The section of dynastic history which refers to the construction, indicates that the anicut was built by the successor of King Harihara II (A.D. 1379-1399), called 'Ajarao'. Sewell (1900:51-52) recognised that the reign of this unknown King coincided with those of two known successors of Harihara II. He therefore ascribed the events in the account to the reigns of Bukka II (A.D. 1399-1406) and Devarāya I (A.D. 1406-1413). Sewell's hypothesis was culled by subsequent writers, but, in most cases, incorrectly. Francis (1904:91,271), Longhurst (1917:56) and Devakunjari (1970:47) dated the account to the reign of Bukka II, whilst Menon (1987:76) thought Devarāya I was solely responsible. Only Sastri (1950:i.124) correctly followed Sewell's argument, stating that the events in the account probably dated to the reigns of Bukka II and Devarāya I.

With regard to the identification of the anicut-fed canal described by Nuniz, Sewell ambiguously suggests:

> ... his great work was the construction of a huge dam in the Tungabhadrā river, and the formation of an aqueduct fifteen miles long

> from the river into the city. If this be the same channel that to the present day supplies the fields which occupy so much of the site of the old city, it is a most extraordinary work. For several miles this channel is cut out of the solid rock at the base of the hills, and is one of the most remarkable irrigation works to be seen in India (1900:51-52).

Two canals, the Hiriya and the Rāya, fit his description. Subsequent scholars took Sewell's description to mean the Hiriya Canal, unquestioningly and without further investigation (Francis 1904:91, 271; Longhurst 1917:56; Devakunjari 1970:47; Menon 1987:76). Only one short section of the Hiriya canal is cut from solid rock, west of the village of Kṛṣṇapuram, whereas several long sections of the Rāya canal are cut through a range west of the 'Urban Core'. For the reasons given above, we have suggested that Sewell is referring to the Rāya canal (see section 5.4.2).

The Setu aqueduct, which carried water from the Anegundi canal to the island of Virūpāpuragadda, received some attention in the literature. The aqueduct was first reported by Suranarain (1909:30). Michell and Filliozat (1981:50, 59) and Deloche (1984:26) provided descriptions, but did not explain the relationship of the aqueduct to the Anegundi canal. Purandare, who also catalogued the wells and tanks of Anegundi fort (1986:239-263), provided an account of the Anegundi anicut at Sānapur (1986:259-260). Morrison included an account of the Setu aqueduct in her regional survey of Vijayanagara (1988:7-8).

A brief notice of the Hiriya canal and the complex water features of the irrigated valley appeared in Fritz *et al.* (1984:10). Deloche described the Matanga bridges and suggested that they may be contemporaneous with the ruined bridge over the Tungabhadrā (1984: 25-26). Menon briefly mentioned the canals in the irrigated valley (1987:76).

The investigation of the agricultural water supply features on the ground was complicated by two factors. Firstly, the completion of the Tungabhadrā Dam submerged the old anicuts at Koragallu, Vallabhāpuram, Rāmanagadda, and Kūruvagadda. It is a great pity that no rescue archaeological survey was carried out at the time (and that Allchin, who recorded a number of prehistoric sites since submerged by the Dam, did not visit these anicuts). Secondly, before the construction of the Dam, the Madras Presidency Public Works Department carried out a number of modifications in order to maintain the efficient working of the Medieval tanks, anicuts and canals. These modifications, which included two lengthenings of the Basvanna canal, are described in detail in sections 4.2.1 and 4.2.2.

5.4.1 How did the agricultural supply system operate?

Before discussing the operation of the agricultural supply system, it is necessary to recall the changes that have occurred to the hydrological environment since the Vijayanagara period. The construction of the Tungabhadrā Dam has raised the water table. Today, wells remain full throughout the year and can be used to provide perennial water for wet cultivation. Water is available in areas where no cultivation had been possible using the irrigational system of the Vijayanagara period. This gives the impression that the area is less dry than it would otherwise be. Before the construction of the Dam, the Vijayanagara anicut-fed canals and catchment drainage tanks provided the only reliable source of water for cultivation away from the low-lying areas near to the river. Although wells now remain full throughout the year, in our view, their supply was not sufficient or reliable enough to permit wet-farming before the 1950s.

Irrigational features of the Vijayanagara period have been incorporated into the Tungabhadrā Dam Project. In particular, the Rāya and Basavanna canals were abutted to the Dam at Mālāpuram. Several medieval features were superseded or abandoned in favour of more modern devices connected to the Dam. A good example is the Mallappannagudi tank, which was made redundant by the construction of the Tungabhadrā Right Bank High Level Canal.

Agricultural water supply or irrigation may be defined as the provision of water to agricultural land raised by artificial means to encourage the growth of cultivated plants, shrubs and trees. Artificial watering has two interconnected purposes. Firstly, it is a method of minimalising climatic or seasonal variability. The seasonal unpredictability of the semi-arid environment necessitated the development of an effective method of perennial watering to sustain settled habitation. A settled population, dependent on agriculture required protection against the vagaries of the climate. During the Vijayanagara period, there were at least seven major famines in South India; A.D. 1335 (Sastri 1950:i.53, 74), A.D. 1390 (Sewell 1932:204-205; Sastri 1950: ii.110), A.D. 1396-1408 (Sewell 1932:205), A.D. 1412 (Briggs 1908:ii.347; Ramaswami 1923:230; Sewell 1932:211), A.D. 1422 (Sewell 1932:214), A.D. 1474 (Sewell 1932:227; Sastri 1950:ii, 114) and A.D. 1540 (Rāmaswami 1932:232); Sewell 1932:248). Perennial watering provided a means of insuring against climatic uncertainty. Secondly, irrigation could be used to exploit and increase the agricultural potential of a given environment. It made possible:

1. The intensification of agricultural production, i.e., multi-cropping.
2. The realisation of genetic potential of plant crops under optimal conditions.
3. The extension of areas already under cultivation, particularly tracts away from rivers.

Thus a limited range of resources (land and population) could be exploited to greater advantage by investment in irrigational projects. Brown (1912:8) has indicated that in India, artificial irrigation increased agricultural output by as much as 30 per cent.

The catchment tanks of Vijayanagara were formed by the construction of earthen embankments across natural drainages and are typical of those built throughout South India in the medieval period. Earthen bunds of this sort are gravity dams, which resist the pressure of the water they hold by their weight alone. Unlike modern earthen dams, they are homogeneous in section and without a puddle wall or core. Puddle cores are used to prevent seepage or infiltration. This is the process by which water penetrates into the mass of the embankment, thereby reducing the frictional resistance and adhesion of the earthwork (Brown 1912:75). However, it has been suggested in the context of Indian earthen dams, that the introduction of a puddle wall destroys the homogeneity of the earthwork (Strange 1927:182-183). Earthen dams are quite capable of withstanding the pressure of the water they hold up, with limited seepage (Strange 1927:159-163). The foundations must be able to carry the bund with only slight and uniform compression when wetted by seepage and should not include any porous layers (Strange 1927:183-185). The failure of the Rāya bund may be attributed to the latter.

Surface run-off collected upstream of the bund, which was revetted with boulders and stone blocks. This was necessary to protect it from wave action resulting from wind acting on the surface of the reservoir (Strange 1927:192-195). The construction of revetting also stabilised the earthwork and protected it from guttering (Brown 1912:73). The downstream faces of bunds were also sometimes revetted. Although the revetting was not grouted, a bund could be effectively staunched by the collection of silt (Strange 1927:139-140).

Tanks are subject to a number of storage losses. Evaporation is probably the most important of these (Buckely 1905:63-65; Strange 1927:139-140; Balek 1983:240-241). Its rate depends upon water temperature, wind speed and aridity (Strange 1927:23). Severe losses also result from percolation and seepage (Strange 1927:19-21).

Large quantities of silt are brought down into tanks with the surface run-off (Pharoah 1855:84; Francis 1904:88). Once in the tank, silt (including fertile matter) tends to settle out, raising the bed. If the silt is not cleared, the tank is eventually rendered useless. British engineers fitted under-sluices to new earthen dams, in order to allow the flushing out of silt (Strange 1927:138-139). Under-sluices were not fitted to medieval bunds. We know from inscriptions that provisions were made for the regular removal of silt from tanks. This suggests that the problem of silting was appreciated. Land and money were often granted to provide labourers with boats and carts for silt clearance (Venkayya 1906:205-208; Mahalingam 1951:49; Gopal 1985:13-14,158).

However, no means of excluding silt, other than manual labour, was found. The digging out of silt was an effective method of keeping tanks functional, but it demanded continued adherence to the terms of the grants. By the nineteenth century, many tanks had become badly silted and the grants, which had been established for the purpose of digging them out, long forgotten. This suggests that some measure of political order was necessary to ensure that agreements, even those beneficial to the community, were implemented.

Tanks were fitted with outlets which allowed water to be safely passed through the bund, so that it could be utilised for the purpose of irrigation. Medieval outlets take the form of culverts under the dam and can operate under a variable head (PPC, printed 27th January, 1869:44). At Vijayanagara, pipes were let through the fabric of the bund at bed level. A masonry frame, located above the upstream end of the pipe, facilitated the regulation of water entering the outlet by the vertical movement of a board attached to a rod. Water passed through the bund into an open, masonry tank on the downstream side. A number of channels took off from the tank, the supply to which could be regulated using stone slabs. Transport losses through evaporation and absorption were reduced, as the irrigable area was usually close to the sources of supply. The tanks were a rich supply of fish, traditionally caught from a coracle. To ensure that water did not rise above a safe height, a waste weir was sometimes provided on one flank of the bund. At Vijayanagara, we find only one type of waste weir, an open spillway without gates.

The supply of the Kamalāpuram tank was supplemented artificially by the Rāya kāluve. The canal was also used for irrigation en route. This method of boosting the supply to tanks was not new. Catchment drainage tanks fed by channels were built in south India from at least the seventh century A.D. onwards. A tank called Parameśvara-taṭāka constructed by the Pallava King Parameśvaravarman, 14.5 km. northwest of Kāñcīpuram, was provided with an additional supply by a feeder channel from the river Pālār (Venkayya 1906:203). The tank is dated to the second half of the seventh century A.D. The Uyyakkoṇḍāṉ channel, which takes off from the Kāverī near Trichinopoly, empties into a tank at Valavandankoṭṭai (Venkayya 1906:209). It was completed by the Coḷa King Kulottuṅga in A.D. 1206.

We have found only three contemporary descriptions of tank building in the Vijayanagara period. Two are provided by Paes and Nuniz, who recorded the construction of the ill-fated Rāya bund southwest of Hospet (see section 4.7). Paes' account suggests that the work, which was paid for and presided over by the king, was portioned out amongst his nāyakas (Sewell 1900:245). As many as twenty thousand people may have worked on the project (Sewell 1900:245). There is no indication how long the work lasted. Both accounts state that a number of men were decapitated as a sacrifice. Paes records that sixty men were killed in this way (Sewell 1900:245). According to Nuniz, these men were prisoners (Sewell 1900:365). There is further evidence for the association of human sacrifice with hydraulic projects, particularly tank building. Excavations in the 'Great Tank' (IVc/1) revealed a vaulted chamber beneath the floor (ARASI 1904-5:30; ASMC 1904-5:36). Inside, the skeletal remains of a single man were found. The excavators concluded that the remains represented a human sacrifice, carried out when the tank was constructed (see section 3.2.2). It is interesting that a local tradition indicates that Mudda, a minister of Kṛṣṇadeva Rāya, was buried beneath the Rāya bund (Francis 1904:90-91) (see section 4.7).

More detailed information is provided by a third source, the important Porumāmiḷḷa tank inscription (EI XIV:97-109). The epigraph, which is dated A.D. 1369, refers to the building of a large tank by Bhāskara Bhavadūra, the son of Bukka I, in Badvel taluk, Cuddapah district. According to the inscription, one thousand labourers worked on the bund every day for two years. One hundred carts were used to transport the stone from a quarry to the construction site. The stone was used to construct the upstream rivetting (bhramā-bhitti) and four sluice outlets (chatur-bhramā-jala-gati). The tank is 4.25 km. in length and has a maximum height of 10 m. This makes it bigger than the largest of the tanks examined in this investigation, the Kamalāpuram tank, which has a bund 2 km. in length. However, the inscription gives a valuable guide to the length of time and the number of workmen employed in the construction of large tanks.

The inscription includes a valuable list of instructions for the building of tanks. It states that the following were needed:

> ... a brahman learned in hydrology (pāthas-śāstra), and ground adorned with hard clay, a river with sweet water (and) yojanas distant (from its source), the hill parts of which are in contact with it (i.e. the tank), between these (portions of the hill) a dam (built) of a compact stone wall, not too long (but) firm, two extremes (pointing) away from fruit-giving land outside, the bed extensive and deep, and a quarry containing straight and long stones, the neighboring fields, rich in fruit (and) level, a water course (i.e. sluices) having strong eddies on account of the position of the mountain, and a gang of men (skilled in the art of) its construction... (EI XIV:108).

Problems likely to be encountered are also mentioned:

> ... water oozing from dam, saline soil, (situation) at the boundary of two kingdoms, elevation in the middle (of the tank) bed, scanty supply of water and extensive land (to be irrigated, and scanty ground and excess water ... (EI XIV:108).

This suggests that the Medieval engineers understood the problems of seepage and salt efflorescence (see above).

Kelsall noted that in Bellary district, a specific group, the Vodda Vandlu, were recognised for their skill in constructing bunds (1872:83-84). In Madras Presidency, the Telugu Oddes or Wudders were responsible for tank digging (Thurston 1913:201). The Wudders used the spade as their caste insignia.

The practice of constructing large catchment drainage tanks to irrigate agricultural lands in and around large urban centres, may be seen at other early historic and medieval capitals. A network of tanks was built to supply the cities of Anurādhāpura and Pollonaruwa in Sri Lanka from the third century B.C. to the twelfth century A.D. (Brohier 1935; Leach 1959; Fernando 1980). Information relating to their construction is contained in the Sinhalese chronicle, the Mahāvaṃśa (*c.* sixth century A.D.).

South of the modern city of Delhi, Muhammad-bin Tughlak (A.D. 1325-1351) constructed the citadel of 'Ādilābād and in A.D. 1328, Jahānpanāh ('The Shelter of the World') (Hearn 1906:104, 194-195; Husain 1963:585-586). 'Ādilābād is joined to Tughlakabad by a fortified bund designed to trap a natural drainage which flowed along the south side of the city (Waddington 1946:60, pl. XI). A series of sluices were let into the north end of the bund (Waddington 1946: pl. XIII). The tomb of Ghiyās-ud-din was located on an artificial island in the reservoir and was reached from Tughlakabad by a causeway. Two further bunds were located to the south and southwest of 'Ādilābād.'

The city of Jahānpanāh joined Sīrī, originally built by Alā-ud-dīn Khalji (A.D. 1296-1326), to Old Delhi (Hussain 1963:615). Just east of the village of Khirki, a complex, fourteen-gated sluice is let into the south wall of Jahānpanāh. The sluice, which is 61 m. in length and is composed of two tiers of seven outlets, controlled the flow of water from the tank (Hearn 1906:110; Husain 1963:616). The sluice is known locally as the Satpulah and was probably built in A.D. 1326 (Hearn 1906:111; Husain 1963:585-586). The city wall at this point acts as a bund, trapping the waters of a stream which flows through the city of Jahānpanāh and eventually discharges into the Yamunā river south of Humāyun's tomb.

At Fatehpur Sīkrī, a masonry bund called the Terah Morī was built by Akbar to dam the waters of the Khārī Nadī (Husain 1970:9-10, 12,13-14, 29; Rizvi and Flynn 1975:8, 115-116,129). The bund, which was fitted with regulating sluices, inundated the land to the north of the city, thus obviating the need for a defensive wall on that side. An earlier tank had been constructed on the same site by Bābur (Rizvi and Fynn 1975:11). Abul Fazl records in the *Akbarnāma,* that the bund burst in July, A.D. 1582, drowning a number of people taking part in Akbar's birthday celebrations downstream of the tank (Hussain 1970:29). The tank must have been repaired, as the *Tuzuk-i-Jahangiri* indicates that it was approximately 23 km. in circumference when

Jahangir held an entertainment on the bund in A.D. 1619 (Husain 1970:29-30). By the mid-nineteenth century, the bund had been breached and its bed turned over to agriculture (Husain 1970:30; Rizvi and Fynn 1975:115).

At Bījāpur, a number of large catchment drainage tanks were constructed during the Medieval period (Cousens 1916:21; Rotzer 1984:151-172). These are now dry. The Begam Talāo was built to the south of Bījāpur, to supply the city with domestic water through a complex pipe system (Cousens 1916:121). Six bunds, labelled B1 to B6 by Rotzer (1984), are located to the west and northwest of the city. B1 to B4 trap small drainages, whilst B5 and B6 are built in series on a large drainage called Rāmling. With the exception of B6 (the Hauḍ-i-Shāhpūr), the bunds appear to have been constructed entirely between the sixteenth and seventeenth centuries A.D. However, Rotzer (1984:155, 162-163) has proposed that parts of B6, which was probably completed by 'Ali 'Ādil Shāh (A.D. 1557-1580), were begun before the Muslim period. Tanks B1 to B6 served two purposes. Firstly, they provided water for intensive, wet agriculture in the environs of the city. Secondly, they furnished Bījāpur and its suburb, Shāhpūr, with domestic water. A further, small catchment tank (B7) was built to the east, between Bijāpur and the village of 'Aīnpūr'.

The Bījāpur bunds, which were massively built to compensate for the permeability of the soil, rest on bedrock. It appears that some attempt was made to provide them with an impermeable upstream face. Their heavy revetting was grouted and thickly covered with lime-mortar, which may have been mixed with ash to increase its impermeability (Rotzer 1984:153,170). This precaution is not found at Vijayanagara. The Bījāpur tanks are provided with elaborate outlets, from which water is drained off at several levels and fed into aqueducts and channels (Rotzer 1984:153-155). The outlets consist of openings, let into the upper faces of a small flight of steps, which supply water at a variable head. They are totally different in function and design to those outlets found at Vijayanagara and we suggest that they may have been influenced by hydraulic technology from the Middle East.

Anicuts were used at Vijayanagara to secure a perennial supply from the Tungabhadrā. Unlike catchment drainage tanks, anicut-fed canals had the benefit of carrying fertile silt to the fields. The purpose of the anicut or diversion weir, which is essentially a solid obstruction placed in the path of the river-flow, is to raise the water level in order to force a supply into a canal (Brown 1912:106-107). The obstruction creates an afflux, which back up the water upstream of the weir (Brown 1912:112; Strange 1927:53-54). Water is taken off in a canal upstream of the anicut when the level of supply is such that it can, by gravity flow, command the area to be irrigated. Anicuts were built in the medieval period on a number of perennial rivers in South India. The best known is probably the 'Grand Anicut' at Śrīrangam on the Kāverī (Walch 1896:29; Buckely 1905:101, 109-110,113; Brown 1912:30,111; Ludden 1985; 21; RIIC:i, 9).

It is likely that the indigenous engineers judged the relative merits of locations for anicuts, using experiences gained through trial and error. For a diversion weir to be effective, it had to be built on a river with a good perennial supply. The river was not normally dammed across its whole width. Anicuts built across entire streams tend to trap silt and quickly become choked (Strachey 1987:76). Permanent weirs were only built where a foundation of rock was available. The medieval builders never managed to devise a successful method of constructing foundations in rivers with sandy beds (PPC, printed 27th January, 1869:45). The Tungabhadrā is ideally suited to the construction of anicuts, as it is relatively easy to join one bank of the river to one of the numerous low islands which divide the river-flow into channels. There is no evidence for the failure of a permanent anicut at Vijayanagara. However, there is some evidence for failure elsewhere in South India. The Harihara stone inscription of A.D. 1410, indicates that a weir on the river Haridrā was breached soon after its completion (Venkayya 1904:210; Mahaligam 1951:55; Kuppuswamy 1976;135-145).

The creation of an obstruction in the river flow tends to cause silting (Brown 1912:112; Strange 1927:59-61). Modern diversion weirs are usually fitted with under-sluices, which allow the silt collected upstream to be washed out when the canal is closed (Buckly 1905:109; Brown 1912:174). The Vijayanagara anicuts were not fitted with under-sluices and the resulting collection of silt reduced the amount of water held upstream of the anicut. Strange (1927:210-212) noted that irrigators often requested that old anicuts be raised to compensate for the deposition of silt and thus enlarge the storage capacity of the backwater pool. All of the Vijayanagara anicuts have been raised in modern times for this reason. One benefit of silt deposition is that it staunches the upstream face of the anicut and increases its impermeability (Francis 1904:91; Brown 1912:114; Strange 1927:254-255).

The Kūruvagadda, Koragallu, Vallabhāpuram and Hosūru weirs were constructed of loose stone. The fabric of the weirs was probably stabilised by the use of clay as a bonding material. Walch (1896:29) notes that in Tamil Nadu traditional diversion weirs were constructed of loose stone in mud, grouted and covered with a layer of chunam. The Hiriya weirs were composed of dressed masonry bonded with metal clamps. The practice of binding crest stones was found in several places in South India by Buckley (1905:109).

The use of a temporary weir crest enabled cultivators to increase the amount of water in the backwater pool. A greater supply could thus be secured during the important winter months when the water level in the river was low (Brown 1912:112). During the monsoon, the crest could be lowered during the first floods, allowing the silt brought down with the initial run-off to be passed over the weir. When the first flood was over, the crest could be put up again to hold an increased supply. In this way, it was possible to discourage some bed silt from entering the canal. Weir crests are usually composed of vertical needles supported by horizontal wales (Strange 1927:82-85). At least one of the Hiriya anicuts was provided with holes for the fitting of wooden weir crests. Buckley (1905:114) noted the existence of similar holes in the rocks near the sites of anicuts in Nasik district, Maharashtra. The remains of stone needles are also extant on the Hosūru anicut and the waste weir of the Kamalāpuram tank.

No head regulators (which are used to shut out flood water from the canal, to control the admission of silt into the canals and to regulate the discharge passing into canals) were fitted at Vijayanagara. In consequence, it was not possible to prevent changes in the velocity of the canal, upon which silt deposition depends. It is possible to discourage bed silt from entering a canal if water is admitted to the canal at a high level (Buckley 1905;39). However, when only the top layer of water is allowed into the canal, bed silt deposition is precipitated in front of the weir and eventually blocks the entrance to the canal. This appears to be what happened at Vijayanagara.

Water was carried by gravity flow from the anicuts and tanks to the agricultural zones. It was kept at a height which commanded the land to be irrigated. The canals were constructed in excavation in flat ground, but were also built against ranges in excavation and embankment when it was necessary to cross irregular terrain. The latter form is best illustrated by the contour-hugging canals used to maintain high command whilst flowing through deep valleys. Occasionally, canals were excavated in solid rock. Natural features were used to good effect by the builders, who led canals around large boulders. Where necessary, natural boulders were accurately split to allow the canal to flow freely. Walch describes the traditional method by which granite can be cut very precisely, using a controlled wood fire and hammers on the surface of the rock (1895:272). His description refers to the Bangalore area, where thin sheets or slabs of granite were produced for construction. Newbold (1842:113-128) also described this process. The canals are too narrow and irregular to be used for the transportation of people or goods. At Setu, an aqueduct of heavy masonry was constructed to transport a subsidiary branch of the Anegundi canal across the river to the island of Virūpāpuragadda. Although none of the canals observed in the field were originally provided with headworks, some were fitted with escape channels at the head, to dispose of surplus discharge.

Rapids would appear to have been constructed where a canal has to overcome a change in gradient. They allow the supply to be taken to a lower level, without an increase in velocity (which would cause erosion and scouring), by breaking the force of the water. There are several examples of this on the Hiriya canal. There are some indications that part of the Hiriya canal were plastered to protect them from erosion in areas where turbulence occurs. Rapids could also be used to compensate for an overly steep gradient, resulting from an error in levelling when the canal was constructed.

Canals are subject to losses through evaporation, seepage and absorption (Buckley 1905:67; Balek 1983:241; Farrington 1985:287). Evaporation losses are determined by water temperature, winds and aridity (Strange 1927:23). Seepage and absorption occur through porous parts of a canal, usually where it is cut through the earth and lacks a lining (Strange 1927:19-21). The extent of the losses depend on the nature of the soil and the dimensions of the channel. Losses decrease as the saturation time increases and, interestingly, they are greater when the canal is in excavation rather than embankment (Strange 1927:24-27). Brown (1912:35) has suggested that transport losses between the canal head and field can account for between 30 and 70 per cent of the supply. In the Punjab, the quantity of water applied to the land is only 28 per cent of that supplied at the head of the canal (Strange 1927:27-30).

Losses through seepage and absorption can be significantly reduced by lining a canal with an impervious material. However, British engineers thought that a cheaper and more permanent solution was to widen, rather than line, canals when the water supply was abundant (Strange 1927:30-32), as it is in the case of the Tungabhadrā. A widening of the canal would reduce velocity and thus inevitably cause silt deposition. It is possible, therefore, that the builders of the Vijayanagara canals compensated for heavy transport losses by increasing the discharge of the canal.

The velocity of a canal (i.e. the mean speed of water expressed in m. sec.$^{-1}$, flowing past a given point) is determined by its cross-section, gradient or longitudinal slope, as well as the type of material from which the canal is constructed. Farrington (1985:291-293) provides details of the Manning formula, which can be used to calculate velocity for canals of uniform flow. Theoretically, if silting is to be prevented, the velocity of flow in an anicut-fed canal should be the same as that of the take-off point on the river (Brown 1912:173). As silt deposition is precipitated by changes in velocity, it is desirable to maintain a constant velocity throughout the canal, so that suspended silt may be carried through to the outlet or fields (Strange 1927:237-239). This is, of course, difficult to achieve in reality. Changes in direction and gradient can result in flow becoming unstable and the creation of turbulence and eddies.

In all canals, it is desirable to attain an equilibrium at which neither scouring nor deposition takes place (Brown 1912:175; Strange 1927:235-236). Above this, flow becomes turbulent, scouring the bed and eroding the banks of the canal (Brown 1912:175; Strange 1927:237-239). Below it, the flow is laminar and silt deposition takes place (Brown 1912:173,175). This encourages the growth of weeds and alters the gradient of the canal by raising the bed (Brown 1912:173). Farrington has used the Froude formula to calculate the stability of flow in some excavated Peruvian canals (1985:293-302). The Vijayanagara canals appear to have reached equilibrium by self-adjustment, either by silt deposition, or by erosion and scouring. This process has been discussed in other contexts by Strange (1927:237-239) and Farrington (1985:290). The main Vijayanagara canals were lined with stone blocks to stabilise the banks and prevent erosion. The original linings were mostly vertical. The beds were not lined. The canals are generally characterised by a liberality of design, allowing for maximum floods and mistakes in the original construction. It should be noted that the deposition of silt in canals can have the beneficial effect of increasing impermeability (Buckley 1905:66; Strange 1927:30-32).

Silt may be divided into three types (Buckley 1905:30,34-37; Brown 1912:173; Strange 1927:407-409):

1. Minerals in solution (fertile and inorganic).
2. Suspension silt (fertile and organic).
3. Bed load (sterile, inorganic silt).

The bed load is detrimental to the efficient working of canals and has no fertilising value. It is rolled along the bottom of the canal by the forward velocity of the water current (Brown

1912:174; Strange 1927:409-412). Suspension silt and minerals in solution are also swept along by the current, but their deposition is largely determined by changes of velocity within the flow. Brown (1912:176) has noted that the proportion by weight of solid matter to liquid may be as great as 1 to 30 when Indian rivers are in flood. Similar proportions are likely to be found in some anicut-fed canals.

Biswas (1970:89) suggests:

> The Romans, like the Egyptians before them, were aware that in order to flow, water requires a downward slope. The slopes of aqueducts were more closely related to topographical conditions than to hydraulic considerations. The bottom slopes of the same aqueduct frequently varied considerably, from about 1 in 2000 to 1 in 250. It is highly unlikely that Roman mensors, or liberators, or architectons had any idea of reconciling a particular cross-section area with a definite slope in such a manner as to produce a desired discharge. All they seem to have done was to construct a part of the aqueduct, and if the resulting discharge was too little for their liking, they may have either increased the area (less likely) or just increased the slope (more probable). Thus in all probability, the slope was fixed by a process of trial and error.

In our opinion, a similar situation prevailed at Vijayanagara. The canals were constructed in sections by indigenous engineers, who used empirical knowledge and rule of thumb. This information was probably passed on by word of mouth. Mistakes resulting from this approximate method of construction could be rectified by the addition of supplementary features. Rapids could be built to reduce the velocity and the gradient steepened to increase it. Unlike modern canals, which are designed to supply a specific and calculated discharge, those at Vijayanagara were probably designed to supply as much water as possible. Thus, the area of land that could be turned over to intensive wet cultivation using water from the canal was determined by the canal's discharge, and not the other way round.

Little evidence concerning the engineers who built canals in the medieval period has survived. We have found only two inscriptional references. The first is the Pratāpa Bukka Rāya maṇḍala inscription of A.D. 1388. It states:

> In order that all the subjects might be in happiness, water being the life of all living things, Vīra Pratāpa Bukka Rāya in his court gave an order to the emperor (or master) of ten sciences, the hydraulic engineer (jala-sūtra) Siṅgāya-bhaṭṭa, that he must bring the Henne river to Penugoṇḍe, and that Siṅgāya-bhaṭṭa, conducting a channel to the Siruvera tank, gave to the channel the name Pratāpa Bukka Rāya maṇḍala channel, and had this sasana written.... In the sciences of hydraulics, in divination or telling omens from sounds (of birds, lizards, etc.), in medical treatment with mercury (or perhaps alchemy?), in speaking the truth, Rudraya's (son) Siṅgāri, what learned man is there in the world to equal you (EC, X, Gb6).

The second, contained in the short inscription on the right bank of the Hirya canal, refers to Kārakunayya splitting the Nadi rock (see section 4.5.6). To us, it seems likely that this person was a hydraulic engineer engaged in the construction of the canal in the sixteenth century A.D.

It has been mentioned above, that the Vijayanagara canals were not provided with regulating works at their head. It appears, therefore, that they were kept in continuous flow, and that their outlets, which controlled the supply taken by the irrigators, were opened and closed in turns. The outlets were used, not only to distribute the water, but also to control the water level in the canal. No original outlets have survived. We suggest that some may have been constructed from timber and needed periodic replacement. Modern capstan-operated outlets were fitted in the nineteenth and twentieth centuries by British engineers.

No evidence for the original distributaries and field outlets survives on the ground. However, we may infer that the devices used today are not essentially different from those originally employed. Water is carried from the canal to the field in unlined distributaries, the design of which appears to depend upon the agricultural regime. The amount of water entering the fields from the distributaries is regulated by small control devices. These range from notches broken through the earthen banks of the field to simple, stone sluices. Water enters the field in one corner and flows along each of the connected furrows. Excess water is allowed out of the field through an outlet in the opposite corner to the inlet. In the case of crops which require heavier watering (such as bananas), the flow is trapped in the furrows by earthen partitions built across their width to form water-filled troughs.

The supply of water to the crops in the field is subject to a number of problems. Water is lost through evaporation, percolation through the soil profile and surface run-off (Farrington 1985:287). The amount of water and the method by which it is supplied to the fields must be carefully gauged. Individual crops have specific and various water requirements (Cotton 1854:170):

1. Rice, peanuts and other crops require constant and intensive watering for between four and six months.
2. Dry grains, such as cholum, require small quantities of water in the dry season.
3. Crops such as sugar cane and chillies require watering for extended periods of time, in the case of cane, ten to eleven months.
4. Vegetables in garden plots require watering throughout the year.

Over-watering can result in water-logging and salt efflorescence, the bringing up of salts by capillary action (Brown 1912;182; Balek 1983:243-244). In addition, standing water can quickly become stagnant and insanitary (Strange 1927:21-22). There is no evidence for the use of field drainage to discharge excess water in Vijayanagara. Brown (1912:185) noted a similar lack of drainage facilities in other irrigation systems and suggested that this may result from a belief that surplus water would be absorbed or evaporate.

The agricultural supply of Vijayanagara was a complex combination of elements working together as a single, integrated whole. The main purpose was the supply of perennial water to the rich agricultural lands to the southwest and west of the 'Urban Core'. In addition, an extensive tract between the 'Urban Core' and the river, the 'Irrigated Valley', was also watered. In our view, the core of the system was the large tank at Kamalāpuram and the Rāya canal, which provided the tank with a perennial supply. The canal was fed by the Kūruvagadda anicut. We suggest that the tank and canal were constructed at the beginning of the fifteenth century A.D. by either Bukka II or Devarāya I (see section 5.4.2). The tank irrigated the agricultural land to southwest of the 'Urban Core' and also supplied the 'Royal Centre' with a supply of water for domestic use. The canal ensured that the lands between Mālāpuram and the tank were supplied with perennial water and also stopped the tank from running dry.

The Kamalāpuram tank served three purposes. Firstly, it irrigated the tracts to the north. Secondly, it provided the Hiriya canal with an additional supply, passing excess water into the canal as it flowed through the irrigated valley. Thirdly, water from the channel which linked the Kamalāpuram tank to the Hiriya canal filled short section of defensive moat around the 'Urban Core'.

Wells and smaller catchment drainage tanks, such as Mallappannagudi kere, supplied areas which could not be reached by the anicut-fed canal or the Kamalāpuram tank. There are indications that a large, catchment tank was located inside the wall of the 'Urban Core' to the northeast of Kamalāpuram village. This tank is likely to have supplied water features lying to the east of the 'Royal Centre' (especially the 'Octagonal Bath') and irrigated gardens inside the 'Urban Core.'

The Vallabhāpuram, Hosūru and Hiriya anicuts and their related canal system represent later additions to the agricultural supply system. The anicuts fed a network of perennial canals on the right bank of the river. To us it seems likely that they were constructed by Kṛṣṇadeva Rāya in the sixteenth century A.D. The Rāmanagadda weir was probably also built at this time. We have suggested that it may have been the failure of the Rāya tank to boost the supply to the existing Rāya canal that led to the building of this secondary feature, which increased the amount of water flowing down to the Kūruvagadda anicut (see section 4.7). The Koragallu and Anegundi anicuts were also built at this time to irrigate agricultural lands on the left bank.

An important tank was located to the southwest of Hospet. In the sixteenth century, Kṛṣṇadeva Rāya tried to dam the drainage which flows down the valley of Sandur to the river at Mālāpuram. Although the bund was completed, the tank never held water. In our view, it was intended to boost the supply of the Rāya canal (see above).

The seven anicuts which were built on the Tungabhadrā to supply the metropolitan area with a perennial agricultural supply worked in different ways. The Vallabhāpuram and Koragally anicuts, which formed a pair in parallel, blocked the whole width of the river, channelling water into canals on both sides of the river. The Hosūru, Hiriya and Anegundi weirs operated as single devices, trapping a single channel of the river between an island and one of its banks. The Rāmanagadda and Kūruvagadda anicuts worked as a pair in series, the former pushing an increased supply into the river channel that was blocked further downstream by the latter (see above).

On the basis of epigraphical evidence, we suggest that the Hiriya canal supplied the Bhūpati tank in the irrigated valley with a perennial supply, in addition to the tank's collection of surface run-off. The Bhūpati tank was probably built to supply Achyutāpura and was broken down after A.D. 1565 to facilitate cultivation of the tank bed (see above and 4.3.1).

The Hiriya canal terminates north of Bukkasāgra, discharging into the river, after crossing the Kampli channel heading northeast. Although not constructed to supply perennial water to the greater metropolitan area of Vijayanagara, we note the existence of four anicuts and canals northeast of the urban settlement. These were probably built in the sixteenth century A.D. by Kṛṣṇadeva Rāya.

The evidence on the ground suggests that no attempt was made to provide Vijayanagara with an encircling moat. However, in two locations, defensive walls were further fortified with short moats. The first protects one short section of the defensive wall of the 'Urban Core'. It is fed by the channel connecting the Kamalāpuram tank with the Hiriya canal. The second is located just to the north of the northwest corner of the Kamalāpuram tank. It is composed of an irrigation channel, which flows along the foot of the fragmentary remains of an outer defensive wall, and is supplied by the tank's waster weir. Moats are described in Paes' contemporary account of the city (Sewell 1900:253).

Wells were used to irrigate garden plots inside the metropolitan area. Inscriptional evidence suggests that hand operated pulley lifts were used to draw on the supply (Gopal 1985:138,140). Water may also have been drawn using the draglift or mot (Grierson 1885:208-209; Mukerji 1907:135) and counter-balanced water lever (lātha or picotah) (Grierson 1885:206-207; Mukerji 1907:140).

5.4.2. When and by whom was the agricultural supply system built?

The agricultural water features have been dated using epigraphical, literary and physical evidence. A major problem on the ground was the separation of original features from modern replacements and alterations. We have been able, using physical evidence in conjunction with archival data, to establish the recognisable period of canal and bund construction. In particular, the examination of the Basavanna canal on the ground and in public works reports, has enabled us to differentiate between modern (i.e. nineteenth and twentieth century) and Medieval constructional styles (see section 4.2.1 and 4.2.2). Thus, it has proved possible to distinguish original canal linings from modern replacements in the field.

The earliest evidence for what was to become the agricultural exploitation of the Vijayanagara period, is the building of river-fed canal and catchment tanks in the tenth and eleventh centuries. A.D. The epigraphical and physical evidence for Cālukyan water management has been discussed in section 1.4 and 5.2. We do not know where the main settlement or capital of the period was located. The presence of inscriptions and architectural fragments of the tenth and eleventh centuries A.D. in the 'Royal

Centre' and at Hampi and Anegundi strongly suggests that a settlement, or perhaps a minor capital, like those at Kampli and Pottalakere (Danāyakanakere), was situated at Vijayanagara.

It may be conjectured that during the Vijayanagara period, the Cālukyan canals and bunded tanks were working. We suggest that they may have given the Vijayanagara builders the idea of constructing further tanks and canals to furnish the city with an agricultural supply system.

During the early fifteenth century A.D., the city was first provided with agricultural supply features. Two tanks were constructed. We have proposed that the Kamalāpuram tank was built at the beginning of the fifteenth century A.D. This correlates well with the stylistic appearance of the sluice outlets of the tank, which are plain and quite different to the ornate types characteristic of the sixteenth century A.D. On the basis of inscriptional evidence, the construction of the Mallappannagudi tank may be dated to A.D. 1412.

If one accepts that the Kamalāpuram tank was built at the beginning of the fifteenth century A.D., the Kūruvagadda weir, and Rāya canal which supplied it, must be contemporaneous. There is some evidence to support this hypothesis. An important account of the construction of an anicut is provided by Nuniz in his dynastic history. It appears to relate to the building of the Kūruvagadda anicut, which was constructed from rough stone without bonding material. The events have been convincingly ascribed to the reign of either Bukka II or Devarāya I by Sewell (1900:51-52) (see section 5.4). Nuniz account states:

> ... the king, desiring to increase that city and make it the best in the kingdom, determined to bring to it a very large river which was at a distance of five leagues away, believing that it would cause much profit if brought inside the city. And so he did, damming the river with great boulders; and according to the story he threw in a stone so great that it alone made the river follow the king's will. It was dragged thither by a number of elephants of which there are many in the kingdom; and the water so brought he carried through such parts of the city as he pleased. This water proved of such use to the city that it increased his revenue by more than 350,000 pardaos. By means of this water they made improvement about the city, a quantity of orchards and great groves of trees and vineyards, of which this country has many, and plantations of lemons and oranges and roses, and other trees which in this country bear very good fruit. But on this turning of the river they say the king spent all the treasure that had come to him from his father, which was a very great sum (Sewell 1900:301-302).

It is clear that the project was paid for and organised by the king himself. The motive was a desire to increase the agricultural revenue of the city by providing a reliable source of perennial water for irrigation.

In our opinion, the weirs and canals, with the exception of the Rāya kāluve and the Kūruvagadda anicut, were all built in the sixteenth century A.D. by Kṛṣṇadeva Rāya as part of a massive building project, financed by successful military campaigns (see below). The construction of the Vallabhāpuram and Koragallu anicuts, and also the original sections of the Basavanna canal, can be dated on epigraphical evidence to A.D. 1521. An inscription indicates that the Hiriya anicuts and canal were already built by A.D. 1524. We suggest that these were also constructed by Kṛṣṇadeva Rāya. Although no inscription refers to its building the Hosūru anicut is also likely to have been built in this period. The Rāmanagadda and Anegundi weirs are constructed in a different manner and form a separate group. We suggest that they were built in the sixteenth century A.D. Local tradition suggests that a minister of Kṛṣṇadeva Rāya, called Mudda, supervised the building of the Tungabhadrā canals (Mackenzie 1800a: 133; Kelsall 1872:112-113; Francis 1904:90-91).

A number of tanks were built in the sixteenth century A.D. The accounts of Nuniz and Paes allow us to date the construction of the Rāya bund to A.D. 1521. We provisionally date the construction of the Bhupati bund in the irrigated valley, designed to supply Achyutāpura with water, to the reign of Achyuta Rāya (A.D. 1530-1542). At least one tank was also modified in this period. The large tank at Danāyakanakere, which we have suggested was originally built in the eleventh century A.D., was partially rebuilt.

The inscriptional record of the fourteenth to sixteenth century A.D. indicates that powerful nāyakas, wealthy individuals and temples invested in hydraulic building projects. They were undertaken as acts of religious merit. The Porumāmiḷḷa tank inscription (discussed above) states:

> Bhāskara ... heard that the merit attaching to the gift of water was the greatest of all. On the authority of the Vedas: 'Verily all this is water'. And the Śruti says that: 'From water alone is produced food; (and) food is Brahman.' ... A shed for distributing water, a well, a canal, and a lotus tank: the merit of (constructing them) is millions and millions (of times) higher in succession. As the water of a tank serves to nurture both movable and immovable creation on (this) earth, even the lotus-seated (Brahmā) is unable to recount the fruit of merit (attaching) to this (EI XIV:107-108).

Bhāskara was a powerful commander, who ruled from the fort of Udayagiri in Nellor district. He constructed the large tank at Porumāmiḷḷa at his own expense for the purpose of acquiring religious merit. On a practical level, irrigational projects increased agricultural revenue by providing a source of perennial water for wet agriculture.

Our researches at Vijayanagara suggest that a different pattern was followed at the capital. Inscriptions, contemporary accounts and local traditions point to the direct involvement of the king in the financial backing and supervision of irrigational projects. Nuniz specifically states that the king paid for and organised the building of a large anicut, which we have identified as the Kūruvagadda weir (see above). The accounts of Paes and Nuniz indicate that the building of the Rāya bund in A.D. 1521. was supervised by Kṛṣṇadeva Rāya, who presided over the project and was perhaps responsible for its failure. The Vallabhāpuram inscription (Kelsall 1872:231; Sewell 1900:162; Francis 1904:38; Venkayya 1906:210; IGI 1908:437) stated that the Koragallu and Vallabhāpuram anicuts and the Basavanna canal, were constructed by Kṛṣṇadeva Rāya in A.D. 1521. The local tradition concerning Mudda (see above), suggests that the building of large tanks and anicut-fed canals at Vijayanagara was paid for using money from the royal treasury.

There is extensive evidence that small irrigation projects were left to wealthy individuals and were carried out for the purpose of acquiring religious merit and for the benefit of the population. Interestingly, there is no literary or inscriptional evidence at Vijayanagara for religious institutions investing in irrigational projects as they did elsewhere in South India in this period. Large-scale agricultural supply projects appear to have remained as rājakāriya (Leach 1959:20) or the personal responsibility of the king. Without an effective system of perennial irrigation, the city could not have flourished or even survived in the hostile, semi-arid environment of the Deccan. To us it seems likely therefore, that the provision of irrigational facilities was viewed as a basic duty of the king, and one which had to be respected in order to maintain the king's position as provider and protector.

The extension and elaboration of the domestic water features of the 'Royal Centre' appears to correspond with periods in which successful war was waged (see section 5.3.2). It is noteworthy that the extensive irrigational programme carried out by Kṛṣṇadeva Rāya at the city, occurred after the conclusion of his successful, military campaign in the Raichur Doab and further east. Portable wealth, taken during warfare, would have provided kings with an opportunity to increase agricultural revenues, acquire religious merit and express their power and prestige in physical terms at the capital. Warfare may also have provided captives for large scale building projects, such as the construction of the Rāya bund.

5.4.3. Does the technology used represent a development of new ideas or merely a continuation of an existing tradition?

It has been suggested, in section 5.2, that the essential elements of the Vijayanagara agricultural supply system, the tank, anicut-fed canal and well, developed out of an ancient and indigenous tradition of water management in Bellary and Raichur districts. However, the scale of irrigational enterprise at Vijayanagara was greater than in previous periods. This was, in our view, largely the result of the availability of labour and an increased investment in hydraulic projects by individuals, particularly the king. These conditions made possible the use of existing technology in new ways, and in particular facilitated the watering of lands away from the river.

To us, it seems likely that the practice of building river-fed canals in Bellary and Raichur districts, developed out of an ancient tradition of field terracing and tank building (see section 4.2). The earliest canals appear to have been constructed in the later Cālukya period (tenth to eleventh centuries A.D.) to irrigate the low-lying areas close to rivers. Their construction may be seen in the context of increasing population and socio-economic organisation. The development of contour-hugging canals and permanent, stone anicuts in the Vijayanagara period, facilitated the supply of perennial water to agricultural tracts away from rivers. A similar pattern has been suggested for the Tambraparni valley by Ludden (1985:52-55).

5.4.4 To what extent did the success of Vijayanagara as an urban centre depend upon the operation of its agricultural supply system?

The population of Vijayanagara relied directly upon the agricultural water supply system for the production of wet and garden crops in an environment of considerable hostility. Perennial irrigation must have made Vijayanagara into a sort of oasis in the extremely barren hinterland that surrounded it. Before the completion of the Tungabhadrā Dam, it was extremely difficult to cultivate vegetables in the surrounding countryside. This situation cannot have been much different in earlier times.

The core of perennially irrigated lands located inside the greater metropolitan area was defensible. We suggest that they provided sufficient food to sustain the population of the city, whilst lands outside the metropolitan area yielded agricultural revenue. The city was, therefore, agriculturally self-sufficient and could theoretically survive if cut off from its surrounding hinterland. This may reflect a need to ensure that the city could withstand not only protracted sieges, for which there is evidence, but also survive periods of political instability, when agricultural revenues from subordinate nāyakas were not forthcoming.

The maintenance of a large population necessitated the construction of precautions against climatic uncertainties, as well as the expansion of agricultural tracts away from low-lying areas close to the river. In our view, the system was expanded by the building of a number of anicut-fed canals. This helped to accommodate an increasing population in the sixteenth century A.D., by expanding the area under wet cultivation. In contradiction to the established pattern of small-scale investment in hydraulic projects by wealthy individuals and temples, the Vijayanagara kings invested their wealth in the construction of large-scale irrigation projects in the city. These projects appear to have been funded using wealth derived from successful military campaigns. By building irrigational features at the city, the kings fulfilled their responsibility to provide for its inhabitants, acquired religious merit and demonstrated their wealth and power.

Conclusions

The development of water supply was essential for settled habitation in the semi-arid environment of the Deccan. Water was needed for domestic and agricultural purposes. From the fourteenth to the sixteenth centuries A.D., the city of Vijayanagara was furnished with water supply systems which provided for both of these needs. The hydrological environment dictated the type of technology which could be used by the medieval engineers. The irregular, granitic landscape was ideally suited to the construction of bunded tanks and anicuts. The clayey soil was quite capable of holding water and an inexhaustible supply of impervious rock was available to the builders of hydraulic works. One characteristic of Vijayanagara's irrigation and water supply systems is that the hydrological environment was exploited to the maximum advantage using available technology. The hydrological environment was fundamentally altered by the Tungabhadrā Dam Project, completed in the 1950s.

A distinction must be made between the domestic supply facilities constructed for the use of the king and his entourage in the 'Royal Centre', and those in the rest of the city. There is no evidence for the construction of municipal supply features at Vijayanagara. Wells constructed by individuals and the river must have provided the primary source of domestic supply outside the 'Royal Centre'. Water may also have been drawn from catchment tanks and irrigation canals. With the exception of the 'Royal Centre', domestic water facilities in the city were no more complex than those found in villages.

In the 'Royal Centre', knowledge of the siphon and gravity flow was used to construct a system of supply and drainage. Water was led from the Kamalāpuram tank by a single, open channel into the 'Royal Centre'. This transport system was, in our view, constructed in the second quarter of the fifteenth century A.D. There are indications that a second, smaller tank to the northeast of Kamalāpuram village may have also supplied water features inside the Royal Centre'.

Once inside the 'Royal Centre', a network of open channels and sealed earthenware pipelines distributed water to a complex of baths, tanks and fountains. Pipes were produced in three standard sizes according to their purpose. Dust and silt was excluded from the system by a number of devices. The greatest concentration of water features is found in enclosure IV, an area which is associated with the demonstration of ritual power. This area was rebuilt and elaborated throughout the occupation of the 'Royal Centre'.

The Muslim-influenced water features of the 'Royal Centre', which we date to the sixteenth century A.D., are characterised by the use of plastering to hide poor quality construction. Many appear to have been rapidly and crudely built. We suggest that they were constructed during Kṛṣṇadeva Rāya's rebuilding of the city. This work may have been funded by wealth taken during military campaigns.

The design of the domestic water system of the 'Royal Centre' suggests that it did not operate continuously, but only on special occasions, such as religious festivals or important visits. The bulk of water appears to have been taken from a network of wells and rainwater collection tanks. The conspicuous consumption of water provided a pleasant environment, and also demonstrated wealth and power. There appears to have been no sense of responsibility on the part of the king to provide domestic supply features for the population of the city.

There is some evidence for the building of open channels and plastered tanks in South India prior to the Vijayanagara period. Therefore, their use may be viewed as the continuation of an existing tradition. Earthenware pipes were used for drainage in Central and Northern India during the early historic period. However, there is no evidence for their use in South India before Vijayanagara. Earthenware pipes were extensively used in the Muslim forts and capitals of the medieval period to supply high status and municipal features.

In medieval South India, irrigational building projects were undertaken by powerful nāyakas, wealthy individuals and temples as acts of religious merit. At Vijayanagara, a different pattern was followed. The king financed and supervised large-scale hydraulic projects. Smaller projects were paid for by wealthy individuals. There is no evidence for the financing of irrigational projects by temples at Vijayanagara. In our view, it was a basic duty of the king, in his capacity as provider and protector, to facilitate wet cultivation at the capital, thereby demonstrating his wealth and power.

The increased availability of labour, investment by the king and possibly the use of captives for building projects permitted irrigational enterprise on a greater scale than previously possible in the area. The construction of anicut-fed canals allowed the exploitation of agricultural areas away from the river. The creation of a core of defensible, irrigated land made the city agriculturally self-sufficient. This meant that it could withstand sieges and dislocation from its agricultural hinterlands during the breakdown of political order.

The core of the agricultural supply system was the Kamalāpuram tank. Its supply was boosted by the anicut-fed Rāya canal. We suggest that both were constructed at the beginning of the fifteenth century A.D. The tank fulfilled a complex role. It irrigated land to its north, passed excess water into the Hiriya canal and also filled several sections of defensive moat. In the sixteenth century A.D., a massive irrigational building project was undertaken by Kṛṣṇadeva Rāya. The Vallabhāpuram, Hosūru and Hiriya anicuts were built to supply a network of potential canals in the metropolitan area. After the failure of the Rāya bund, which had been intended to boost the supply to the Rāya canal, the Rāmanagadda anicut was constructed to boost the supply to the Kūruvagadda weir. In addition, weirs were built at Koragallu and Sānapur to irrigate the left bank of the river. Achyuta Rāya probably enlarged the system by building the Bhūpati tank to supply the suburb of Achyutāpura. There is no evidence for the pro-vision of an agricultural drainage system.

The technique of building catchment drainage tanks had been perfected long before the fourteenth century A.D. as an effective means of securing a perennial supply for irrigation. At Vijayanagara, bunds were built wherever a good site on a drainage could be found. With the exception of the Rāya bund, tanks were successful. However, they suffered from two shortcomings. Firstly, fertile silt settled out and did not reach the field. Secondly, it was usually only possible to irrigate lands close to the bund. The development of anicut-fed canals and particularly those which followed contours, provided a more abundant supply of water and allowed areas further field to be irrigated. Anicuts had the advantage over tanks that fertile silt was passed into the fields. Storage losses were also avoided. However, anicut-fed canals were expensive and difficult to build. At Vijayanagara, their construction, using empirical knowledge of the afflux and gravity flow, may be attributed to the stimulus of royal sponsorship.

The longevity of the irrigation system of Vijayanagara is remarkable. The achievement of its builders in creating a durable and effective means of irrigating extensive agricultural tracts away from the river is measured by its continued use up to the present day.

With regard to future research, we suggest that a detailed examination of water management at other medieval urban sites, such as Tughlukabad, Fatehpur Sīkrī and Golkoṇḍa should be undertaken. A study of the parallel development of irrigation and water supply in Nepal and Sri Lanka is also necessary. This would permit a more thorough comparison between the different types of hydraulic technology employed in India during the medieval period. This study should include a thorough examination of contemporary literary sources and archival material, particularly public works records. It should also anticipate changes which may have taken place to the hydrological environment as a result of modern irrigation projects.

[illegible] probably in larger measure by building the [illegible]

[illegible] seasonal [illegible]

[illegible] the [illegible] possible [illegible] The development of [illegible] canals and [illegible] provided a more abundant supply of water [illegible] to be irrigated. [illegible]

The majority of the irrigation system of [illegible]

With regard to [illegible] detailed examination of [illegible] passages [illegible] in other medieval [illegible] should be undertaken [illegible] would permit a more [illegible] employed in India during the medieval period. The study should include [illegible] examination of [illegible] material [illegible] as a result of modern irrigation projects.

of water for irrigation in southern India during [illegible] municipal features.

[illegible] irrigation [illegible] projects [illegible] by individuals and temples [illegible]

[illegible]

Bibliography

Administration Reports of the Madras Public Works Department (Irrigation Branch).
1891-2:63
1893-4:74
1894-5:69
1895-6:73
1896-7:78
1897-8:81
1925-6:16
1926-7:17
1927-8:18
1928-9:44
1929-30:47
1930-1:58
1933-4:15
1934-5:17
1935-6:14
1936-7:13
1946-7:34-36

Allchin, F.R.
1954 'The Development of Early Culture in the Raichur District of Hyderabad', Ph.D. Dissertation, University of London.
1963 *The Neolithic Cattle-keepers of South India.* University of Cambridge, Oriental Publication no. 9. First edition. Cambridge.

Annual Report, Archaeological Survey of India.
1903-4:61-64
1904-5:24-30
1905-6:52-53
1907-8:5-6,110, pls. XXVII, XXXII
1908-9:22-25
1911-2: pt.1:7,8,38-39
1912-3:pt.1:5-6
1913-4:8
1914-5:10-11,67-71, pls. XLVIII, XLIXa
1915-6:46
1916-7:5
1917-8: pt.1:13
1918-9: pt.1:9
1919-20:14-15
1920-1:11, 19
1921-2:31-32
1922-3:66-69

1923-4:11, 37-38, pls. IVb, XVI
1924-5:40,47, pl. VIIIb
1925-6:44,140
1926-7:44
1927-8:45,61,pl. XVIIIb
1928-9:45,90, pls. XXXVIIa, XXXVIIIa,b, XLb
1929-30:48
1930-4: pt.1:41,122,127
1930-4: pt.2:pls. LVIIb, LXId, LXIIe
1934-5:21-2
1935-6:25

Annual Report, Archaeological Survey of Mysore
1929:50-58

Annual Report of the Archaeological Department, Hyderabad
1927-8:1-6
1928-9:1-11
1929-30:6-23

Annual Report of the Archaeological Department, Southern Circle
1909-10:3-6,31,47-43,111
1910-1:26,125
1911-2:28,45-46
1912-3:7-8,19-20,22,24-25,44-53
1913-4:3-4,16,20-27,41-43
1914-5:3-4,9-10,14,19-22
1915-6:4-5,8,11,14-17,46, pls. XVIIa,b
1916-7:5-9,12-15,28-30, pls. X-XVI
1917-8:7-8,10,14-15,29, pl. XI
1918-9:4,8,10,13-14
1919-20:4-5,11,15-18,33
1920-1:8-9,13-15,28

Annual Report on South Indian Epigraphy
1892, nos. 266, 272
1904, no. 16
1904, no. 18
1904, no. 25
1922, no. 683
1922, no. 697
1934-5, no. 351

Annual Progress Report of the Archaeological Survey Department, Southern Circle
1905-6:8,22
1906-7:10,38
1907-8:9-13,33
1908-9:15-16

Annual Progress Report of the Archaeological Survey of Madras and Coorg
1903-4:5,36-37,70-81
1904-5:14-18,36

Armillas, P.
1971 'Gardens on swamps'. *Science*, Vol. 174, no. 4010, pp. 653-661.

Asher, C.B.
1985 'Islamic Influence and the Architecture of Vijayanagara'. In Dallapiccola, A.L. (ed.), *Vijayanagara: City and Empire*, pp. 188-195. Stuttgart.

Aziz, K.K.

1985 'Glimpses of Muslim Culture in the Deccan'. In Dallapiccola, A.L. (ed.), *Vijayanagara: City and Empire*, pp. 159-176. Stuttgart.

Balsaubramnya,

1985 'Vīrabhadra Sculptures in Hampi'. In Nagaraja Rao, M.S. (ed.) *Vijayanagara: Progress of Research,* 1983-1984, pp. 133-135. Mysore.

Balek, J.

1983 *Hydrology and Water Resources in Tropical Regions.* Developments in Water Science Series no. 18. Prague

Ballhatchet, K. And Harrison, J. (eds.)

1980 *The City in South Asia.* London.

Biswas, A.K.

1970 *History of Hydrology*. London.

Bose A.,

1942 *Social and Rural Economy of Northern India.* Calcutta.

Bray, F.

1984 *Science and Civilisation in China,* vol. 6, *Biology and Biological Technology. Part II: Agriculture.* Joseph Needham Series. Cambridge.

Bray, W.

1976 'From Predation to Production: The Nature of Agricultural Evolution in Mexico and Peru'. In Sieveking, G. de G., Longworth, I.H., and Wilson, K.E. (eds.), *Problems in Economic and Social Archaeology*. London.

Breckenridge, C.A.

1985 'Social Storage and the Extension of Agriculture in South Asia'. In Dallapiccola, A.L. (ed.), *Vijayanagara: City and Empire,* pp. 41-72. Stuttgart.

Briggs, J.

1908 *History of the Rise of Mahomedan Power in India, Till the Year A.D. 1612.* In four volumes. Second edition. Calcutta.

Brohier, R.L.

1935 *The Ancient Irrigation Works of Ceylon.* In three volumes. Colombo.

Bromehead, C.E.N.

1942 'The Early History of Water Supply'. *Geography Journal,* 99, pp. 142-151, 183-196.

Brown, P.

1942 *Indian Architecture: The Islamic Period.* Second edition. Bombay.

1976 *Indian Architecture: Hindu, Buddhist and Jain.* Seventh reprint. Bombay.

Brown, R.H.

1912 *Irrigation: Its Principles and Practice as a Branch of Engineering.* Second edition. London.

Brown, W.N.

1943 *Chanhu-daro Excavations, 1935-1936.* American Oriental Series, vol. 20. New Haven.

Buckley, R.B.

1905 *The Irrigation Works of India.* Second edition. London.

Chander, S.

1987 'From a Pre-Colonial Order to a Princely State: Hyderabad in Transition', c. 1748-1865 A.D. Ph.D. Dissertation, University of Cambridge.

Clark, G.

1944 'Water in Antiquity'. *Antiquity,* vol. XVIII, no. 69, pp. 1-15.

Clerk, H.E.

1901 *Preliminary Report on the Investigation of Protective Irrigation Works in the Madras Presidency.* Madras.

Costa, P.M.
1983 'Notes on Traditional Hydraulics and Agriculture in Oman'. *World Archaeology,* vol. 14, no. 3, pp. 273-295.

Cottam C.
1980 'City, Town and Village: The Continuum Reconsidered'. In Ballhatchet, K. and Harrison, J. (eds.), *The City in South Asia.* London.

Cotton, A.
1954 *Public Works in India, their Importance; with Suggestions for their Extension and Improvement.* Second edition. London.

Cousens, H.
1916 *Bījāpur and its Archaeological Remains.* Archaeological Survey of India. Imperial Series, vol. XXXVII. Bombay.

Dallapiccola, A.L. (ed.)
1985 *Vijayanagara: City and Empire.* Stuttgart.

Dames, M.L.
1918 *The Book of Duarte Barbosa.* In two volumes. London.

Dani, A.H.
1963 *Indian Palaeography.* Oxford.

Deloche, J.
1984 *The Ancient Bridges of India.* New Delhi.

Dimbleby, G.W.
1967 *Plants and Archaeology.* London.

Devakunjari, D.
1970 *Hampi.* Archaeological Survey of India. Calcutta.

Drower, M.S.
1954 'Water Supply, Irrigation and Agriculture'. In Singer, C., Holmyard, E.J., and Hall, A.R., (eds.), *A History of Technology,* vol. 1, chapter 19, pp. 520-557. Oxford.

Dutt, B.B.
1925 *Town Planning in Ancient India.* Calcutta.

Dutt, R.
1904 *India in the Victorian Age: An Economic History of the People.* London.

Epigraphia Carnatica
V, Hn. 133
V, Cn. 256
X, Gb. 6

Epigraphia Indica
IV:226-269 (inscription no. 38).
VII:119-128 (inscription no. 17).
VIII:36-49 (inscription no. 6).
XIV:97-109 (inscription no. 4).

Erdosy, G.
1985 'Urbanisation and the Evolution of Complex Societies in the Early Historic Ganges Valley'. Ph.D Dissertation, University of Cambridge.

Farrigton. I.S.
1980 'The Archaeology of Irrigation Canals, with Special Reference to Peru'. *World Archaeology,* vol. II, no. 3, pp. 287-305.

Fernando, A.D.N.
1980 'Major Ancient Irrigation Works of Sri Lanka'. *Journal of the Sri Lanka Branch of the Royal Asiatic Soceity,* special number, new series vol. XXII, pp. 1-24.

First Report of the Madras Public Works Commissioners.

1853 'History and Present State of the Maramut Department of Public Works under the Board of Revenue'. In Return to the House of Lords. Dated 6th May 1853. Printed 2nd June 1853, pp. 95-96.

Forbes, R.J.

1954 'Hydraulic Engineering and Sanitation'. In Singer, C., Holmyard, E.J., and Hall, A.R. (eds.), *A History of Technology*, vol. 2, chapter 19, pp. 663-694. Oxford.

Francis, W.

1904 *Gazetteer of the Bellary District.* Madras.

Fritz, J.M.

1983a 'Report of Research, January-March 1983'. In Nagaraja Rao, M.S. (ed.), *Vijayanagara: Progress of Research, 1979-1983,* pp. 43-44. Mysore.

1983b 'The Roads of Vijayanagara: A Preliminary Study'. In Nagaraja Rao, M.S. (ed.), *Vijayanagara: Progress of Research, 1979-1983,* pp. 51-56. Mysore.

1985a 'Features and Layout of Vijayanagara: The Royal Centre'. In Dallapiccola, A.L. (ed.), *Vijayanagara: City and Empire,* pp. 240-256. Stuttgart.

1985b 'Was Vijayanagara a Cosmic City?' In Dallapiccola, A.L. (ed.), *Vijayanagara: City and Empire,* pp. 257-273. Stuttgart.

1985c Report of Research, 1980-1984. In Nagaraja Rao, M.S. (ed.), *Vijayanagara: Progress of Research, 1983-1984,* pp. 69-95. Mysore.

Fritz, J.M. and Michell, G.

1981 'The Vijayanagara Documentation and Research Project: A Progress Report'. In Allchin, B. (ed.), *South Asian Archaeology.* Cambridge.

1985 'Research at Vijayanagra: Report on Field Work', December 1983-February 1984. In Nagaraja Rao, M.S. (ed.), *Vijayanagara: Progress of Research, 1983-1984,* pp. 54-56. Mysore.

Fritz, J.M., Michell, C. and Nagaraja Rao, M.S.

1984 *The Royal Centre at Vijayanagara Preliminary Report.* Department of Architecture and Building, University of Melbourne, monograph series no. 4. Melbourne.

Garbrecht, G.

1980 'The Water Supply System at Tuspa' (Urartu). *World Archaeology,* vol. II, no. 3, pp. 306-312.

Ghosh, A.,

1948 'Taxila' (Sirkap), 1944-1945. *Ancient India,* Bulletin of the Archaeological Survey of India, no. 4, double number, July to January, pp. 41-84.

Gopal, B.R. (ed.)

1985 *Vijayanagara Inscriptions. Volume one.* Mysore.

Goswami, A. (ed.)

1953 *Glimpses of Mughal Architecture.* Calcutta.

Goswami, K.G.

1948 *Excavations at Bangarh, 1938-1941.* Asutosh Museum Memoir no. 1. Calcutta.

Government of Madras, Archaeological Survey.

1902 *Ancient Kindgom of Vijayanagara,* pp. 11-17.

Grierson, G.A.

1885 *Bihar Peasant Life.* Calcutta.

Grover, S.

1981 *The Architecture of India: Islamic, 727-1707 A.D.* New Delhi.

Havell, E.B.

1913 *Indian Architecture.* London.

Hamberley, G. and Swaan, W.

1968 *Cities of Mughal India: Delhi, Agra and Fatehpur Sikri.* London.

Hawley, A.H.
1973 'Ecology and Population'. *Science*, vol. 179, pp. 1196-1201.

Hearn, G.
1906 *The Seven Cities of Delhi*. London.

Heras, H.
1929 *Beginnings of Vijayanagara History*. Madras.
1932 'The Pre-portuguese Remains in Protuguese India'. *Journal of the Bombay Historical Society*, vol. IV, no. 2, pp. 125-180.

Hocart, A.M.
1936 *Kings and Councillors*. Egyptian University Faculty of Arts Publication, no. 12. Cairo.

Hughes Buller, E.
1906 'Gabrbands in Baluchistan'. In *Annual Report, Archaeological Survey of India*. 1903-4, pp. 194-201. Calcutta.

Husain, A.B.M.
1970 *Fatehpur-Sikri and its Architecture*. Dacca.

Husain, A.M.
1963 *Tughluq Dynasty*. London.

Hyderabad Archaeological Series
1922 *The Munirabad Stone Inscription of the 13th year of Tribhuvanamala (Vikramaditya VI)*. No. 5, pp. 1-12. Calcutta.

Imperial Gazetteer of India.
1908 Madras, vol. 1. Provincial Series, pp. 168-170, 437-438. Calcutta.
1909 Hyderabad State Provincial Series, pp. 33-35, 98-99. Calcutta.

Indian Archaeology: A Review
1953:7,35,pl. IIb
1954-5:9,13,16,20,24-25,40,54, pls. IX, XXIa, XXVIII, XXXVIa,b
1955-6:19,60, pl. XXVIIb
1956-7:15,27,37,55,72,pls. XIa,b, XXVIIIb, XXXb, XXXIa, LVa,b
1957-8:8-9,12,25,47,51,74, pls. X, XI, XIIIa,b, XXVIb, LXIIIa,b, LXIVb, LXXI
1958-9:8, 13-14,47,99, pls. VIIa, XIIIa,b, XIVa,b, LIVb, XCIV
1959-60:17, pl. XIIa
1960-1:4,32,33,37, pls.Ia, XLIIIa,b, LIIIa, LXIb
1961-2:10,43-44,51,56,58,74, pls. XVIIIa, LXIIIa,b, LXIVb, LXVa,b, LXVIIb, LXXIXa, LXXXIXb, XCIa,b, CXXIVb
1962-3:27,30, pls. LIVa, LVb, LVIII, LXXXIIIb
1963-4:20, pl. XIV
1964-5:2,7,18,19,43, pls. Ia,b, VII, XIVB, XV, XXXV
1967-8:45, pls. XIII, XIV, XVb, XVII, XXI
1968-9:2, pl. Ia
1970-1:1-2,8,35,94, pls. I, IIa, IIIa, XIXb, LIIIa, XCa,b
1971-2:1-2,2,7,16, pls. IIIa, XVII, XVIII, XXIV
1972-3:1,9,12,26, pls. XIIa, XIXb, XXXIb
1973-4:7, pl. VIIa
1974-5:14-15,16, 39,40,50, pls. XIVa,b, XVa,b, XVIIb, XXXIIIb, XXXIVb, XLIV
1975-6:15,20,30,62, pls. XIVb, XVIIb, XXIIa-XXVI, XLIIa
1976-7:25,30,45,48,49,50-51,52-53, Pls. XIVb, XXXIIb, XXXIXa,b, XLa,b, XLI, XLIIb, XLIIIb, XLIV, XLVIa, La, LXVIa
1977-8:43
1978-9:43-45,57-59,71-72,73-74, pls. XI-XIV, XXa,b, XXVIa,b, XXVIIa,b, XXVIIIa, XXIXa,b

1979-80:33
1980:1:26-28, pls. XIV-XVII
1981-2:25-26

Junker, L.L.
1985 'Morphology, Function, and Style in Traditional Ceramics: A Study of Contemporary Pottery from Bellary District'. In Nagaraja Rao, M.S. (ed.), *Vijayanagara: Progress of Research, 1983-1984,* pp. 144-151. Mysore.

Kamath, S.U.
1977 *Karnataka: A Handbook.* Bangalore.

Kanakasabhai Pillai, V.
1890 'Tamil Historical Text: no. 2. The Kalingattu Parani'. *Indian Antiquary,* vol. XIX, pp. 329-245.

Kelsall, J.
1872 *Manual of the Bellary District.* Madras.
1873 Archaeology of the Belāri District. *Indian Antiquary,* vol. II, pp. 177-180.

King, A.D.
1980 'Colonialism and the Development of the Modern South Asian City: Some Theoretical Considerations'. In Ballhatchet, K. and Harrison, J. (eds.), *The City in South Asia.* London.

Kirke, C.M.S.
1980 'Prehistoric Agriculture in the Belize River Valley'. *World Archaeology,* vol. II, no 3, pp. 281-286.

Krishnaswami Aiyanagar, S.
1936 'The Character and Significance of the Empire of Vijayanagara in Indian History'. In Karnakar, D.P. and Krishnaswami Aiyanagar, S. (eds.), *Vijayanagara Sexcentenary Commemoration Volume,* pp. 1-28. Dharwad.

Kulke, H.
1985 'Maharajas, Mahants and Historians: Reflections on the Historiography of Early Vijayanagara and Sringeri'. In Dallapiccola, A.L. (ed.) *Vijayanagara: City and Empire,* pp. 120-143. Stuttgart.

Kundangar, K.G.
1928 'The Hosahaḷḷi Copper-plate Grant of Harihara II'. *Journal of the Bombay Historical Society,* vol. 1, no. 2, pp. 121-138.

Kuppuswamy, G.R.
1976 'Economic Implications of the Harihar Stone Inscription of Devarāya I'. *Journal of the Karnatak University, Social Sciences,* vol. 12, pp. 135-145.
1981 'Irrigation System in the Karnataka Dynastic Analysis'. *Journal of the Karnatak University, Social Sciences,* vol. 16-17, pp. 75-86.
1983 'Irrigation Facilities Provided by Feudatory Families in South Karnataka'. Reprint from the *Quarterly Journal of the Mythic Society,* vol. LXXXIV, no. 1, pp. 1-9.

Lal, B.B.
1949 'Śiśupālgarh 1948: An Early Historical Forest in Eastern India'. *Ancient India, Bulletin of the Archaeological Survey of India,* no. 5, January, pp. 62-105.
1955 'Excavations of Hastināpura and other Explorations in the Upper Gangā and Sutlej Basins', 1950-1952. *Ancient India, Bulletin of the Archaeological Survey of India,* nos. 10-12, pp. 5-51.

Leach, E.R.
1959 'Hydraulic Society in Ceylon'. *Past and Present,* vol. 15, pp. 2-26.

Longhurst, A.H.
1917 *Hampi Ruins: Described and Illustrated.* Second edition. Calcutta.

Lournados, H.
1980 'Change or Stability? Hydraulics, Hunter-gatherers and Population in Temperate Australia', *World Archaeology,* vol. II, No. 3, pp. 245-264.

Ludden D.
1979 'Patronage and Irrigation in Tamil Nadu'. *The India Economic and Social History Review,* vol. XVI, no. 3, July-September, pp. 347-365.
1985 *Peasant History in South India.* Princeton.

Mackay, E.J.H.
1938 *Further Excavations at Mohenjo-daro, 1927-1931.* Two volumes. New Delhi.

Mackenzie, C.
1800a 'Account of the Present State of the Anagoondy Country'. In *Mackenzie General,* vol. X, 13, pp. 127-135 (India Office Library and Records).
1800b *Plan of the Ancient City of Beejnagar* (scale 1 inch to 450 yards). In the National Archives, New Delhi (F189/3).
1800c *Beejnagaur (scale 2? to 100 yards).* In the Mackenzie Collection, no. WD 2646 (India Office Library and Records).
1800d *Sketch Plan of the Ruins of Beejnagur (scale 1 inch to 450 yards).* In the Mackenzie Collection, no. WD 2650 (India Office Library and Records).
1800e *Plan of a Singular Structure remaining at Beejanaggur.* In the Mackenzie Collection, no. WD 2651 (India Office Library and Records).

Madras Survey Maps
1800 *Hampi Ruins (scale 12 inch to 1 mile).*
1882 *Ruins of Hampi (scale 6 inch to 1 mile).*
1900 *Map of the City of Vijayanagara and Surrounding Country (scale 4 inch to 1 mile).*

Mahalingam, T.V.
1951 *Economic life in the Vijayanagara Empire.* Madras.
1955 *South India Polity.* Madras.
1975 *Administration and Social Life under the Vijayanagara Empire.* In two volumes. Madras.

Major, R.H.
1857 *India in the Fifteenth Century.* London.

Maloney, C.
1975 'Archaeology in South Asia: Accomplishments and Prospects'. In Stein, B. (ed.). *Essays on South Asia,* Asain Studies at Hawaii series, no. 15. Hawaii.

Manjunathaiah, T.M. and Parameshwar Kumar, M.
1985 'Sculptures on the Hillock South of Matanga'. In Nagaraja Rao, M.S. (ed.), *Vijayanagara: Progress of Research, 1983-1984,* pp. 136-137. Mysore.

Marshall, J.
1931 *Mohenjo-daro and the Indus Civilisation.* In three volumes. London.

Mate, M.S.
1970 'Early Historic Fortifications in the Ganga Valley'. *Purātattva, Bulletin of the Indian Archaeological Society,* no. 3, pp. 58-69.

McCrindle, J.W.
1926 *Ancient India as described by Megasthenes and Arrian.* Calcutta.

Mehta, R.N.
1963 'Ancient bunds in Sabarkāṇṭha district', Gujarat. *Journal of the Oriental Institute, M.S. University of Baroda,* vol. XII, June, no. 4, pp. 359-365.
1979 *Medieval Archaeology.* New Delhi.

Menon, P.
1986 'Agrarian Economy of the Carnatic in the Sixteenth and Seventeenth Centuries A.D'. Ph.D. Dissertation, University of Aligarh.
1987 'Vijayanagara: Striking Waterworks'. *Frontline,* December 12th-25th, pp. 76-78.

Merklinger, E.S.

1981 *Indian Architecture: The Deccan, 1347-1686 A.D.* London.

Michell, G.

1977 *The Hindu Temple.* London.

1982 'Vijayanagara: City of Victory'. *History Today*, vol. XXXII, pp. 38-42.

1983a 'Report on Field Work, January-March 1983'. In Nagaraja Rao, M.S. (ed.), *Vijayangara: Progress of Research, 1979-1983,* pp. 41-42. Mysore.

1983b 'A Small Dated Temple at Vijayanagara'. In Nagaraja Rao, M.S. (ed.), *Vijayanagara: Progress of Research, 1979-1883,* pp. 45-49. Mysore.

1985a 'A Never Forgotten City'. In Dallapiccola, M. (ed.), *Vijayanagara: City and Empire,* pp. 196-207. Stuttgart. Also in Nagaraja Rao, M.S. (ed.), *Vijayanagara: Progress of Research, 1983-1984,* pp. 152-163. Mysore .

1985b 'Architecture of the Muslim Quarters at Vijayanagara'. In Nagaraja Rao, M.S. (ed.). *Vijayanagara: Progress of Reasearch, 1983-1984,* pp. 101-118. Mysore.

Michell, G. And Filliozat, F.

1981 *Splendours of the Vijayanagara Empire: Hampi.* Marg Publications. Bombay.

Miller, R.

1980 'Water Use in Syria and Palestine from the Neolithic to the Bronze Age'. *World Archaeology*, vol. II, no. 3, pp. 331-341.

Moreton, W.E.

1853 'Memoranda on Irrigation by Rajbuhas'. Roorkee College Press. Bound with other articles in a volume marked, Professional Papers, Roorkee College (SOAS Library).

Morrison, K.D.

1988 'The Vijayanagara Metropolitan Survey: Prelimnary Investigation'. In Nagaraja Rao, M.S. (ed.), *Vijayanagara: Progress of Research, 1985-1987.* Mysore.

Mukerji, N.G.

1907 *Handbook of Indian Agriculture.* second edition. Calcutta.

Nagaraja Rao, M.S. (ed.)

1983 *Vijayanagara: Progress of Research, 1979-1983.* Mysore.

1985 *Vijayanagara: Progress of Research, 1983-1984.* Mysore.

Nagaraja Rao, M.S.

1983a 'Nomenclature of a Royal Road at Vijayanagara'. In Nagaraja Rao, M.S. (ed.), *Vijayanagara: Progess of Research, 1979-1983,* pp. 57-59. Mysore.

1983b 'Ahmadkhan's Dharmaśāla'. In Nagaraja Rao, M.S. (ed.), *Vijayanagara: Progess of Research, 1979-1983,* pp. 64-65. Mysore.

1985 'Research at Vijayanagara, 1979-1983'. In Dallapiccola, A.L. (ed.). *Vijayanagara: City and Empire,* pp. 208-215. Stuttgart.

Nagaraja Rao, N.S. and Patil, C.S.

1985 'Epigraphical References to City Gates and Watch Towers of Vijayanagara'. In Nagaraja Rao, M.S. (ed.), *Vijayanagara: Progress of Research, 1983-1984,* pp. 96-100. Mysore.

Nath, R.

1978 *History of Sultanate Architecture.* New Delhi.

National Insititute of Design, Ahmedabad

1983 *Fatehpur Sikri: Integrated Development Plans.* For the Department of Tourism, Government of India. New Delhi.

Newbold, Lieut.

1842 'On the Process prevailing among the Hindu, and formerly among the Egyptians, of quarrying and polishing Granite'. *Journal of the Royal Asiantic Society,* no. VII, pp. 113-128.

1845 'Notes, Chiefly Geological, across the Peninsular of Southern India'. *Journal of the Asiatic Society of Bengal,* vol. XIV, part 2, pp. 517-519.

Oates, D. and Oates, J.

1976 'Early Irrigation Agriculture in Mesopotamia'. In Sieveking, G. de G., Longworth, I.H. and Wilson, K.E. (eds.), *Problems in Economic and Social Anthropology,* pp. 109-136. London.

O'Brien, M.J. Lewarch, D.E., Mason, R.D. and Neely, J.A.

1980 'Functional Analysis of Water Control Features at Monte Alban', Oaxaca, Mexico. *World Archaeology,* vol. II, no 3, pp. 342-355.

Panda, S.K.

1985 'The Kṛṣṇa-Godavari Delta: A Bone of Contention Between the Gajapatis of Orissa and the Rāyas of Vijayanagara'. In Dallapiccola, A.L. (ed.), *Vijayanagara: City and Empire,* pp. 88-96. Stuttgart.

Parliamentary Papers, Returns to the House of Commons

1851 *Public Works Report, Bombay Presidency.* Dated 4th March, 1850. Printed 1st August, 1851, pp. 198-201.

1853 *Particular of some of the more important works executed under the Madras Civil Engineers Department.* Third Division. Dated 15th April, 1853. Printed 20th August, 1853, pp. 127.

1854 *Public Works Report, Madras Presidency.* Dated 21st February, 1854. Printed 8th-10th May, 1854, pp. 68-69.

1870 *Proceedings of the Madras Government, Public Works Department.* Dated 27th January, 1869. Printed 27th July, 1870, pp. 38-45. Bound in a volume entitled, East India (Irrigated Works), vol. V (SOAS Library).

Patil, C.S.

1983a 'Krishna Temple'. In Nagaraja Rao, M.S. (ed.), *Vijayanagara: Progress of Research, 1979-1983,* pp. 60-63. Mysore.

1983b 'Door Guardians of Vijayanagara'. In Nagaraja Rao, M.S., (ed.), *Vijayanagara: Progress of Research, 1979-1983,* pp. 66.67. Mysore.

1985a 'Palace Architecture at Vijayanagara: The New Excavation'. In Dallapiccola, A.L. (ed.), *Vijayanagara: City and Empire,* pp. 229-239. Stuttgart.

1985b 'Palace Architecture of Vijayanagara'. In Nagaraja Rao, M.S. (ed.), *Vijayanagara: Progress of Research, 1983-1984,* pp. 119-132. Mysore.

1985c 'Sculptures at Koṭilinga'. In Nagaraja Rao, M.S. (ed.), *Vijayanagara: Progress of Research, 1983-1984,* pp. 138-143. Mysore.

Pharoah and Company

1855 *A Gazetteer of Southern India with the Tenasserim Provinces and Singapore,* pp. 83-86, 98-100, 108-109. Madras.

Pichumuthu, C.S.

1980 *Physical Geography of India.* Fourth edition. New Delhi.

Purandare, S.

1986 'The History and Archaeology of Anegundi'. Ph.D. Disseration, University of Poona.

Rajasekara, S.

1985a 'Inscription at Vijayanagara'. In Dallapiccola, A.L. (ed.), *Vijayanagara: City and Empire,* pp. 101-119. Stuttgart.

1985b *The Map Approach to Vijayanagara History.* Dharwad.

Ramanayya, N.V.

1933 *Vijayanagara: Origins of the City and the Empire.* Madras.

Ramaswami, P.N.

1923 'Early History of Indian Famines'. *Indian Antiquary,* vol. LII, pp. 227-242.

Rangacharya, V. (ed.)

1915 *A Topographical List of Inscriptions in the Madras Presidency, Collected till 1915 with Notes and References.* In three volumes. Madras.

Rao, S.R.

1963 'Excavation at Rangpur and other Explorations in Gujarat'. *Ancient India, Bulletin of the Archaeological Survey of India,* no. 18-19, pp. 5-207.

Rea, A.

1886 'Vijayanagara I'. *Christian College Magazine,* Madras, December, pp. 428-436.

1887 'Vijayanagara II'. *Christian College Magazine,* Madras, January, pp. 502-509.

Report of the Indian Irrigation Commission (1901-1903)

1903 In four parts. London.

Richards, F.J.

1930 'Race Drift in South India'. *Indian Antiquary,* vol. LIX, pp. 211-218, 229-230.

1933 'Geographical Factors in Indian Archaeology'. *Indian Antiquary,* vol. LXII, pp. 235-243.

Rizvi, S.A.A. and Flynn, V.J.A.

1975 *Fatehpur-Sikri.* Bombay.

Robins F.W.

1946 *The Story of Water Supply.* London.

Rotzer, K.

1984 'Bījāur. Alimentation en eau d'une ville Musalmane du Dekkan aux XIV^{e}XVIe Sicles'. *Bulletin de L'-École Française D' Extreme Orient,* no. 17, pp. 125-195.

Rouwse, H. and Ince, S.

1957 *History of Hydraulics.* Iowa.

Rowland, B.

1977 *The Art and Architecture of India: Hindu, Buddhist and Jain.* Second, integrated edition. London.

Saletore, B.A.

1934 *Social and Political Life in the Vijayanagara Empire.* Madras.

1936 'Theories Concerning of the origin of Vijayanagar'. In Karnakar, D.P. and Krishnaswami Aiyangar, S. (eds.), *Vijayanagara Sexcentenary Commemoration Volume,* pp. 139-159. Dharwad.

Saletore, R.N.

1937a 'Town-planning in the Vijayanagara Empire I'. *Karnataka Historical Review,* vol. IV, pp. 43-50.

1937b 'Town-planning in the Vijayanagara Empire II'. *Karnataka Historical Review,* vol. V, pp. 5-9.

1982 *Vijayanagara Art.* New Delhi.

Sastri, H.K.

1908 'The First Vijayanagara Dynasty: Its Viceroys and Ministers'. In *Annual Report, Archaeological Survey of India.* 1907-8, pp. 235-254. Calcutta.

1912 'The Third Vijayanagara Dynasty: Its Viceroys and Minister'. In *Annual Report, Archaeological Survey of India,* 1911-12, pp. 177-197. Calcutta.

Sastri, K.A.N.

1950 *History of India.* In three volumes. Madras.

Saunders, N.

1985 'The Civilising Influence of Agriculture'. *New Scientist,* 13th June, pp. 16-18.

Schimmel, A.

1985 'Deccani Art and Culture'. In Dallapiccola, A.L. (ed.), *Vijayanagara: City and Empire,* pp. 177-187. Stuttgart.

Selections From the Records of the Madras Government (New Series).

XXVII (1856):60

XXIX (1856):46

XXXVIII (1857):174-175

LIV (1858):342-343
LIX (1959):11

Sewell, R.
1900 *A Forgotten Empire.* London.
1932 *The Historical Inscriptions of Southern India.* Madras.

Shama Sastry, R.
1917 'A Few Inscriptions of the Ancient Kings of Anegundi'. *The Quarterly Journal of the Mythic Soceity (Bangalore)*, vol. VII, July, no. 4, pp. 285-291.

Shamasastry, R.
1951 *Kauṭilya's Arthaśāstra.* Fourth edition. Mysore.

Sharma, G.R.
1969 'Excavation at Kauśāmbī, 1949-1950'. *Memoirs of the Archaeological Survey of India*, no. 74. New Delhi.

Sherratt, A.
1980 'Water, Soil and Seasonality in Early Cereal Cultivation'. *World Archaeology*, vol. 11, no. 3, pp. 313-330.

Sherwani, H.K.
1953 *The Bahmanis of the Deccan: An Objective Study.* Hyderabad.

Singer, C., Holmyard, E.J. and Hall, A.R. (eds.)
1954 *A History of Technology.* Oxford.

Sinha, B.P. And Narain, L.A.
1970 *Pāṭaliputra Excavation, 1955-1956.* Patna.

Sinopoli, C.
1983 'Earthenware Pottery of Vijayanagara: Some Observations'. In Nagaraja Rao, M.S. (ed.), *Vijayanagara: Progress of Research, 1979-1983*, pp. 68-74. Dharwad.
1985a 'Earthenware Pottery of Vijayanagara: Documentation and Interpretation'. In Dallapiccoḷa, A.L. (ed.), *Vijayanagara: City and Empire*, pp. 216-228. Stuttgart.
1985b 'Vijayanagara Ceramic Analysis: 1983-1984 Progress Report'. In Nagaraja Rao, M.S. (ed.), *Vijayanagara: Progress of Research, 1983-1984*, pp. 57-68. Dharwad.

Sircar, D.C.
1965 *Selected Inscriptions, vol. 1, from the sixth century B.C. to the sixth century A.D.* Second edition. Calcutta.

Stein, B. (ed.)
1975 'Essays on South Asia'. Asian Studies at Hawaii Series no. 5. Hawaii.

Stein, B.
1980 *Peasant State and Society in Medieval South India.* New Delhi.
1985a 'The Problematical "Kingdom of Vijayanagara"'. In Dallapiccola, A.L. (ed.), *Vijayanagara: City and Empire*, pp. 1-4. Stuttgart.
1985b 'Vijayanagara and the Transition to Patrimonial Systems'. In Dallapiccola, A.L. (ed.), *Vijayanagara: City and Empire*, pp. 73-87. Stuttgart.

Strachey, R.
1870 'The Extension of Irrigation in Guzerat, Candeish, and the Deccan'. Memoradum no. 9. In *Return to the House of Commons.* Dated 17th April, 1867. Printed 27th July, 1870. pp. 73-79. Bound in a volume entitled, *East India (Irrigation Works)*, vol. V. (SOAS Library).

Strange, W.L.
1923 *Indian Engineering Relating to Irrigation, Water Supply of Towns, Roads and Buildings.* London.
1928 *Indian Storage Reservoirs with Earthen Dams.* London.

Suranarain Row, B.
1909 *A Short Description of the City of Vijayanagara.* Madras.

Survey of India Map Series
1928 No. 57 A/8 (scale 1:63,360)
1929 No. 57 A/7 (scale 1:63,360)
1947 No. 57 A/8 (scale 1:63,360)
1947 No. 57 A/7 (scale 1:63,360)
1976 No. 57 A/7 (scale 1:50,000)
1976 No. 57 A/8 (scale 1:50,000)
1976 No. 57 A/11 (scale 1:50,000)

Terry, J.
1955 *The Charm of Indo-Islamic Architecture.* London.

Thapar, R.
1981 *A History of India.* London.

Thomas, I.J.
1985 'Cultural Developments in Tamil Nadu During the Vijayanagara Period'. In Dallapiccola, A.L. (ed.), *Vijayanagara: City and Empire,* pp. 5-40. Stuttgart.

Thurston, E.
1913 *The Madras Presidency with Mysore, Coorg and the Associated States.* Cambridge.

Times of India
1973 'New fountains unearthed in Taj'. 25th June.

Van Liere, W.J.
1980 'Traditional Water Management in the Lower Mekong Basin'. *World Archaeology,* vol. 11, no. 3, pp. 265-280.

Vats, M.S.
1940 *Excavations at Harappā, 1920-1922 and 1933-1934.* In two volumes. Calcutta.
1950 'Repairs at Agra and Fatehpur Sikri: 1944-1949'. *Ancient India, Bulletin of the Archaeological Survey of India,* no. 6, January, pp. 91-99.

Venkayya, V.
1906 'Irrigation in South India in Ancient Times'. In *Annual Report, Archaeological Survey of India, 1903-4,* pp. 202-211. Calcutta.

Volwahsen, A.
1970 *Living Architecture: Islamic India.* London.

Waddington, H.
1946 'Ādilābād: A Part of the Fourth Delhi'. *Ancient India, Bulletin of the Archaeological Survey of India,* no. 1, January, pp. 60-76.

Walch, G.T.
1896 *The Engineering Works of the Godavari Delta: A Descriptive and Historical Account.* In two volumes. Madras.
1899 *The Engineering Works of the Kistna Delta: A Descriptive and Historical Account.* In two volumes. Madras.

Warth, H.
1895 'The Quarrying of Granite in India'. *Nature,* vol. LI, pp. 272.

Watson, A.
1964 *The War of the Goldsmith's Daughter.* London.

Wheeler, R.E.M.
1948 'Iran and India in Pre-Islamic Times: A Lecture'. *Ancient India, Bulletin of the Archaeological Survey of India,* no. 4, double number, July to January, pp. 85-103.

Willcocks, W.
1904 *The Assuan Reservoir and Lake Moeris.* Cairo.
1911 *The Irrigation of Mesopotamia.* London.
1930 *Ancient Systems of Irrigation in Bengal.* Calcutta.

Wilson, H.H.

1828 *The Mackenzie Collection: A Descriptive Catalogue of the Oriental Manuscripts and other Articles.* Second Edition. Calcutta.

Wilson, H.M.

1909 *Irrigation Engineering.* Sixth edition. New York.

Winter Jones, J.

1863 *The Travels of Ludovico di Varthema.* London.

Wittfogel. K.A.

1957 *Oriental Despotism.* New Haven.

Index

Achyuta Rāya 7, 24, 25
Achyutāpura 12, 29, 62, 65, 78
agriculture 12, 13, 14, 15, 17, 19, 20, 21-3, 25, 26, 29, 31, 60, 72, 73, 89ff, 92, 99, 103
Amarāvati 10, 52, 53, 70
Anegundi 9, 10, 11, 19, 24, 27-9, 74, 75
Anegundi canal 66-7, 91, 96
anicuts (diversion weirs) 11, 19, 20, 26, 27, 33, 50, 52-3, 56-7, 73, 74, 75, 89ff, 95ff, 100
    Anegundi 66-7, 101
    Hiriya 57-60, 96, 99-100, 101
    Hosūru 56, 96, 99-100, 101
    Kuruvagadda 53, 96
    Rāmanagadda 53, 101
    Vallabhāpuram 52, 53, 96, 99-100, 102
aqueducts 24, 34, 65, 66-7, 91, 95
    Setu 65, 91

Bahmani kingdom 6, 7, 86
Balakṛṣṇa temple 24, 29, 47, 78
Basavanna canal 11, 20-1, 52-6, 91, 101, 102
baths 24, 26, 33-6, 37, 40, 41, 76, 77, 80, 88
Bellary district 9, 14, 20-1, 55, 72-3, 75, 76, 94
Bhojanśāla 43, 44, 76, 77
Bijāpur 7, 9, 29, 81, 86-7, 95
bridges 24, 44, 61, 62-4, 65, 91
Bukka 6, 7, 24, 93
Bukka II 24, 90
bunds *see* earthworks

Cālukyas 10-11, 73ff, 101
canals (*kāluve*) 10, 19, 20-1, 33, 34, 46, 50, 51-7, 57-71, 73, 74, 75, 78, 89ff, 93ff
    at Besnagar 74
Cālukyan 74
    at Rajghat 74
    in Rudradaman's inscription 74
    at Kaveripattinam 74-5
    at Nāgarjunakoṇḍa 75
    in Kharavela inscription 74
    in Megasthenes' account 74
    capstans 49, 50, 54, 98-9
Chandraśekhara temple 33, 34, 36
channels/ conduits 33, 34, 35, 37-8, 39, 40, 41, 42, 43, 44, 46, 47, 49, 50, 51-2, 64, 67, 71, 73, 74, 76, 77, 78, 79, 81, 84, 85, 86, 88, 90, 99

Danāyakanakere 11, 73, 75, 89, 101
Delhi Sultanate 5-6, 87, 94
desilting 36, 38-9, 42, 43, 69, 79, 80, 92-3, 96, 97-8
Devarāya I 90
diversion weirs *see* anicuts
drains 38, 40, 41, 47, 50, 53, 83, 84, 85, 86, 89, 95
drinking water 23, 79

earthworks (bunds) 25, 33, 35, 43, 48-9, 50, 60, 61, 67, 73, 74, 75, 92, 94, 101
    Mallappannagudi 70-1, 90
    Matanga Parvatam 61-2, 65-7, 78
    Rayas 67-70, 93

Fatehpur Sikri 81, 87-8, 94-5
Faṭhu' lāh Shīrāzī 81
fountains 41, 42, 43, 76, 77, 78, 82, 83

Gauripuram Vanka 53

Harihara 6, 7, 24, 95-6
Harihara II 90
Hemakuntam hill 9, 24, 46, 76, 77
Hiriya (Turtha) canal 11, 19, 24, 46, 47, 56, 57-66, 78, 89-91, 98, 99, 100, 101
Hospet (Nāgalāpura) 7, 10, 19, 21, 26, 27, 29, 30, 32, 48, 52-6, 67, 70
Hosuru canal 56-7
Hoysalas 6, 24-5, 79

Indus civilization 84
inscriptions 5, 9-12, 16, 24, 27, 45, 46, 48, 58, 59, 62, 65, 71, 73, 74, 83, 90, 93, 94, 95-6, 98, 100
iron fittings/clamps 37, 66, 85
iron production 30-1

Kadirāmpuram 20, 26, 29
Kālaghatta canal 56-7
Kamalāpuram 8, 20, 26, 27, 31, 46, 48, 52, 53-4, 55, 56, 61
Kamalāpuram tank 33, 48-51, 52, 54, 55, 56, 61, 70, 71, 76, 78-9, 89, 90, 93, 94, 96, 99, 100
Kampli 6, 9, 11, 26, 32, 59, 74, 101

Kṛṣṇadeva Rāya 7, 10, 27, 69, 83, 89, 90, 93, 100, 101, 101, 102

land tenure 12, 13
Lokapāvana tank 47

Madhavācarya Vidyāraṇya 6
Madura 6, 7
Mālāpuram 11, 20, 33, 52, 53, 54, 55, 72, 91, 100
Mallappannagudi 12, 27, 29, 55, 70-1, 77, 92
Manmatha tank 24, 77, 78
masonry 21, 25, 34, 36, 37, 38, 39, 40, 41, 43, 44, 45, 46, 48-9, 50, 51, 53, 55, 58, 59, 60, 64-5, 66, 67, 68, 73, 75, 83, 86, 92, 94, 96

Octagonal Bath 35, 36, 43, 77, 82, 83
Octagonal Fountain 41, 42, 43, 76, 77, 78, 82, 83

Pataliputra 74, 85
Paṭṭananda Ellammā temple 34
pipes 35, 36, 40, 41, 43, 44, 45, 49, 71, 77, 79-81, 84, 86, 88, 90
plaster, mortar, linings 33, 34, 35, 36, 37, 38, 40, 41, 42, 43, 44, 49, 50, 51, 55, 56, 58, 59, 60, 61, 66, 71, 73, 79, 81, 83, 85, 95, 96
population of urban centre 4-5, 7, 8, 14, 15, 25, 29
Portuguese 7, 14

qunats 86, 87
Queens Bath 33, 34, 35, 36, 43, 76, 77, 81, 82, 83

Raichur doab 6, 7, 14, 72-3, 75, 76
Rāma Rāya 7
Rāya canal 20-1, 48, 51, 52-6, 57, 90, 91, 93, 100, 101
revenue (state respources) 4, 5, 14, 15, 16, 17, 101
Royal Centre 24, 25-6, 28, 30, 32, 33-45, 49, 51, 52, 72, 76-89, 101

Sagar 6
Sandur hills and valley 19, 53, 67, 69, 70
Sāyaṇa 6
'segmentary state' 3-5, 15, 16-18
siphon 79
sluices 38, 42, 64-5, 67-9, 71, 92-4, 95, 96
soil 21
spillways (waste weirs) 50-1, 58-9, 93, 95, 96
springs 47
Stepped Tank 26, 38, 39, 79

tanks (*kere*) 11, 12, 20, 23, 26, 33, 34, 35, 36, 38-9, 40, 42, 44, 45-6, 73, 78-81, 84, 86, 87, 93
  catchment drainage 19, 20, 35, 45-6, 73, 75, 86, 92
  control 43, 44, 79, 80-1
  Great Tank 38-9, 41, 76, 77, 80, 82, 83, 93
  irrigation 48-51, 62, 67-9, 73-5, 89, 101
  Kaveripattinam 74
  rain water 30, 35, 45, 46, 75, 77, 81, 87
  step 26, 38, 39, 45
  temple 30, 47, 76, 77, 78, 83, 86, 88-9
taxation 5, 12, 14, 16, 17
technology of water supply 72-8, 84-8, 103
Tungabhadrā Dam and irrigation scheme 11, 20-1, 27, 33, 40, 46, 47, 52, 53, 54, 55, 58, 67, 71, 72, 79, 91-2
Tungabhadrā river 6, 10, 12, 19, 20, 23, 24, 26-7, 48ff, 73, 74, 91, 100
Turtha anicut 11, 57-60

Urban Core (of Vijayanagara) 19, 23, 25-6, 27-30, 35-6, 45, 49, 51, 52, 61, 99, 100
urban industries 30-1

Vallabhāpuram anicut 11, 52, 53, 54, 102
Vijayanagara city 2, 5, 6, 19, 29
  defences 22, 24, 25, 49
  end 78
  environs 1, 5, 12, 19, 20, 37, 72
  gardens 23, 29
  lay out 23-31
see also Anegundi, Hemakutam, Hospet, Kamalāpuram, Royal Centre, Urban Core, Virupākṣa temple
Vijayanagara kingdom
  historiography 3-5, 8, 12-18
  nāyakas 5, 13, 14, 16, 29
  organization 3-4, 5, 12-18
  political history 4-7
  sources 5-12, 69-70 (*see also* inscriptions)
Vijayanagara, royal capital 1, 5, 7, 13, 22-4, 29
Virūpākṣa temple 4, 6, 9, 11, 24, 46, 47, 76
Virūpāpuragadda island 24, 66, 67, 91
Viṭṭhala temple 24, 29, 47

Water Pavilion 40, 44-5, 76, 77
Wells 12, 20-1, 26, 30, 39, 40, 41, 44, 46, 73, 75, 77, 78, 79, 81, 83, 86, 88, 91, 100
  step 45, 46, 77
  terracotta ring wells 73, 84, 85

ILLUSTRATIONS

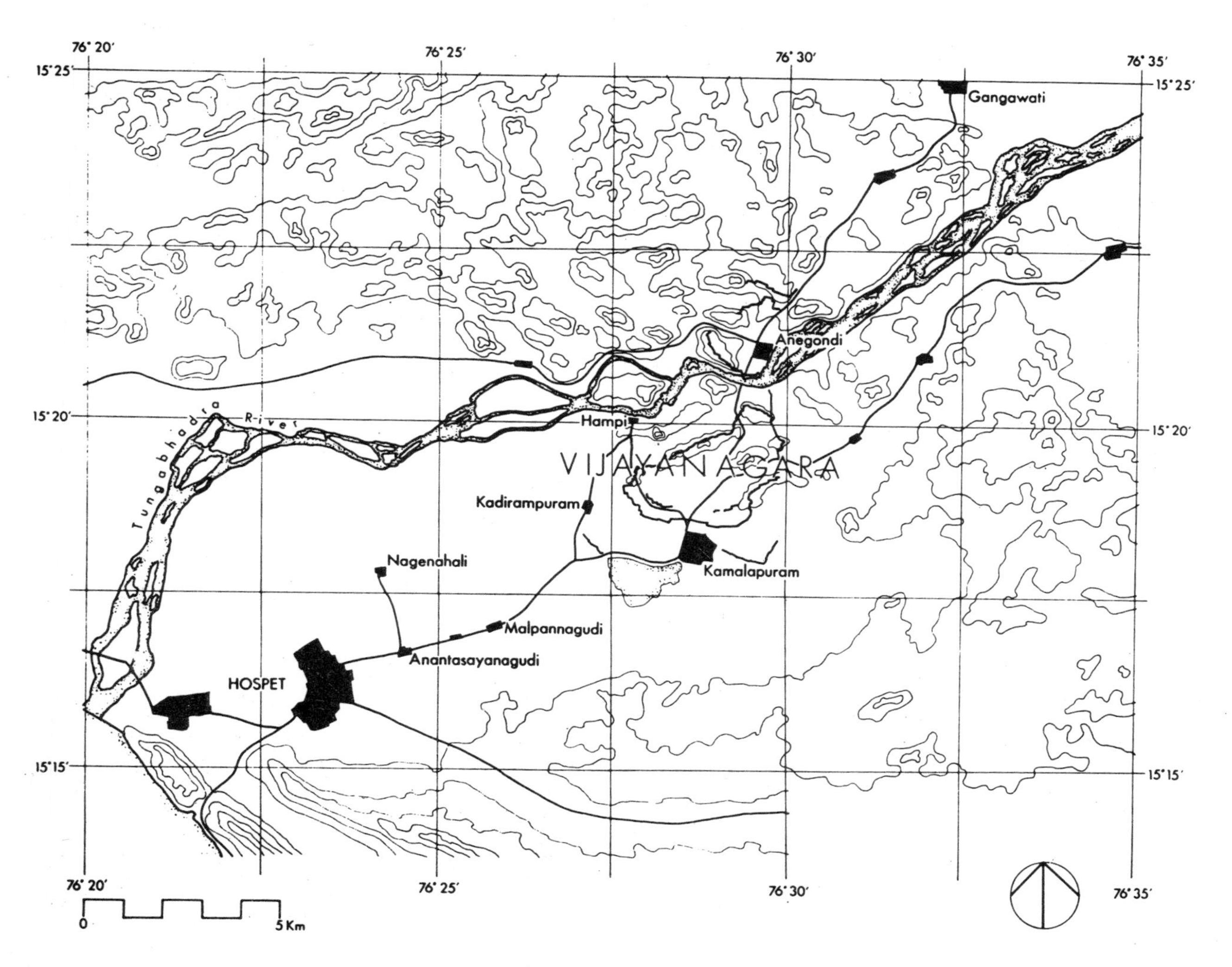

2.1 Map of Vijayanagara its environs (VRP).

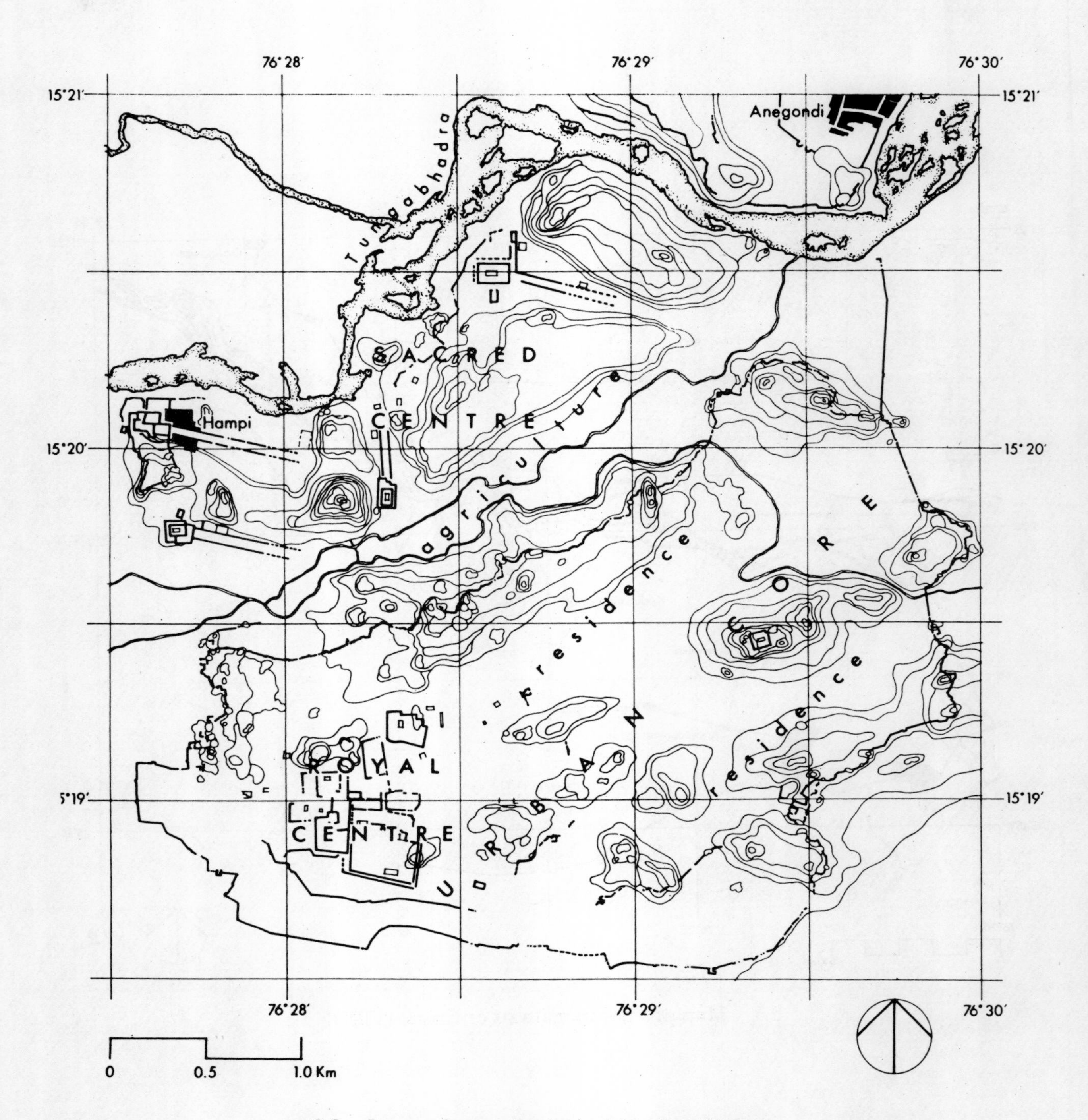

2.2 Internal organisation of the site (VRP).

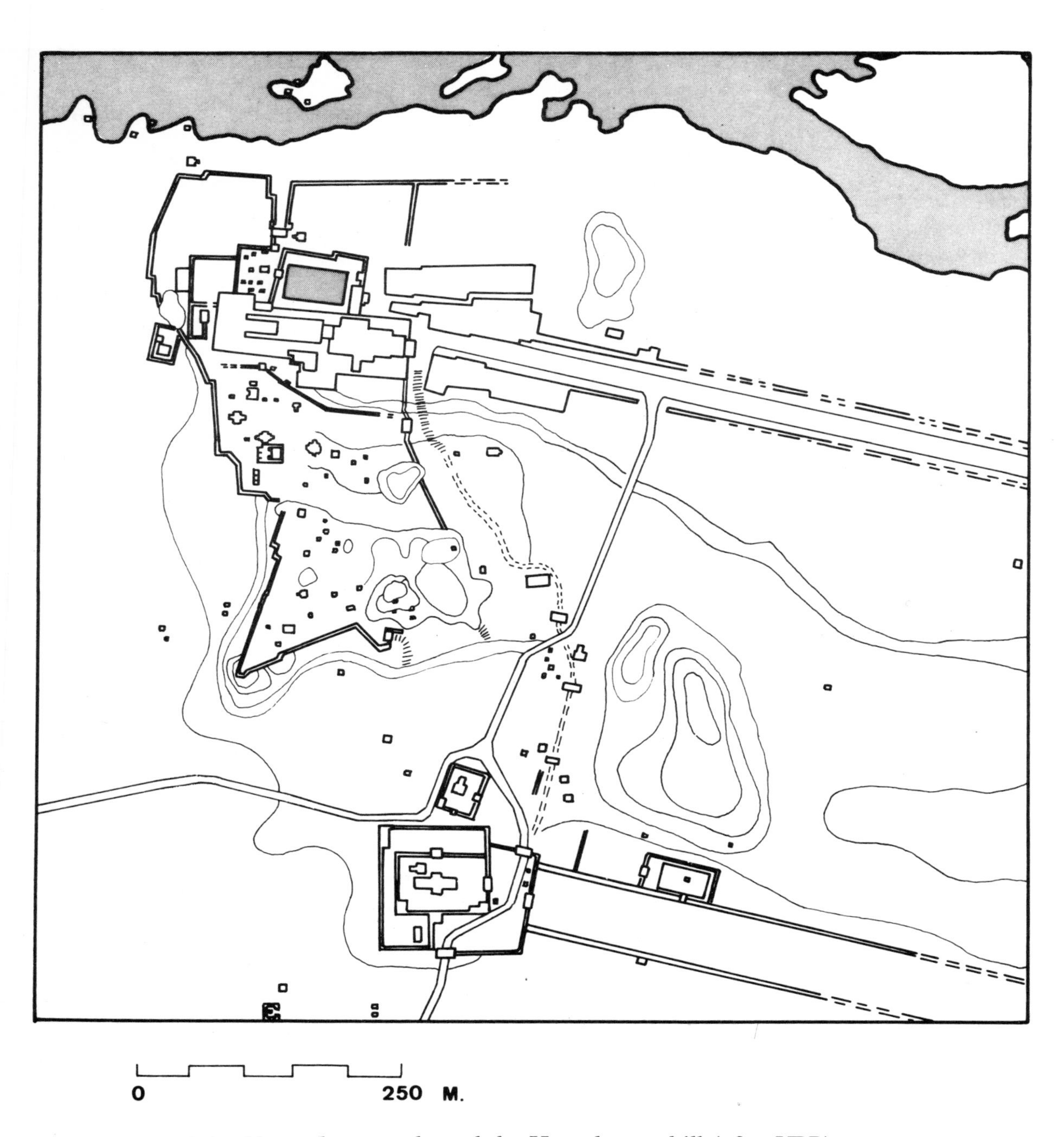

2.3 Vīrūpākṣa temple and the Hemakutam hill (after VRP).

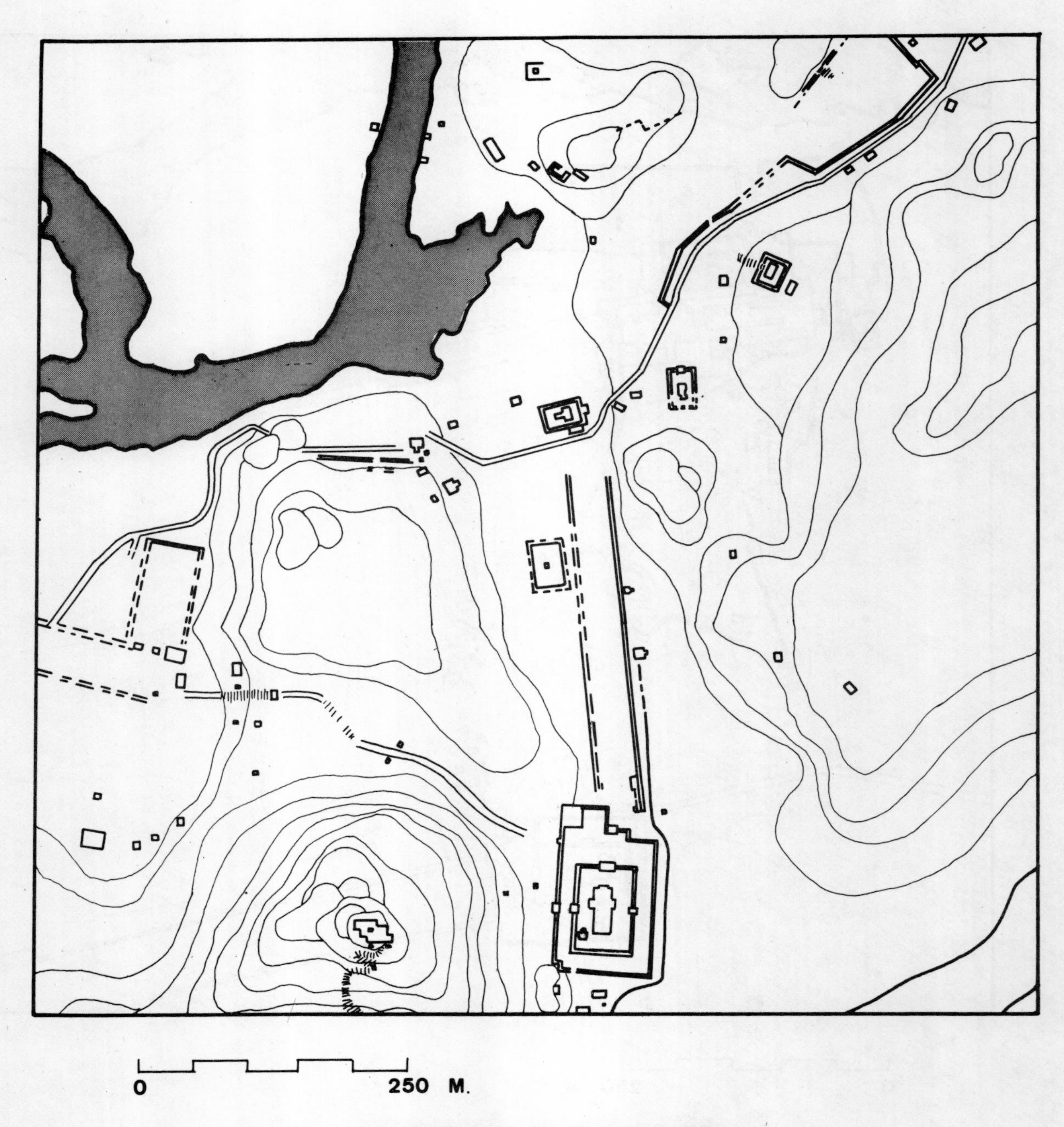

2.4 Achyuta Rāya's temple (after VRP).

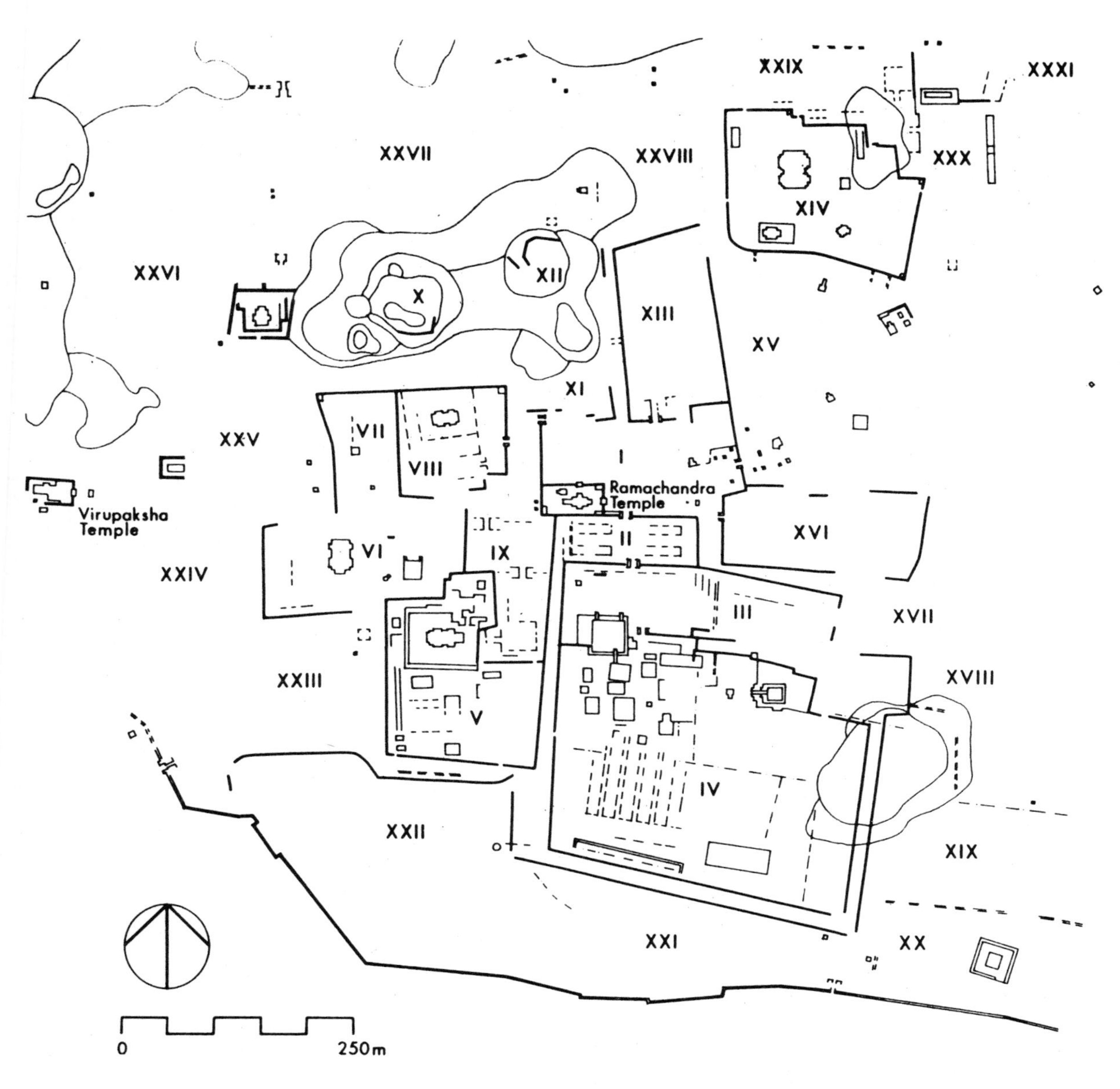

3.1 Map of the 'Royal Centre' (VRP).

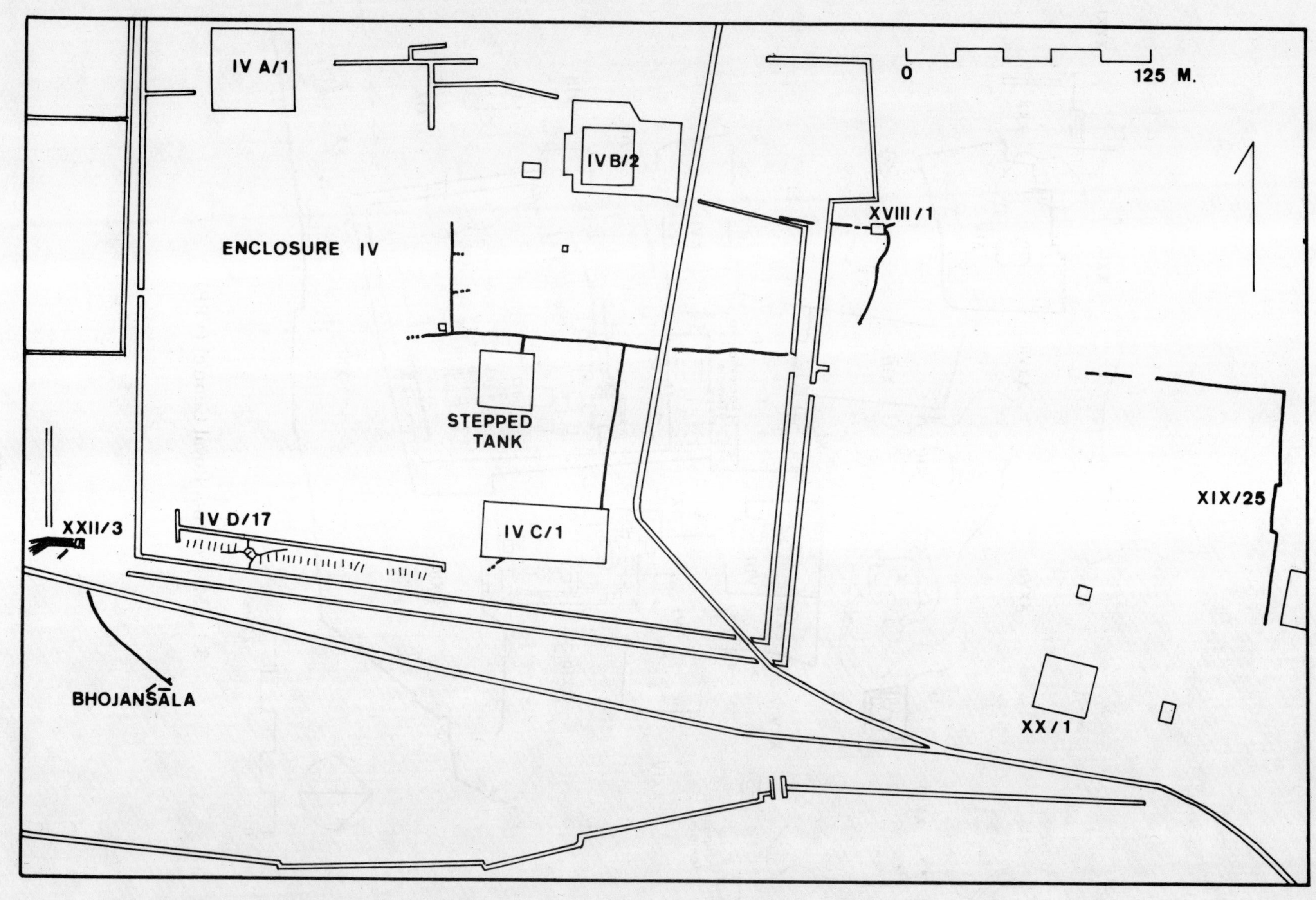

3.2 Map showing water features in enclosure IV and its environs.

3.3 Open channel XIX/25.

3.4 Open channel XIX/25.

3.5. Chunam skirting on the west side of XIX/25.

3.6 Panorama of the disturbed area to the south of Chandraśekhara temple.

3.7 Ruined tank south of Chandraśekhara temple.

3.8 Close-up of the tank's lining.

3.9 Open channel XIX/25; 90° turn to the west, view to north.

3.10 Open channel XIX/25; 90° turn to the east, view to east.

3.11 View to the north along open channel XIX/25; disarticulated sections.

3.12 View to the north along open channel XIX/25; disarticulated sections.

3.13 View to the east along open channel XIX/25; disarticulated sections.

3.14 View to the east along open channel XIX/25; disarticulated sections.

3.15 XIX/25 acting as a boundary wall; view to east.

3.16 XIX/25 acting as a boundary wall; view to east.

3.17 XIX/25 running across toe of hill/ outcrop; view to east.

3.18 XIX/25 running across toe of hill/ outcrop; view to west.

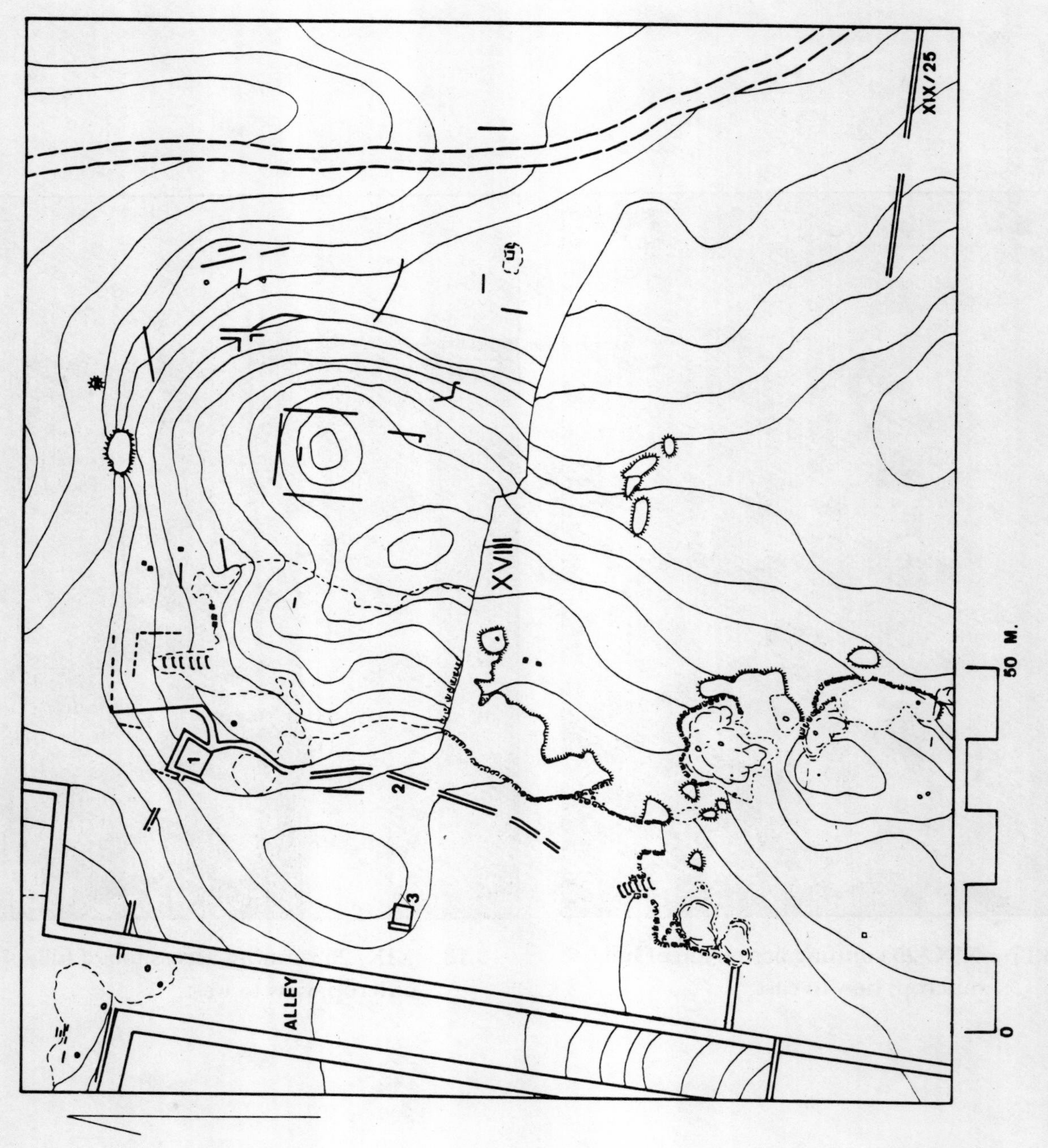

3.19 Map of enclosure XVIII (after VRP).

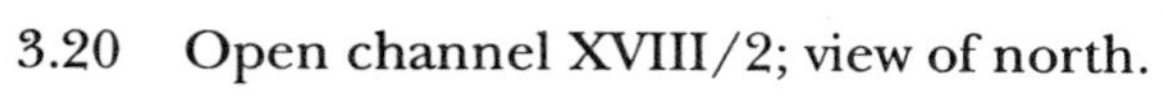

3.20 Open channel XVIII/2; view of north.

3.21 Open channel XVIII/2; view of south.

3.22 Close-up of the chunam lining in XVIII/2.

3.23. XVIII/2 passing through the crevice in a boulder; view to north.

3.24 Close-up of the plaster lining in the crevice.

3.25 XVIII/2, supported by brickwork, running along terrace.

3.26 XVIII/2 in excavation; view to north.

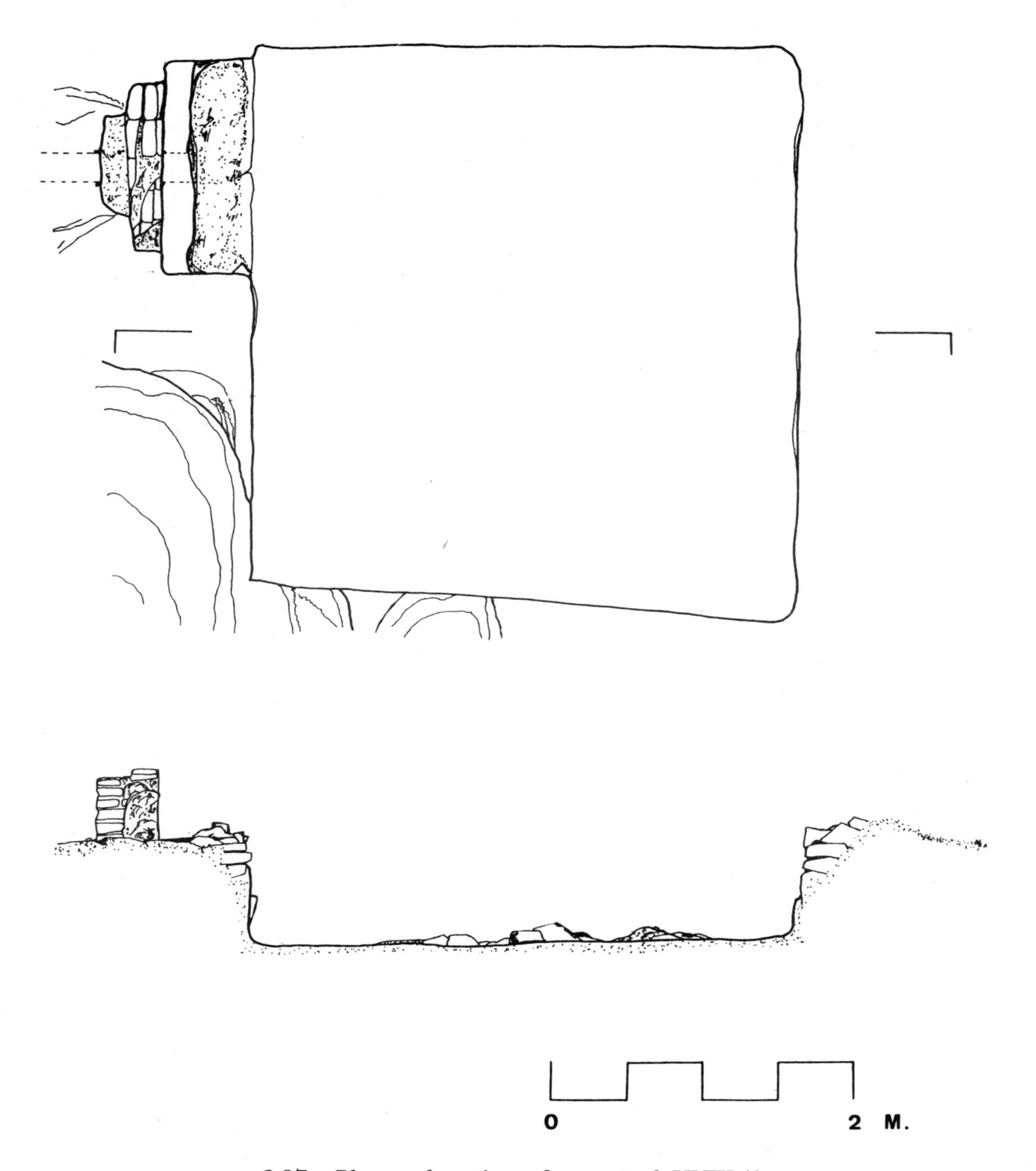

3.27 Plan and section of open tank XVIII/1.

3.28 Open tank XVIII/1; view to northwest.

3.29 Open tank XVIII/1; view to northeast.

3.30 Close-up of the plaster lining in XVIII/1.

3.31 Open tank XVIII/1; the east face of the pipe outlet.

3.32 Open tank XVIII/1; close-up of the east face of the pipe outlet.

3.33 Open tank XVIII/1; the west face of the pipe outlet.

3.34 Open tank XVIII/1; close-up of the west face of the pipe outlet.

3.35 Pipeline leading from XVIII/1 passing through enclosure wall.

3.36 Close-up of the pipeline leading from XVIII/1.

3.37 Pipeline passing through alley between XVIII and IV.

3.38 Pipeline running along the outside of the enclosure wall of IV.

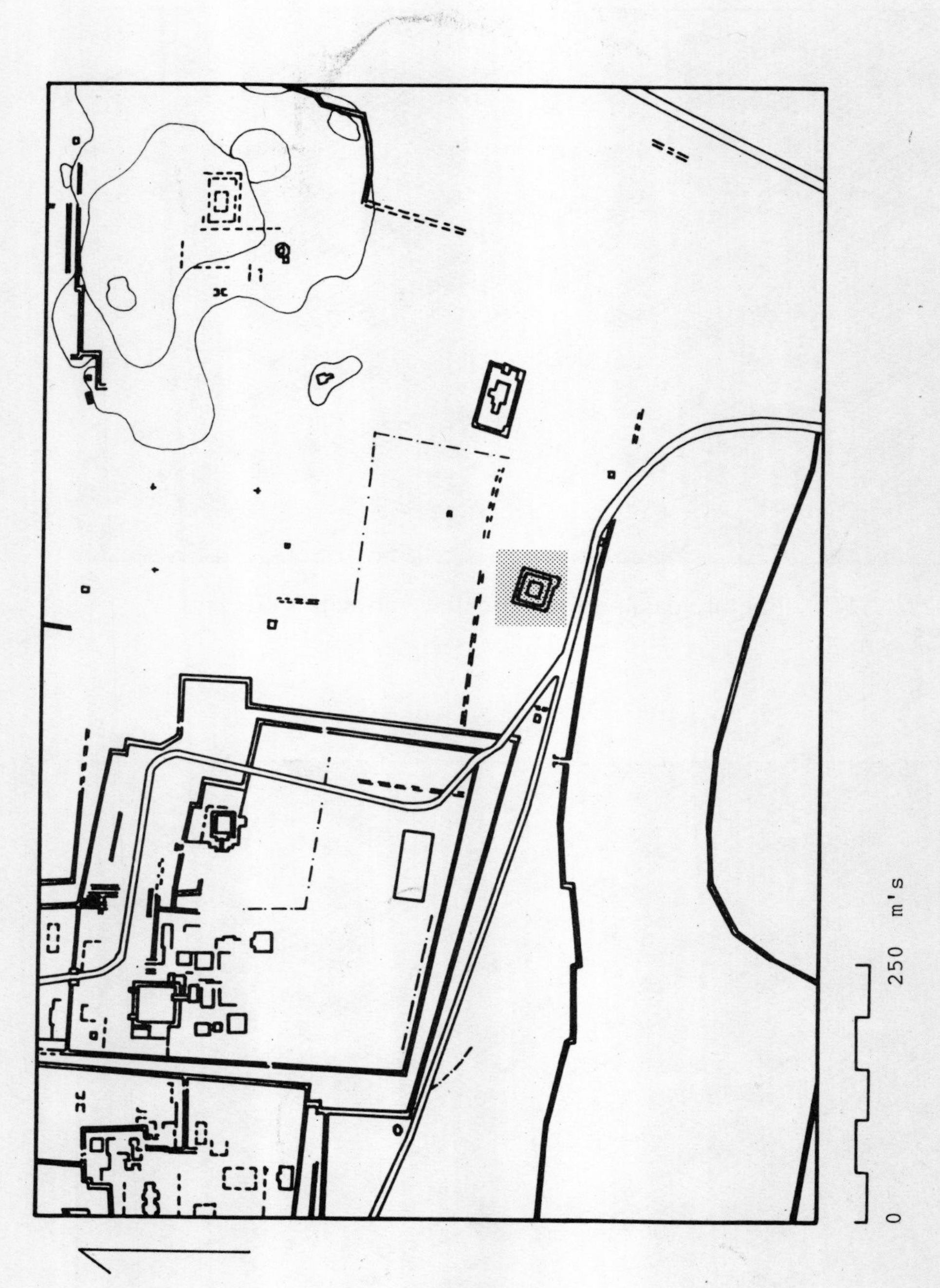

3.39 Map showing location of 'Queen's Bath' (XX/1) (after VRP).

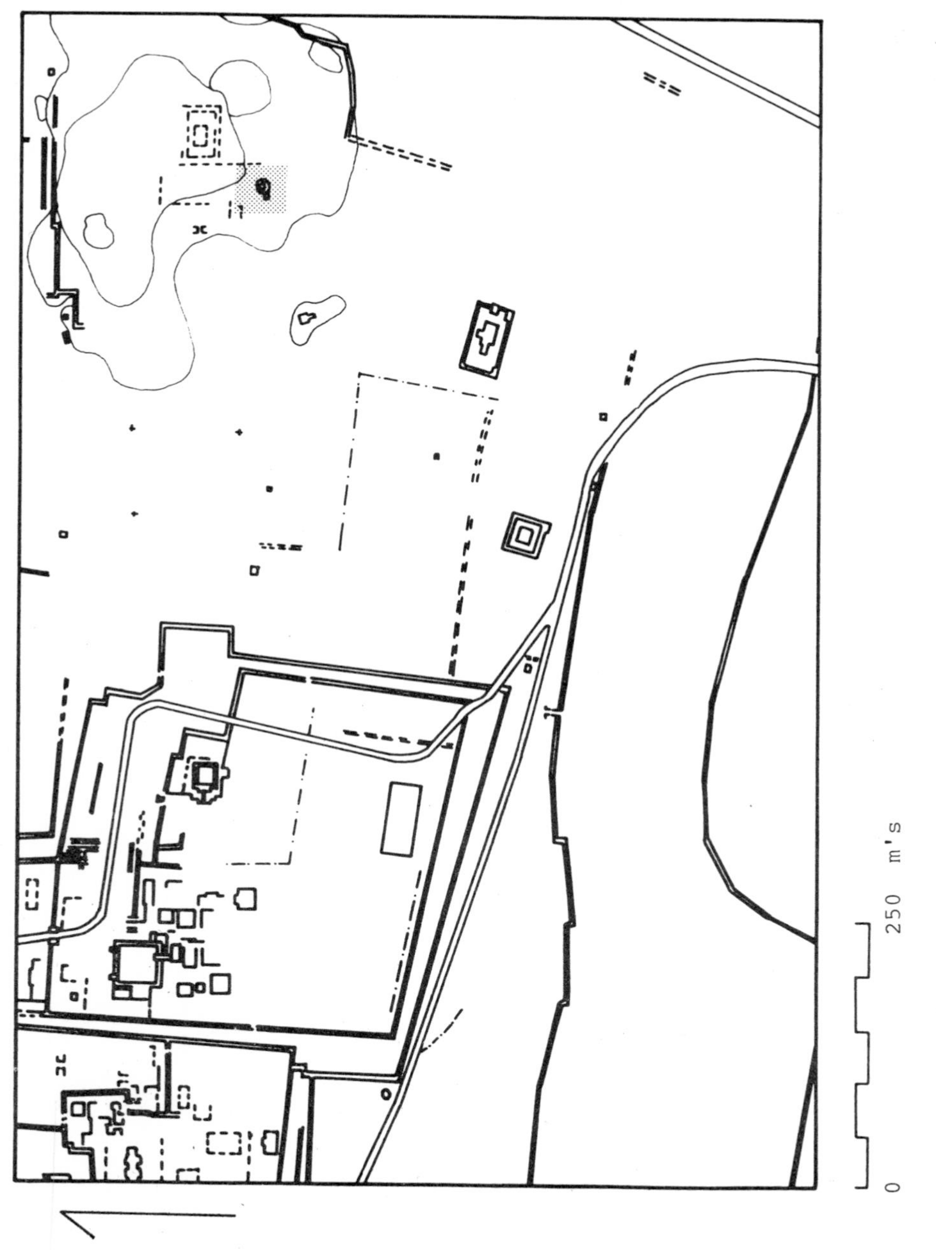

3.40 Map showing location of 'Octagonal Bath' (after VRP).

3.41 Supply spout to 'Queen's Bath' (XX/1).

3.42 'Queen's Bath' (XX/1); view to northwest.

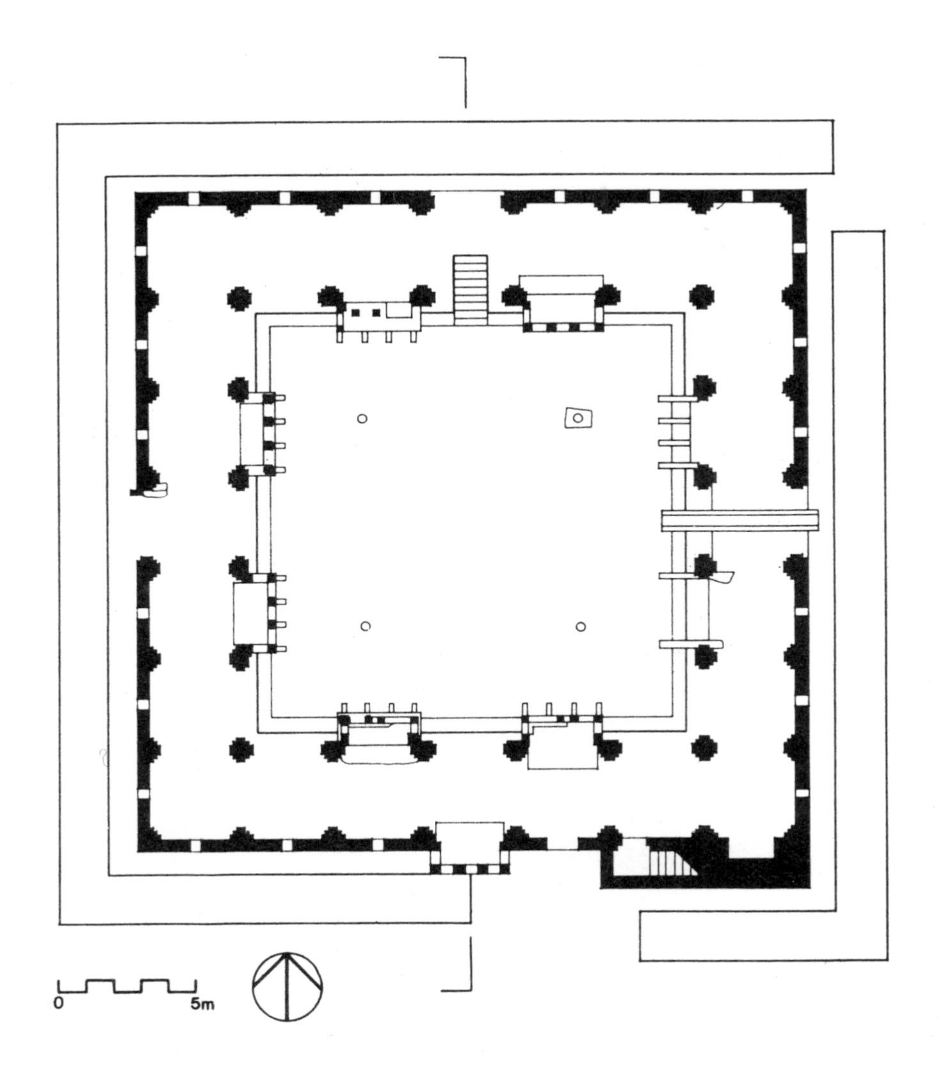

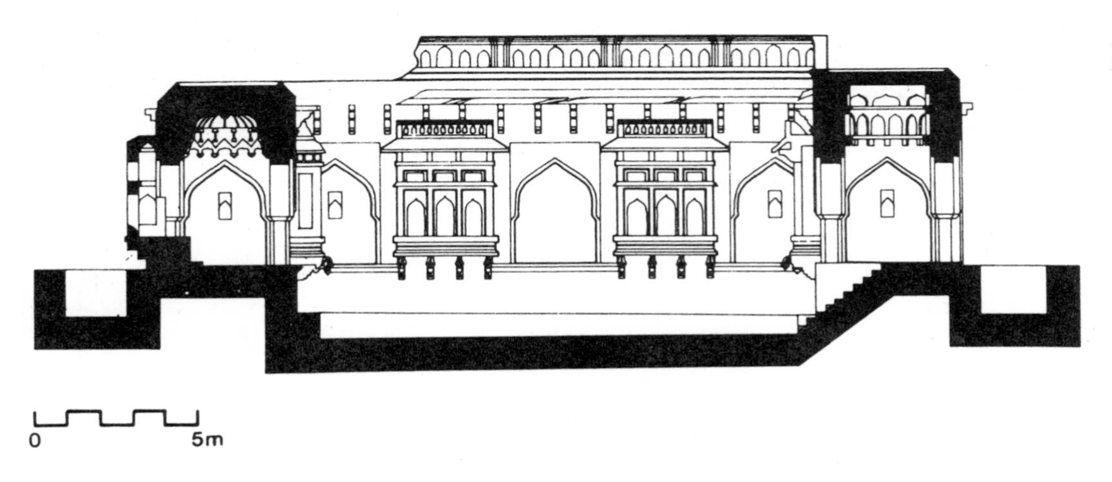

3.43 'Queen's Bath' (XX/1); plan and section (VRP).

3.44 Moat around the 'Queen's Bath' (XX/1).

3.45 'Queen's Bath' (XX/1) outlet.

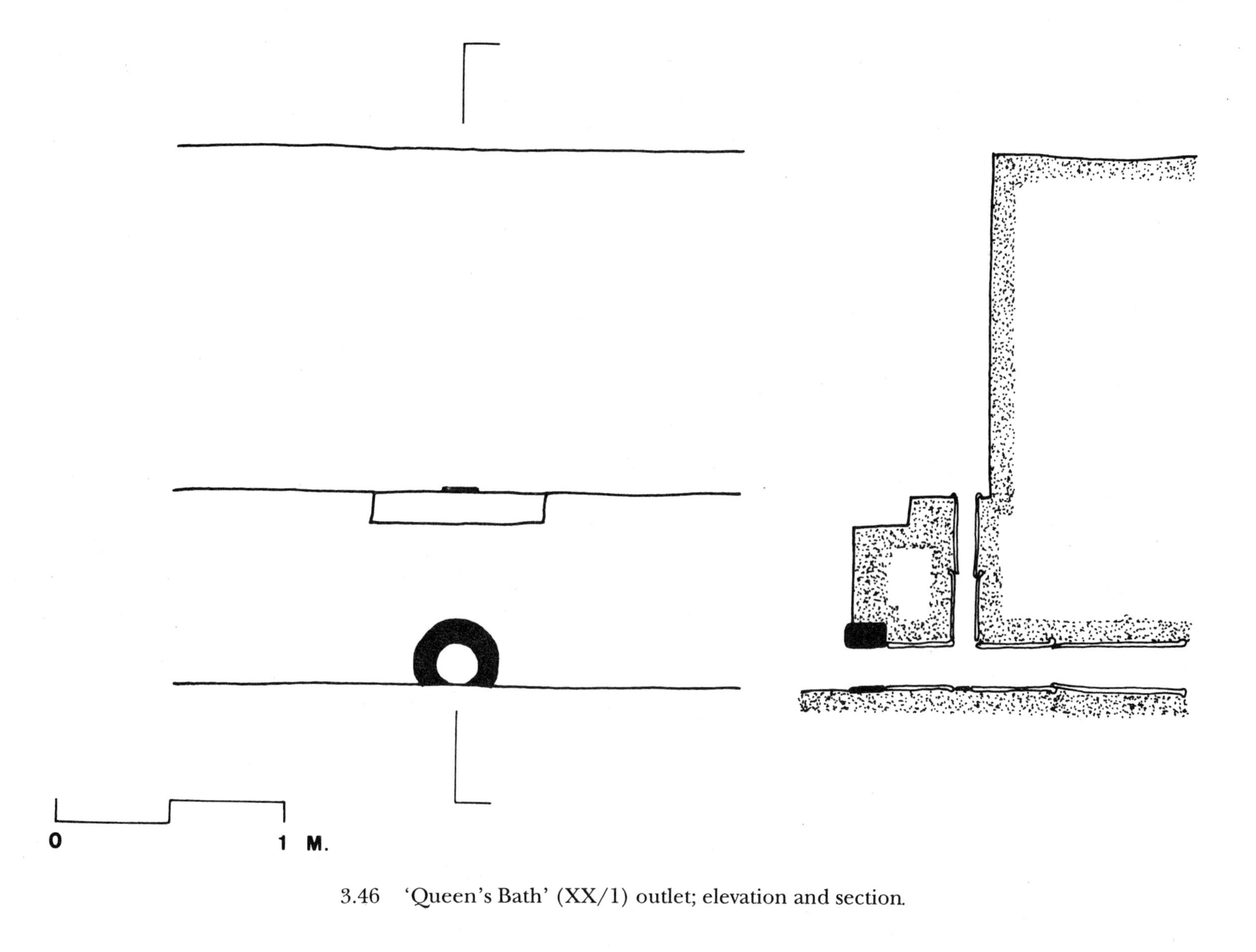

3.46 'Queen's Bath' (XX/1) outlet; elevation and section.

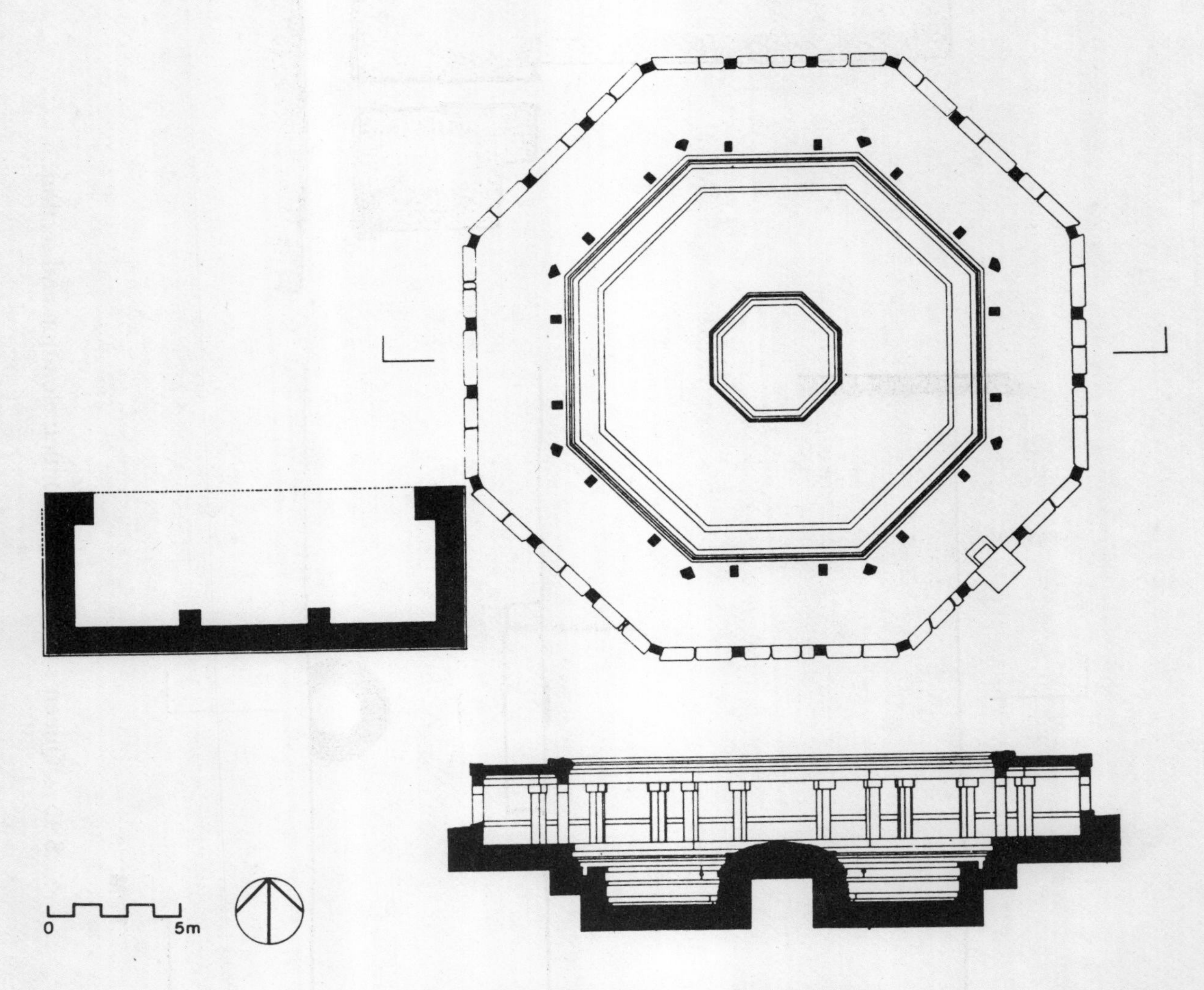

3.47 'Octagonal Bath' plan and section (VRP).

3.48 'Octagonal Bath'; central fountain.

3.49 'Water Tower'.

3.50 Plaster-lined bath (XIX/25); view to west.

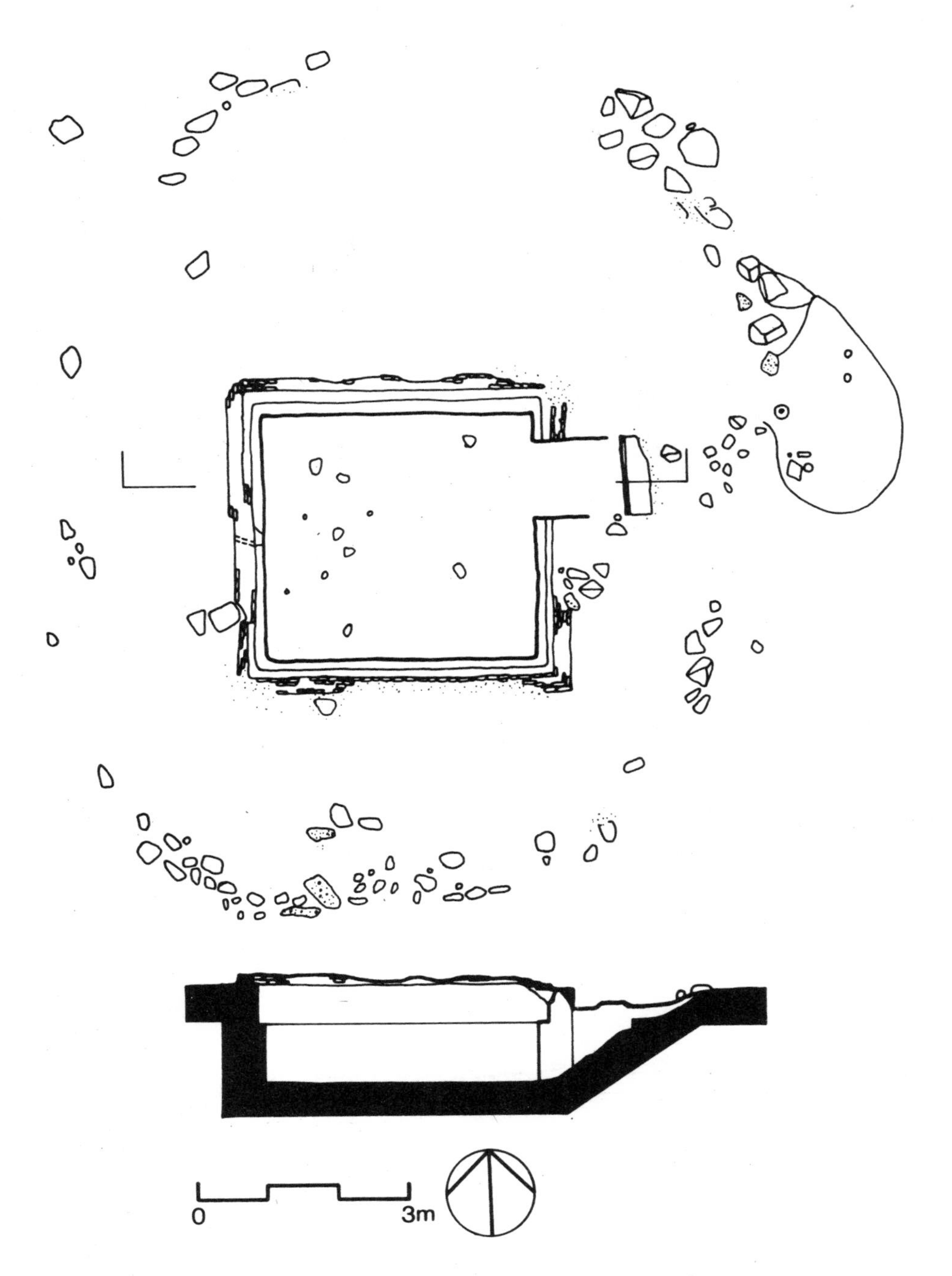

3.51 Plaster-lined bath (XIX/25); plan and section (VRP).

3.52 Raised channel in IV running over outcrop; view to west.

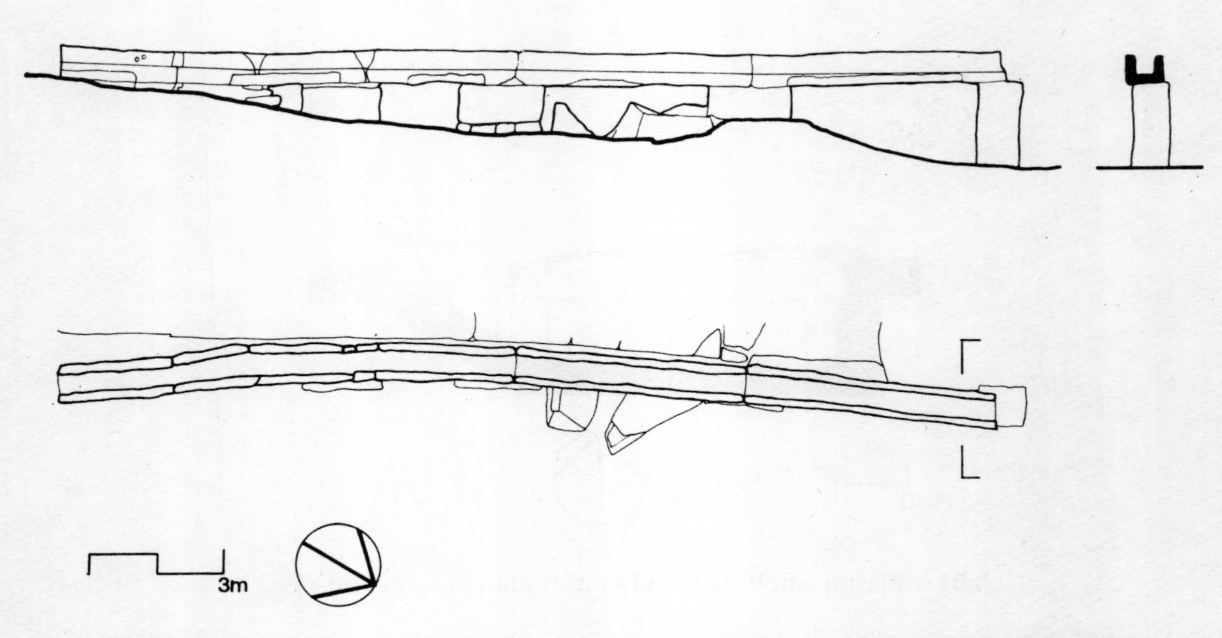

3.53 Raised channel in IV running over outcrop; plan, section and elevation (VRP).

3.54 Raised channel in IV; view to south.

3.55 Raised channel in IV running over outcrop; iron-stained holes.

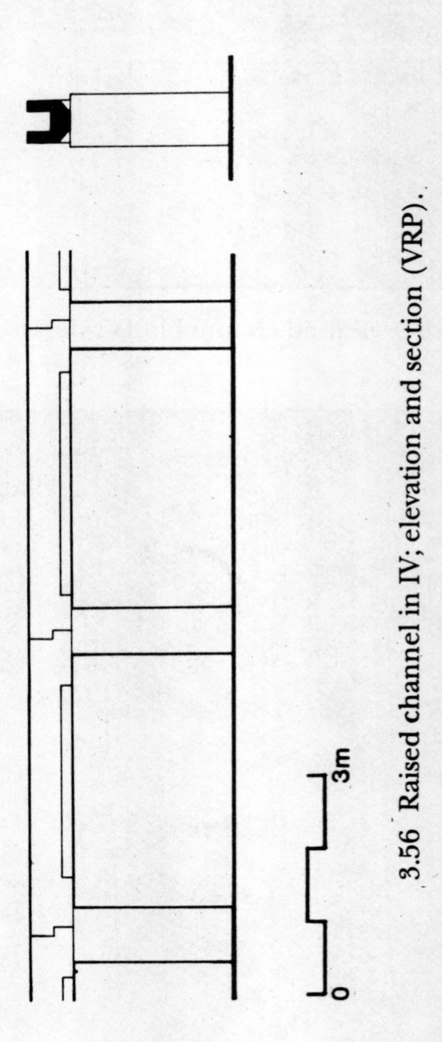

3.56 Raised channel in IV; elevation and section (VRP).

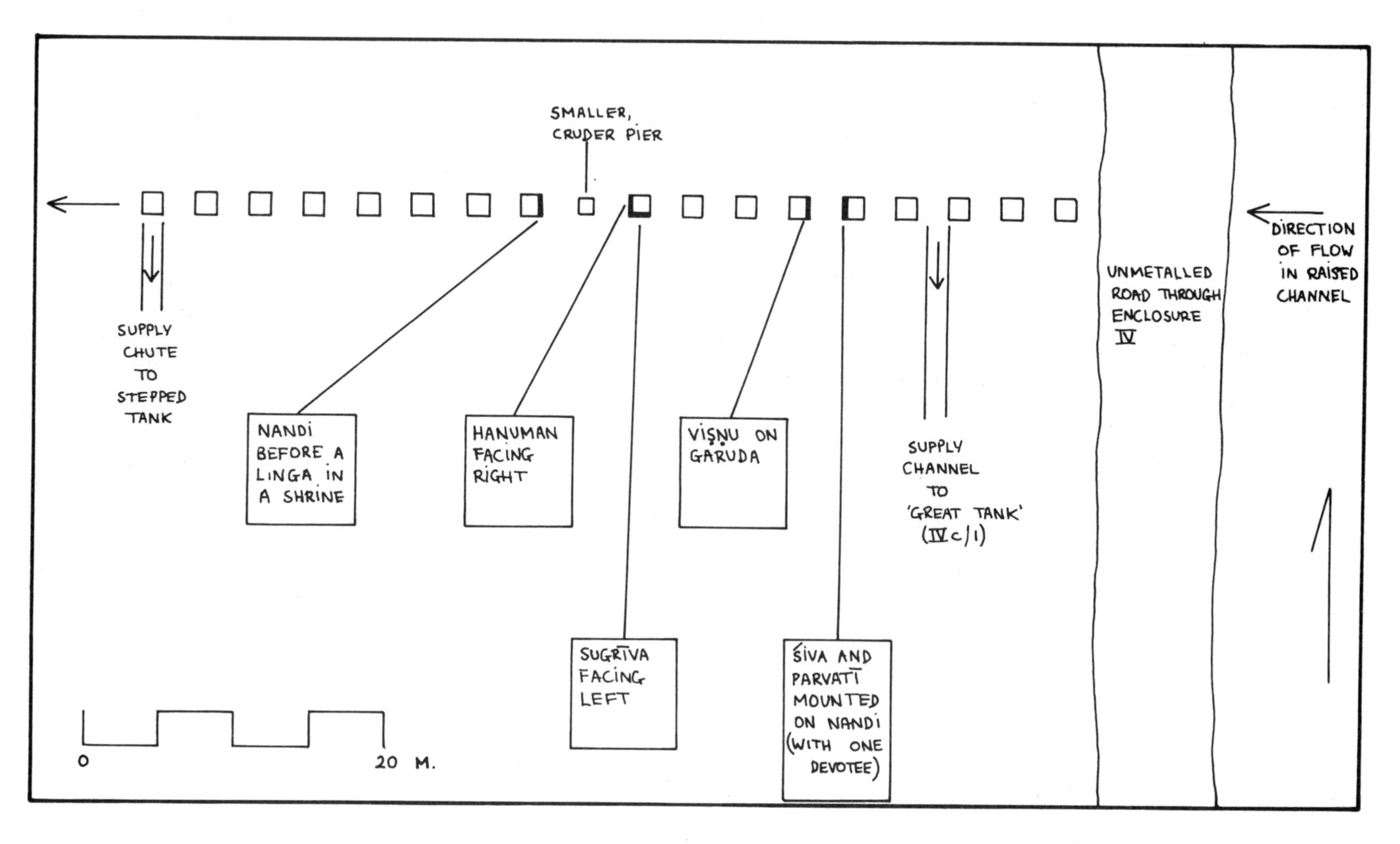

3.57 Diagrammatic sketch-map showing location of icons on the piers of the raised channel.

3.58 Śiva and Parvatī, mounted on Nandi.

3.59 Viṣṇu on Garuda.

3.60 Sugrīva.

3.61 Hanuman.

3.62 Nandi in a shrine.

3.63 Feeder channel to the 'Great Tank' (IVc/1); view to north.

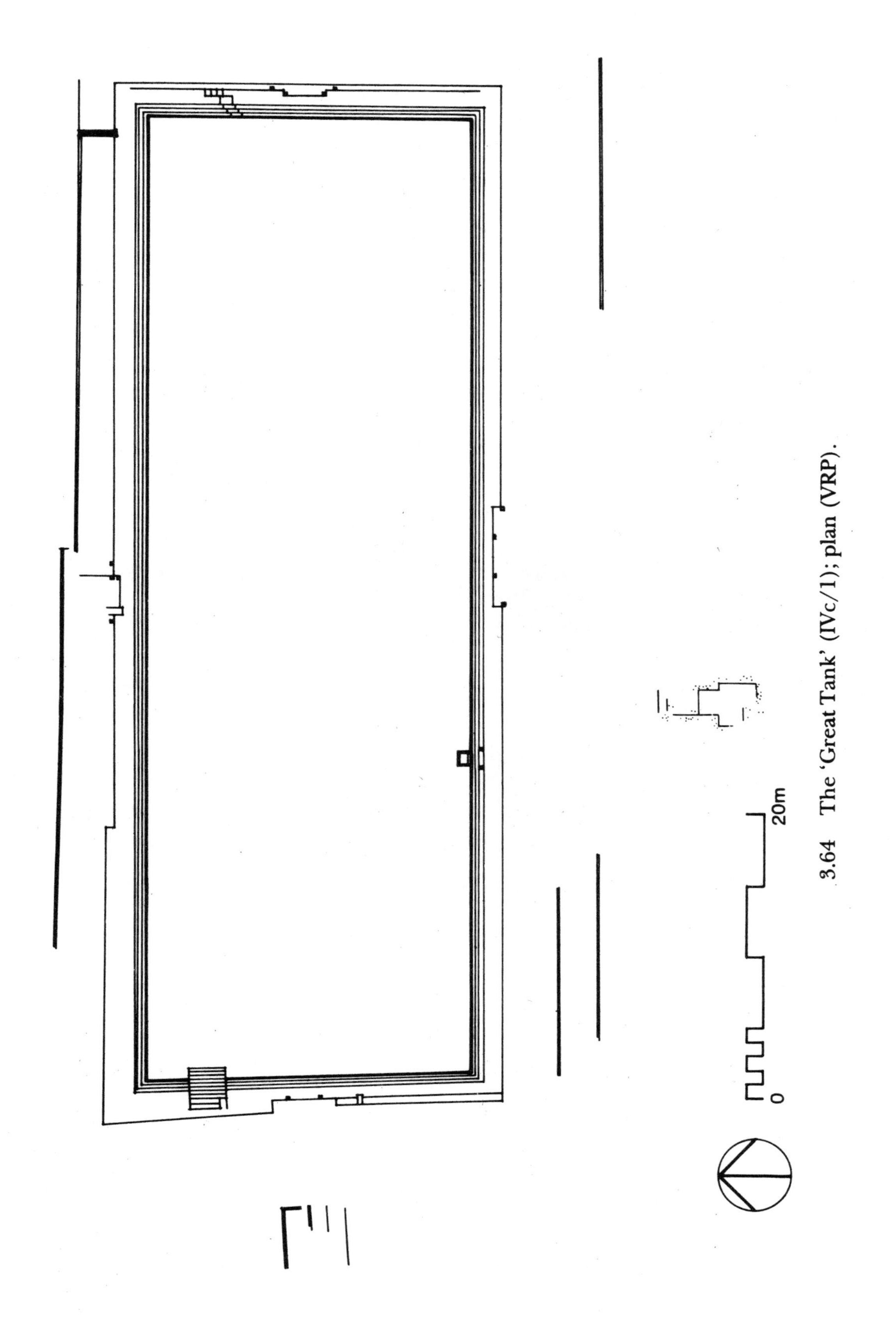

3.64 The 'Great Tank' (IVc/1); plan (VRP).

3.65 The 'Great Tank' (IVc/1); view to west.

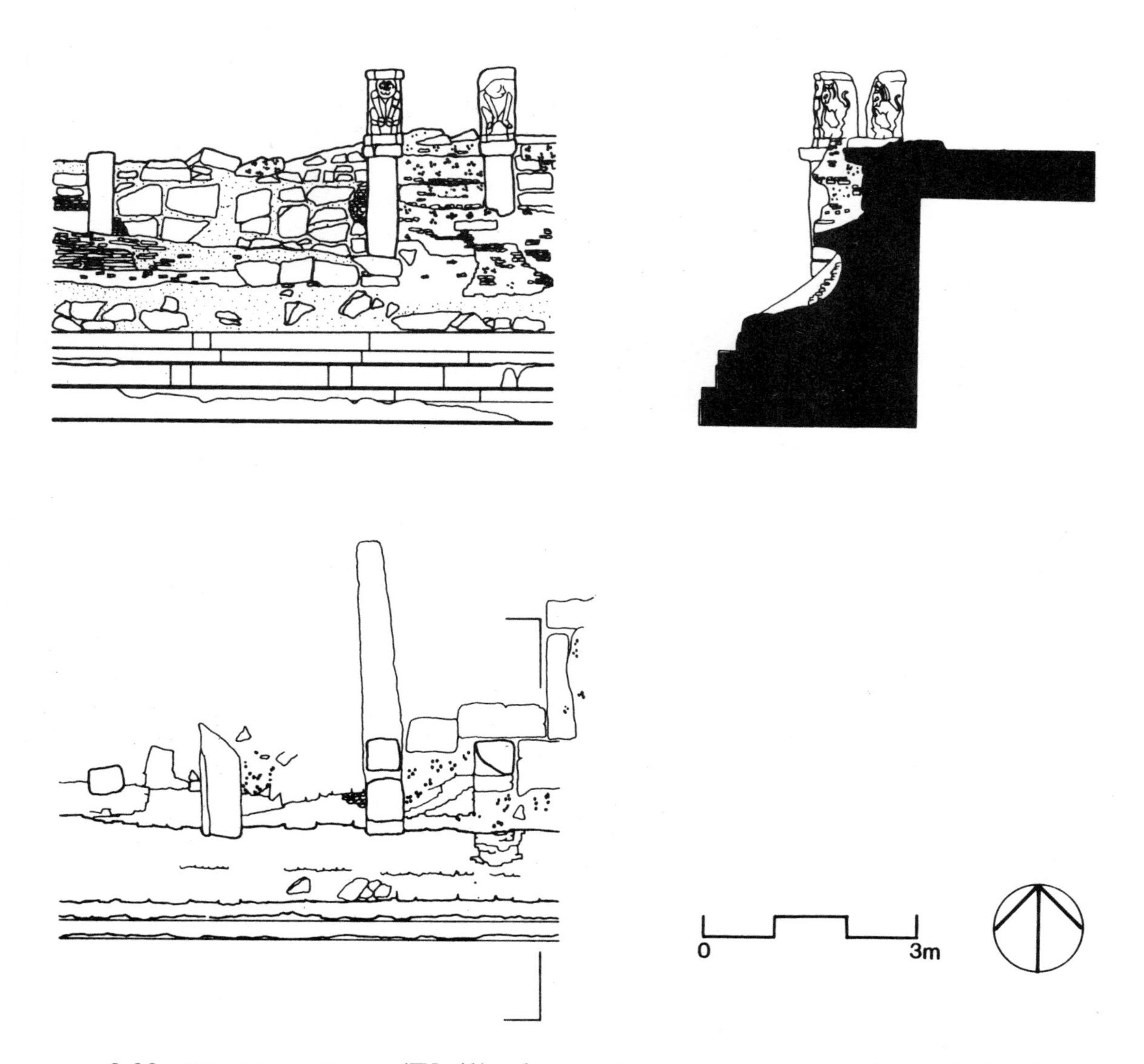

3.66 The 'Great Tank' (IVc/1); plan, section and elevation of side wall (VRP).

3.67 The 'Great Tank' (IVc/1); outlet.

3.68 Supply chute to the 'Stepped Tank'.

3.69 'Stepped Tank'; view to southwest.

3.70 'Stepped Tank'; view to northwest.

3.71 Kannada cypher.

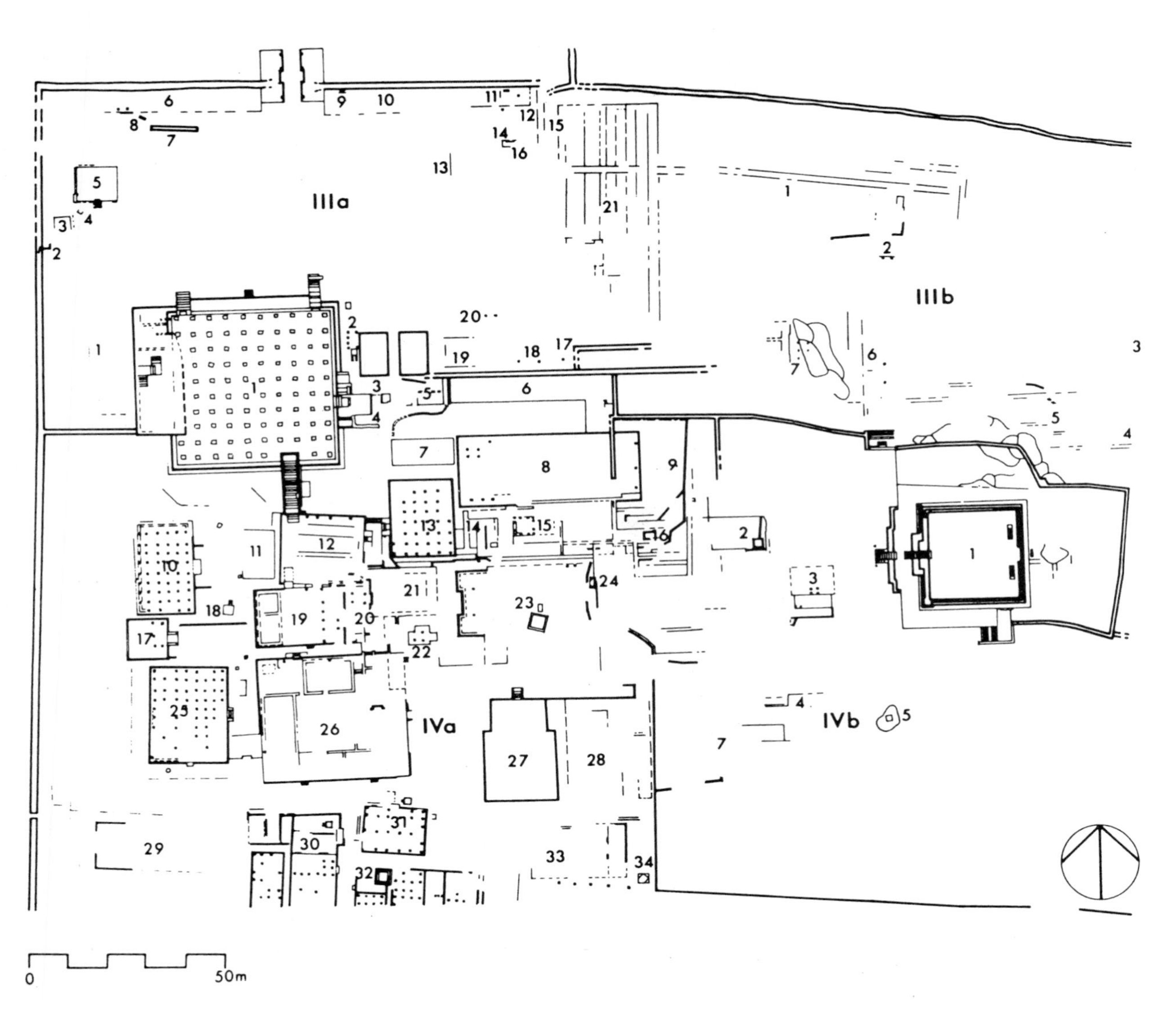

3.72 Plan of enclosure IVa/b (northern sector) (VRP).

3.73 Panorama of enclosure IV; view to west.

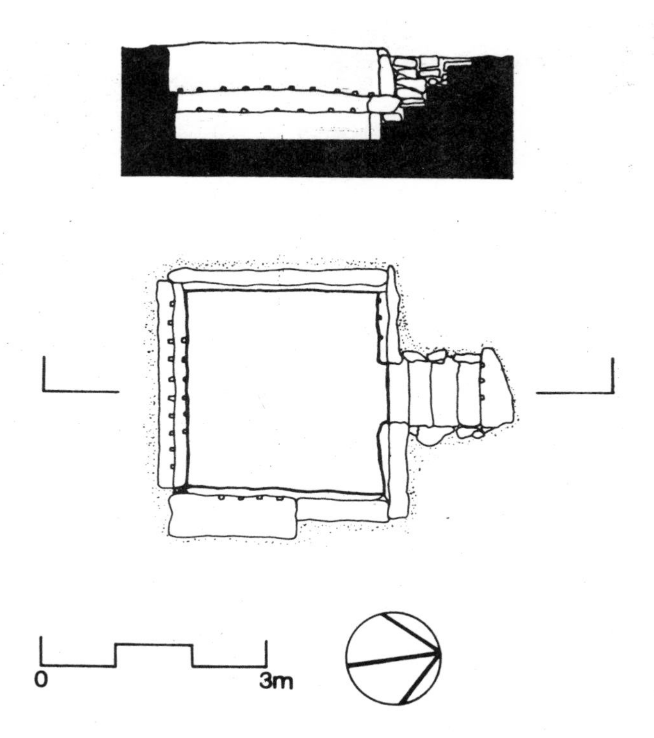

3.74 Plaster-lined, stone tank (IVa/18); plan and section (VRP).

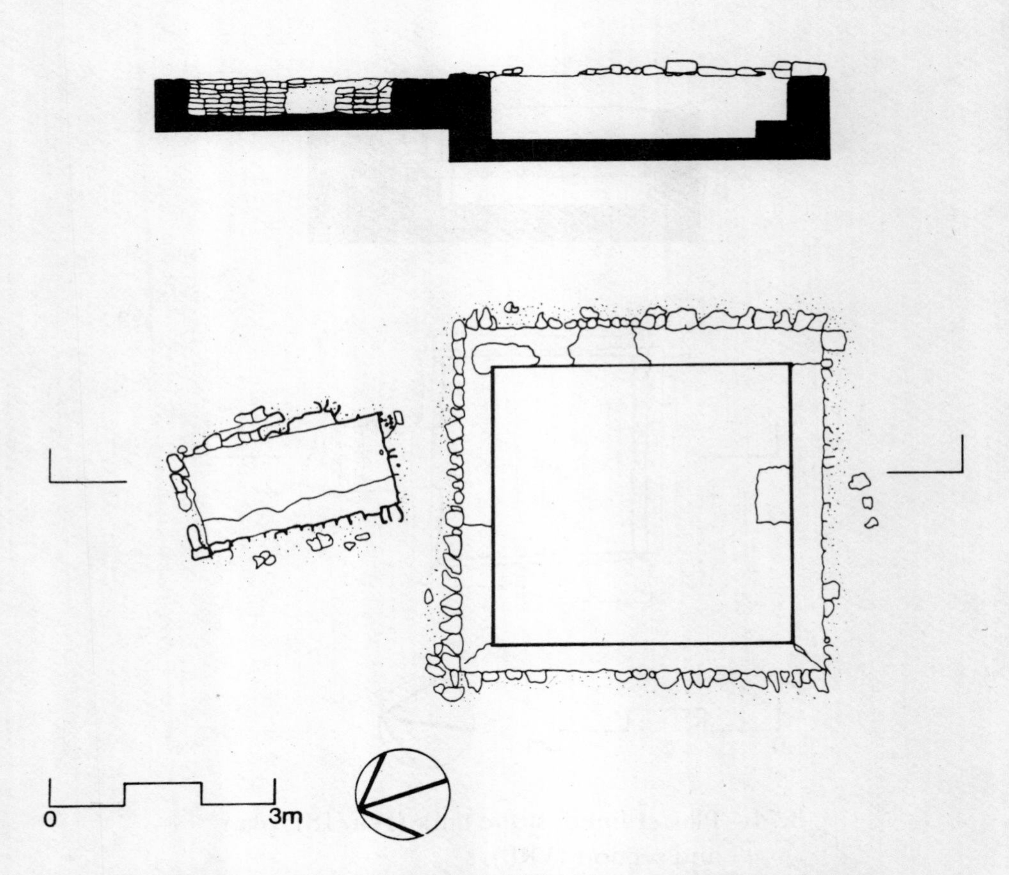

3.75 Plaster-lined tank (IVa/23); plan and section (VRP).

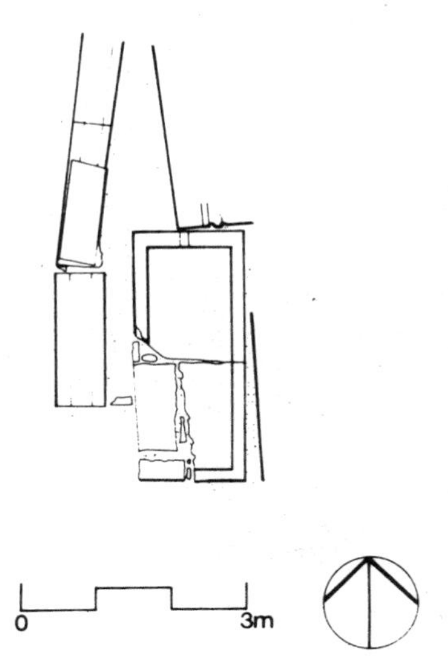

3.76 Stone tank (IVa/24); plan (VRP).

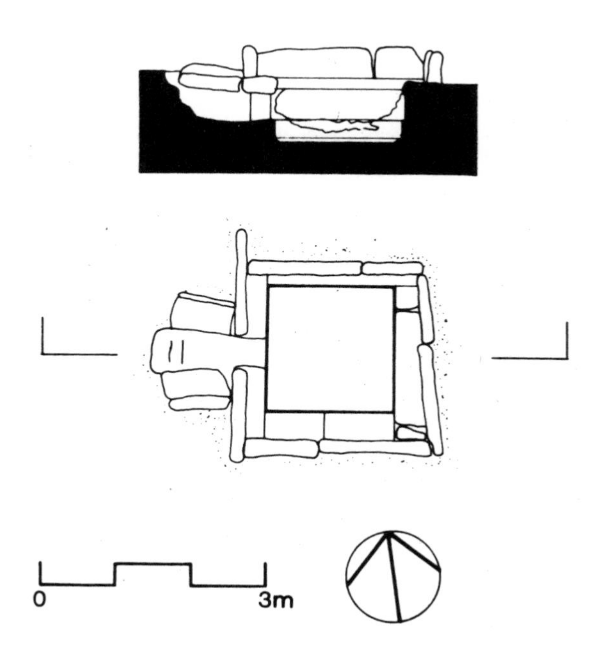

3.77 Plaster-lined tank (IVa/31b); plan and section (VRP).

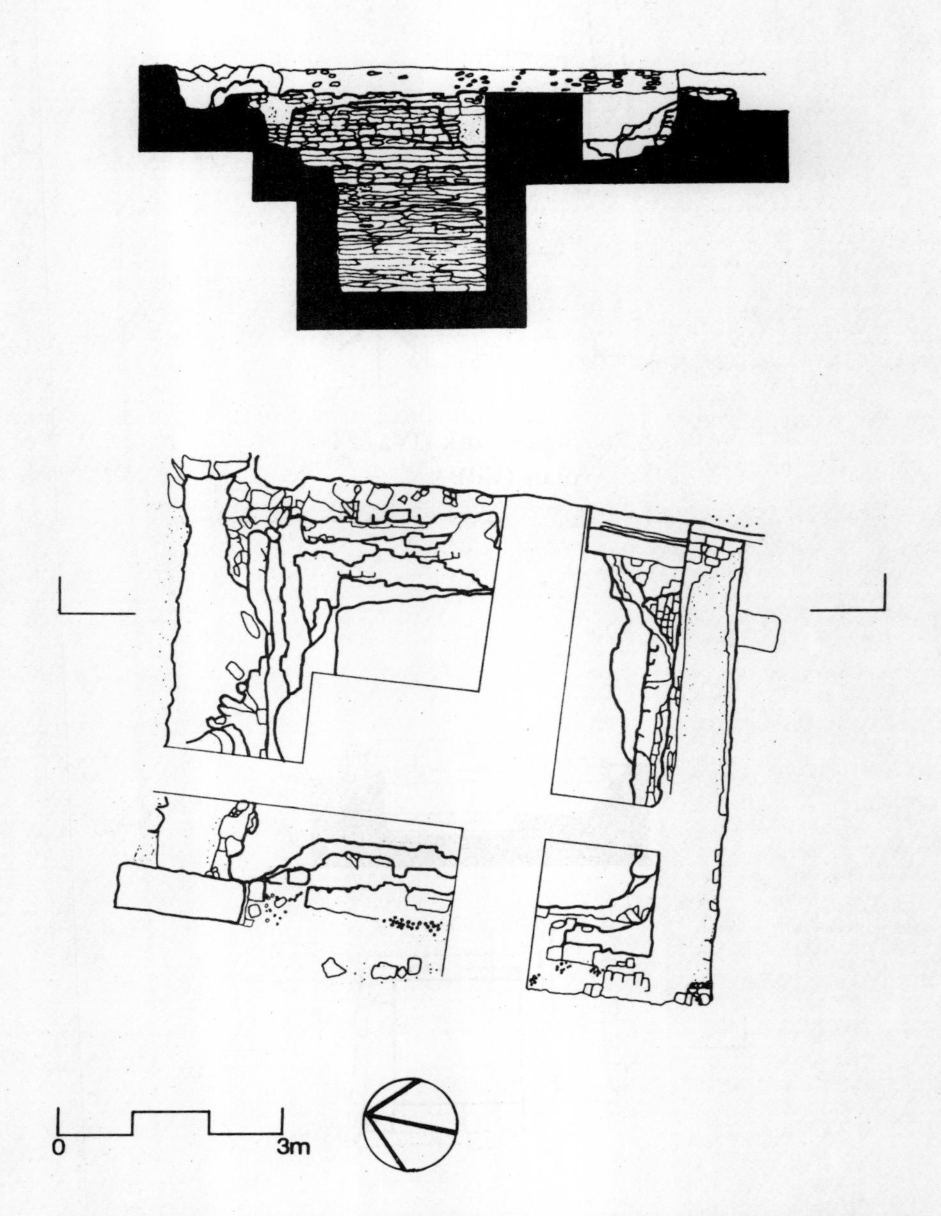

3.78 Plaster-lined brick tank (IVa/32); plan and section (VRP).

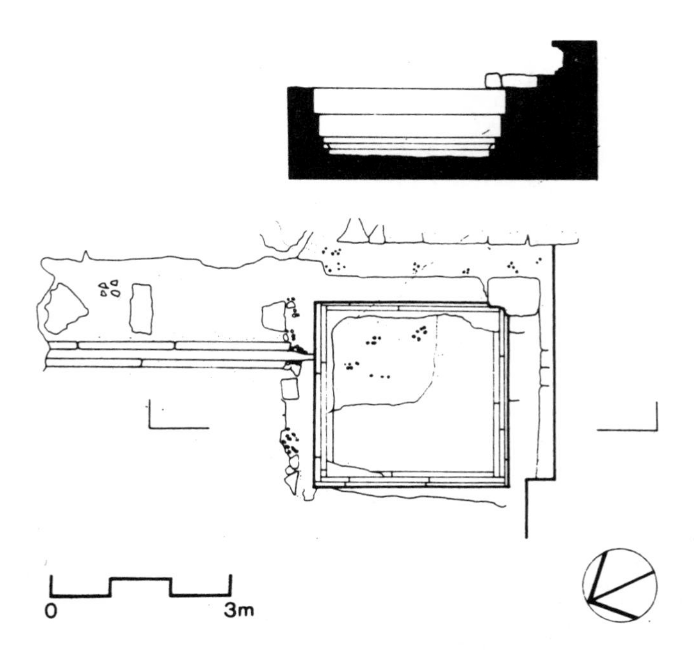

3.79 Stone tank (IVb/2); plan and section (VRP).

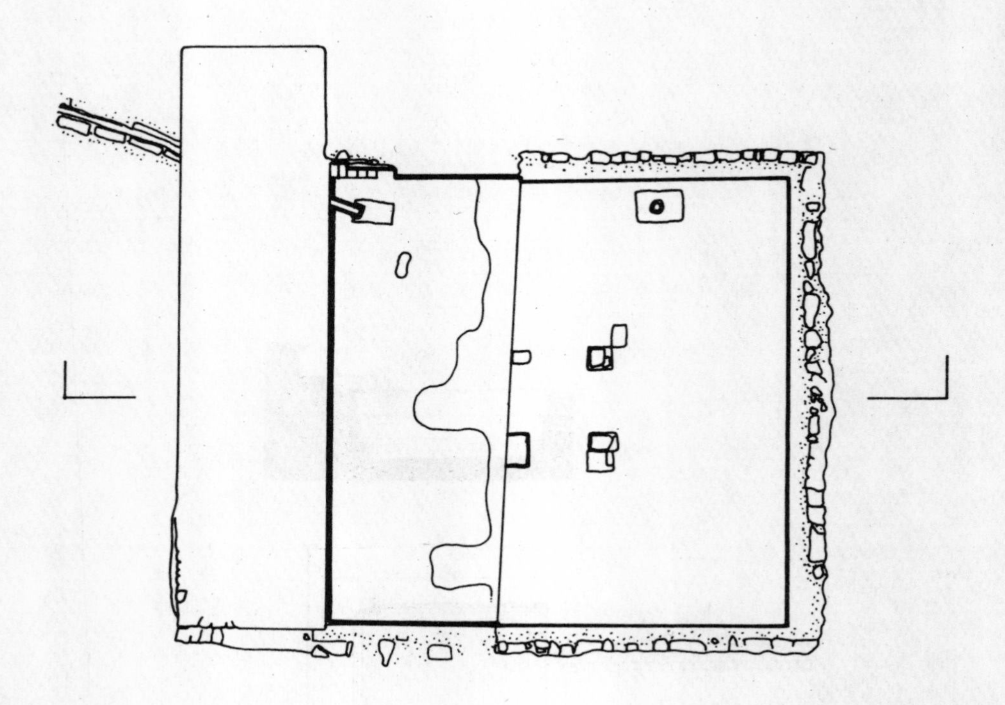

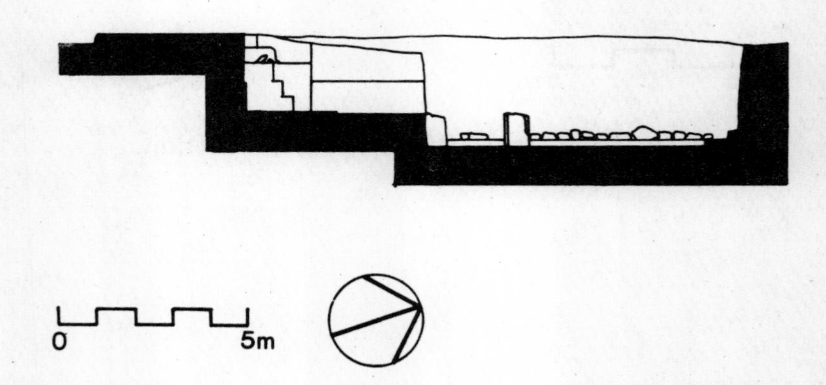

3.80 Plaster-lined tank with central pavilion (IVb/3) plan and section (VRP).

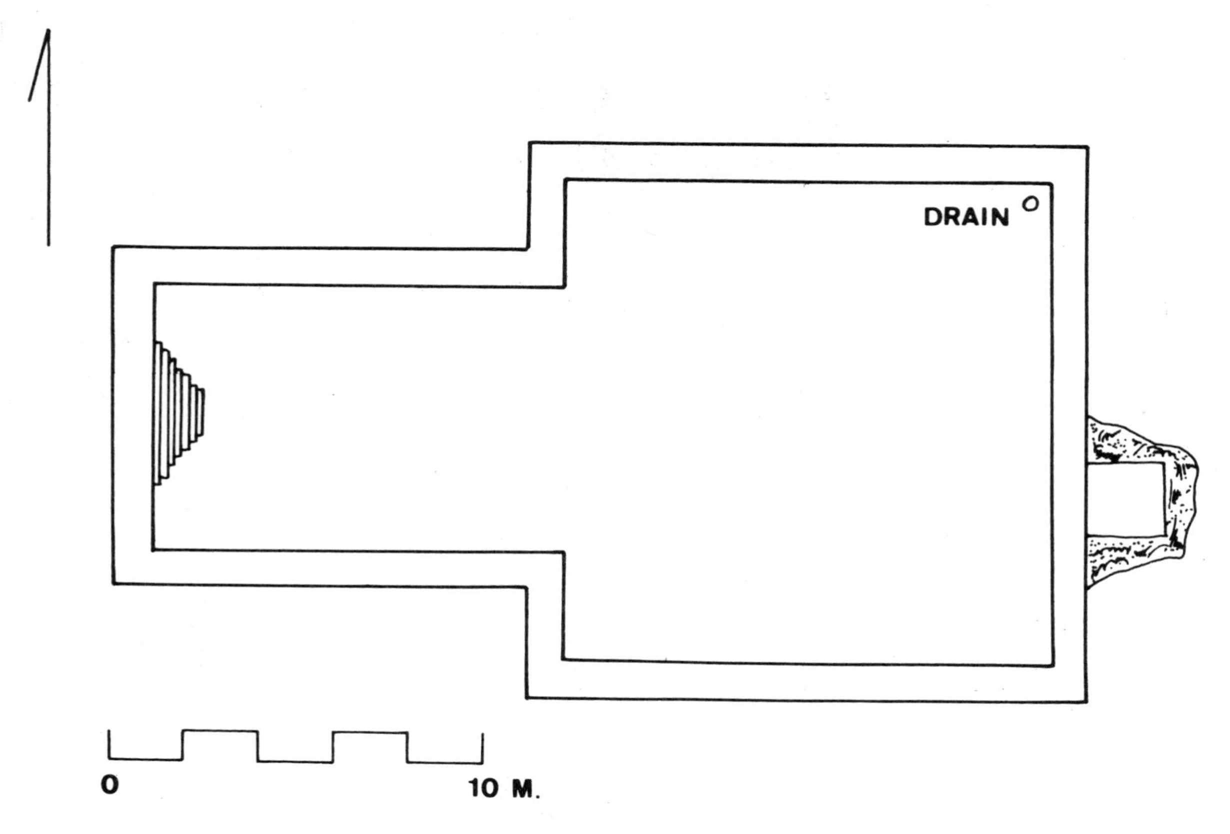

3.81 Stone-lined tank (IVb/4); plan.

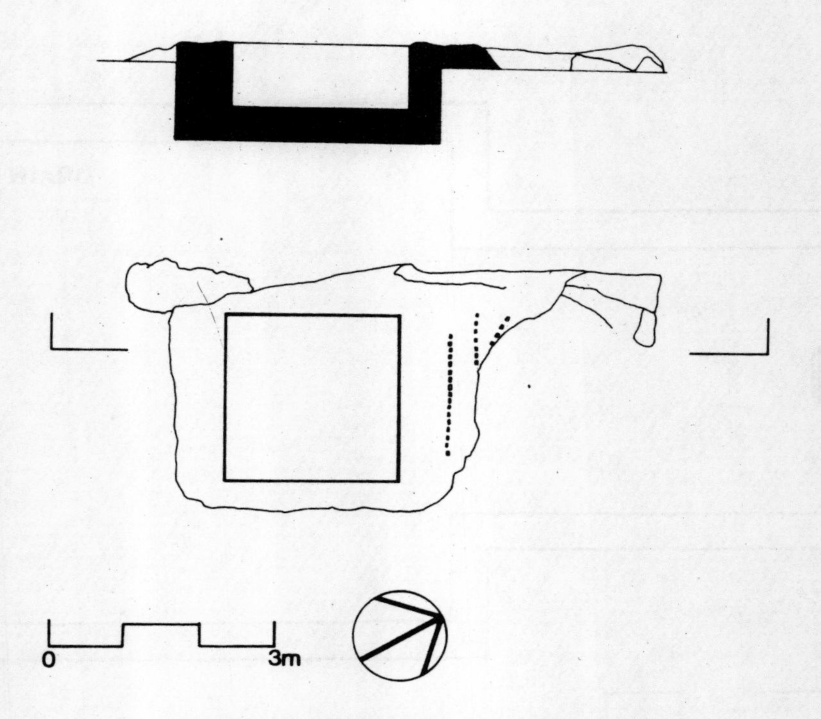

3.82 Rock-cut basin (IVb/5); plan and section (VRP).

3.83 Conduit of type 1.

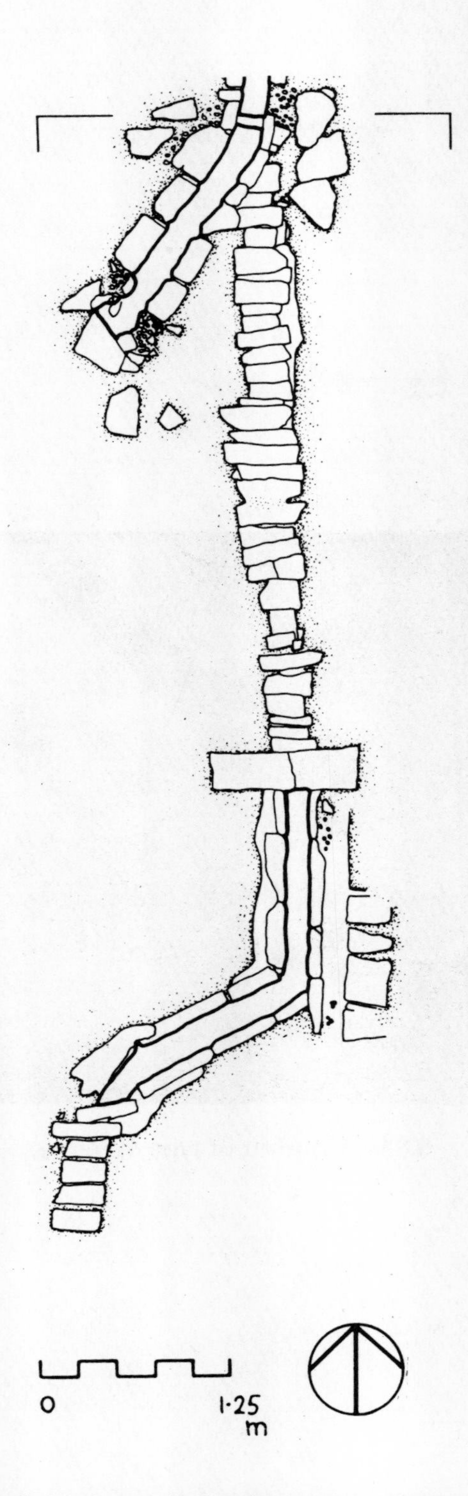

3.84 Conduit of type 1; IVa/9 and 16, plan (VRP).

3.85 Conduit of type 2.

3.86 Conduit of type 3.

3.87 Plaster-lined tank with central pavilion (IVb/3); view to west.

3.88 Plaster-lined tank with central pavilion (IVb/3); outlet.

3.89 Stone-lined tank (IVb/4); view to east.

3.90 Stone-lined tank (IVb/4); outlet.

3.91 Covered channel IVd/17; view to west.

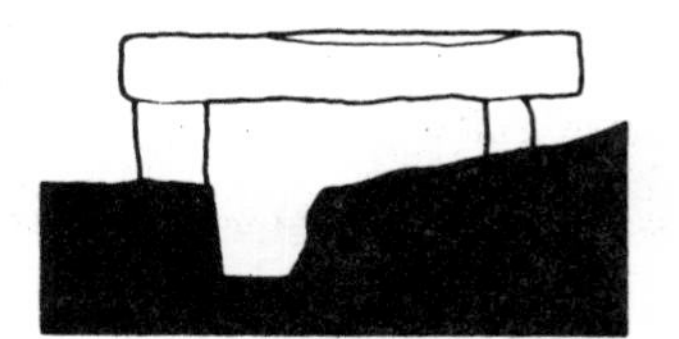

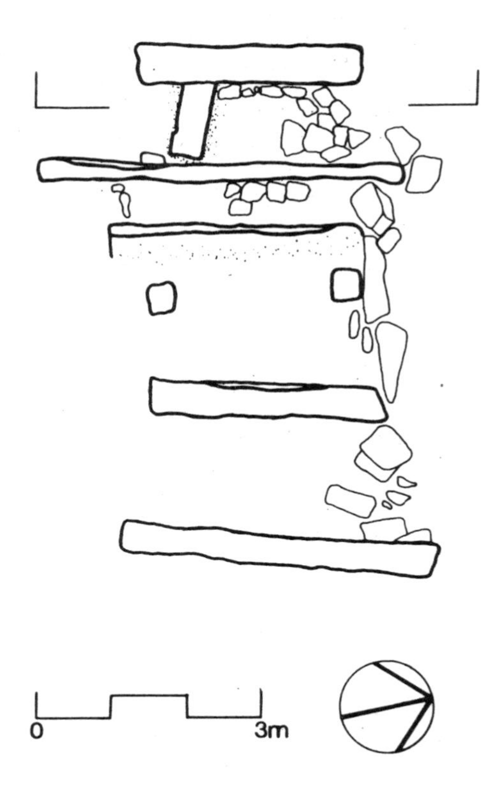

3.92 Covered channel IVd/17; plan and section (YRP).

3.93 Covered channel IVd/17; plan and section (VRP).

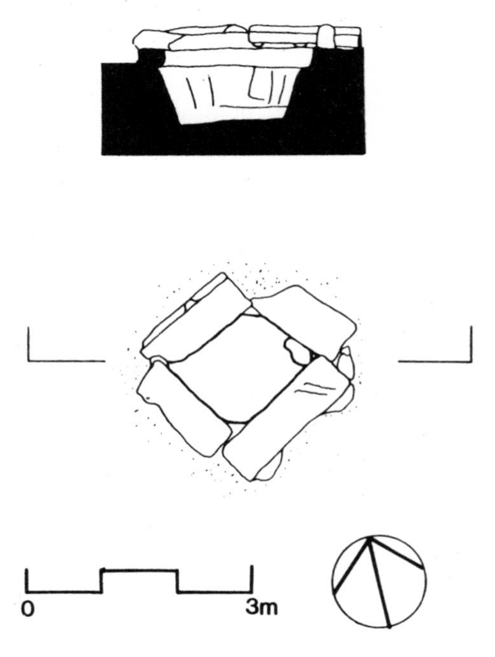

3.94 Well (IVA/10b); plan and section (VRP).

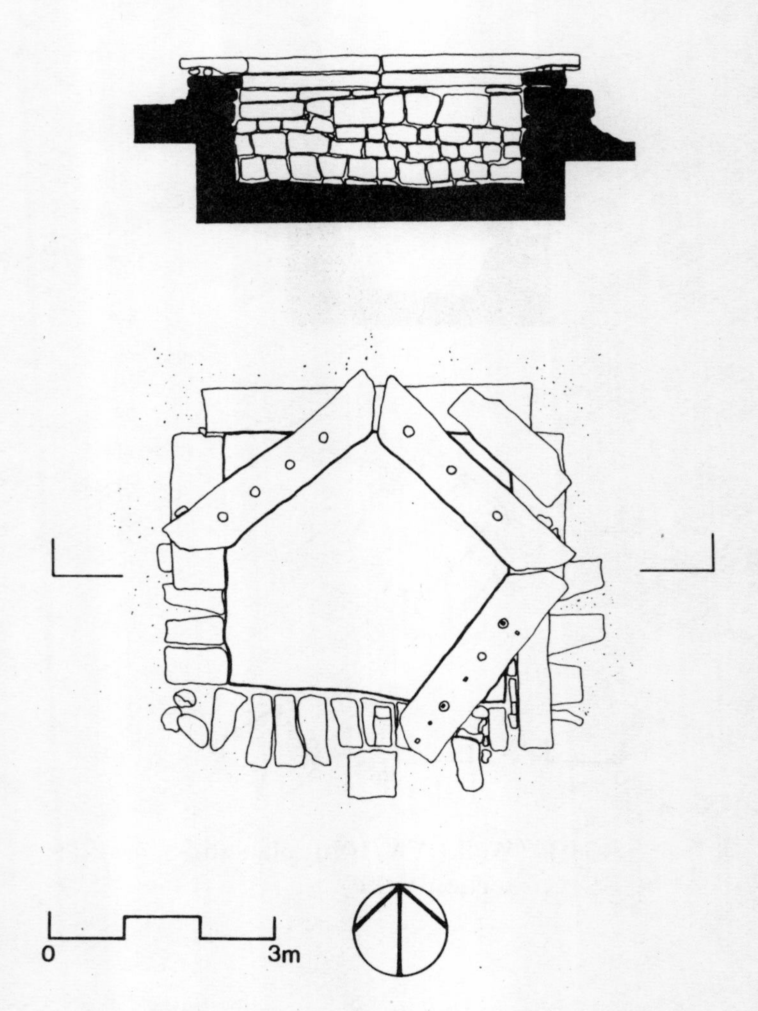

3.95 Well (IVa/34): plan and section (VRP).

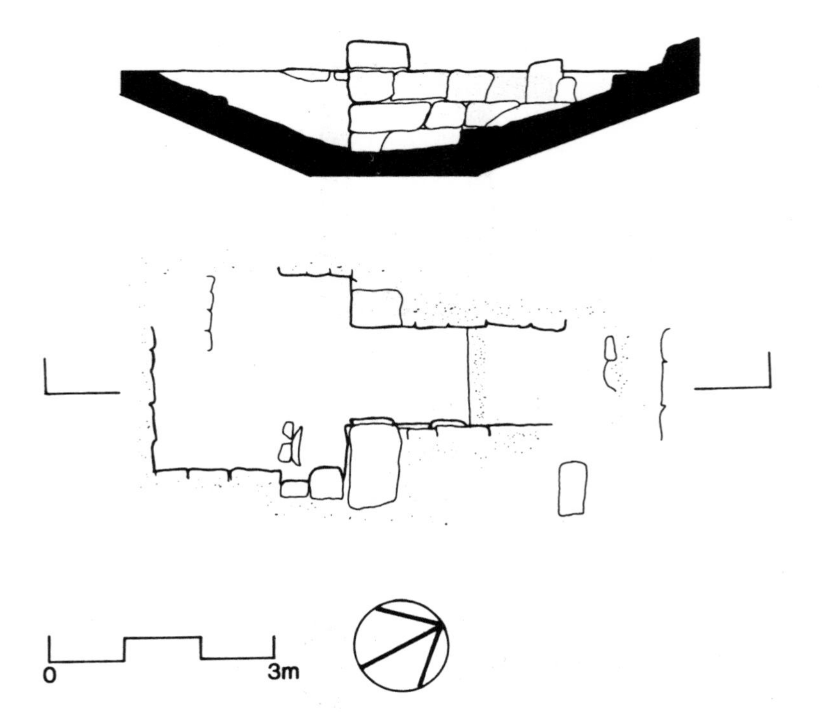

3.96 Well (IVc/3); plan and section (VRP).

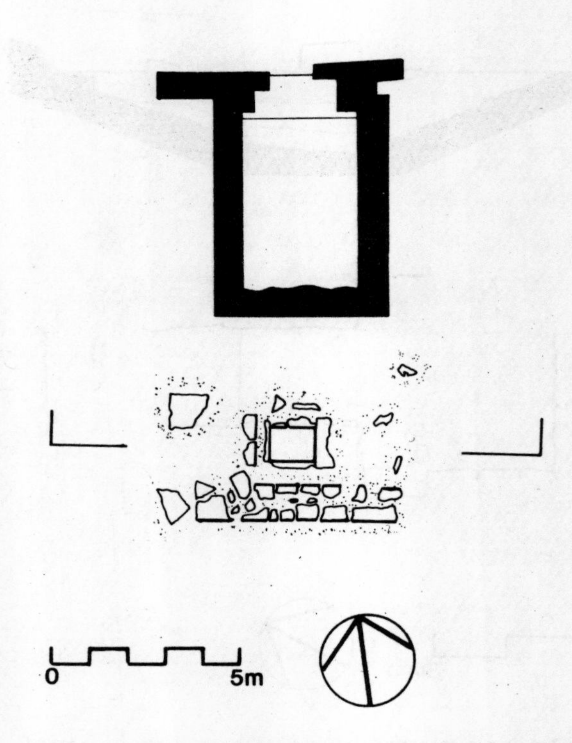

3.97 Well (WA/1); plan and section (VRP).

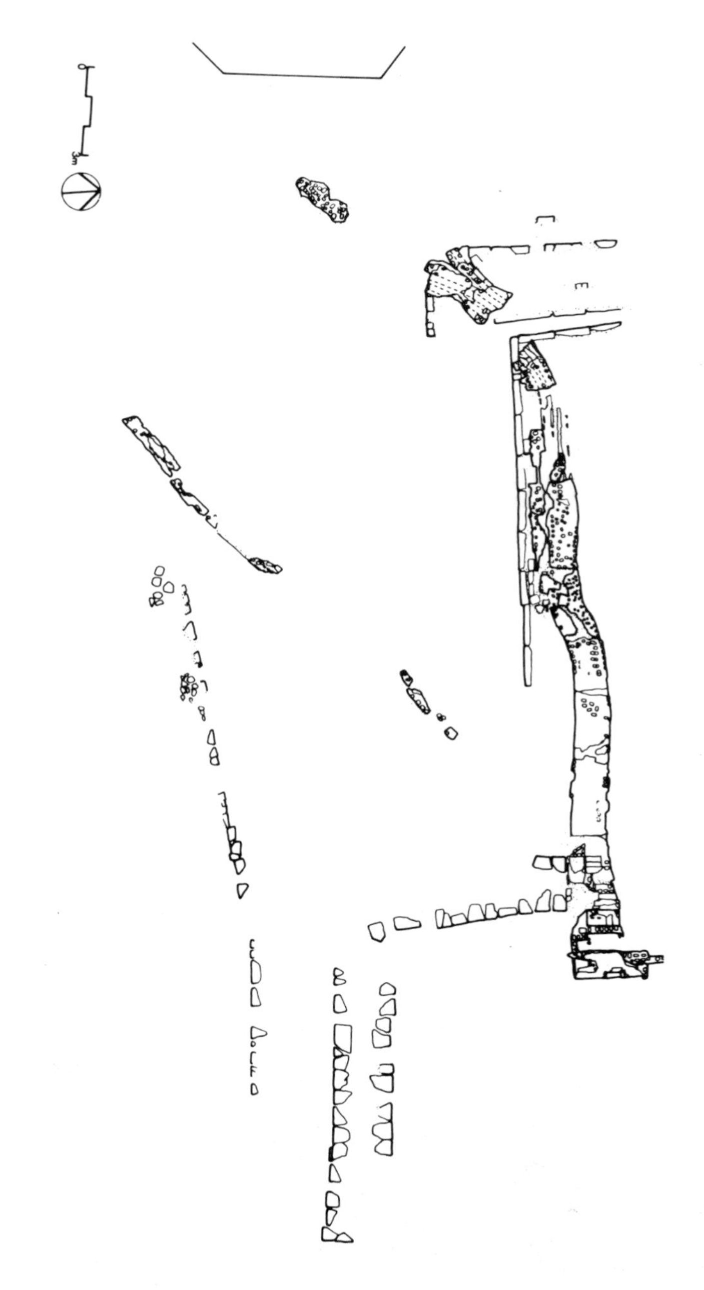

3.98 Complex of water features in the northeast corner of XXII (XXII/2 and 3); plan (VRP).

3.99 East face of XXII/3.

3.100 South face of XXII/3.

3.101 View eastwards along XXII/2.

3.102 Fragment of the dismantled tank.

3.103 Fragments of the dismantled tank built into a dry-stone wall.

3.104 Traces of plaster covering visible on XXII/2.

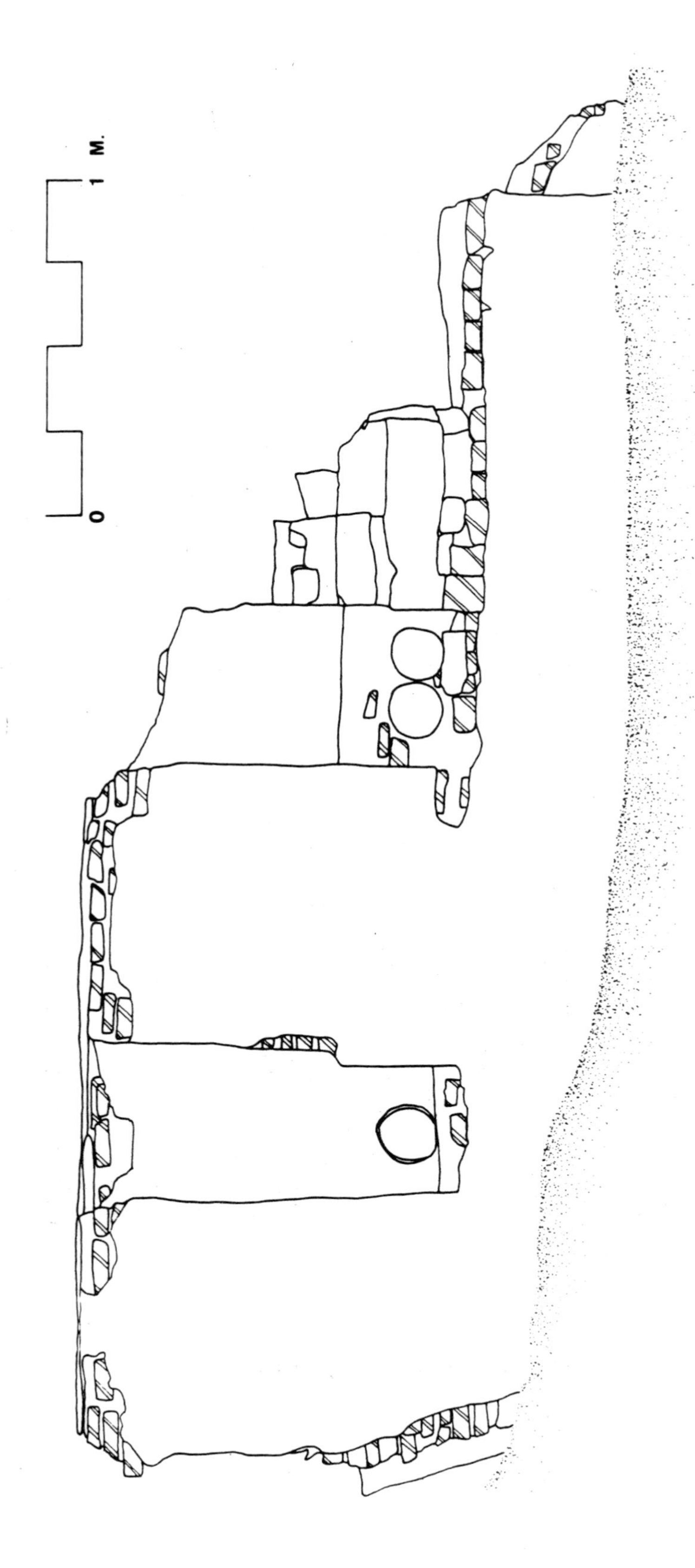

3.105 XXII/3; east elevation (VRP).

3.106 XXII/3; plan (VRP).

3.107 XXII/3; close-up of the single pipe outlet, recessed in notch.

3.108 XXII/3; close-up of the double outlet.

3.109 XXII/3; close-up of a hole for the fitting of a pivotal rod.

3.110 XXII/3; raised tank before excavation.

3.111 XXII/3; raised tank after excavation (VRP).

3.112 XXII/3; raised tank, inlet with notch (VRP).

3.113 XXII/3; three pipe outlets in the raised tank (VRP).

3.114 XXII/3; recess behind the pair of pipe outlets in the raised tank (VRP).

3.115 XXII/3; impression of a fourth outlet in chunam in the raised tank.

3.116 XXII/3; plastered insert in the raised tank.

3.117 Panorama of the west end of XXII/2.

3.118 'Octagonal Fountain' (XXII/1).

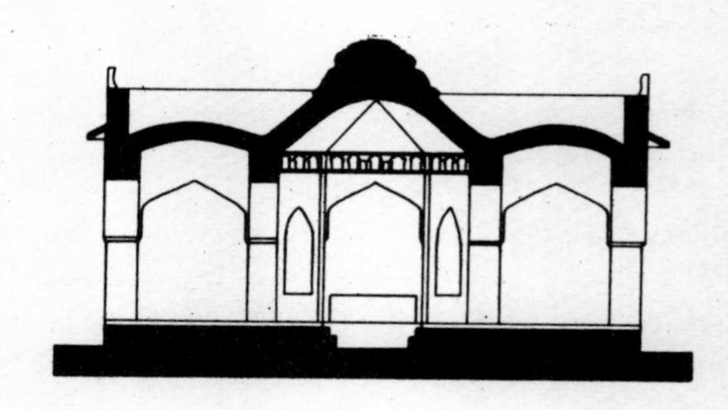

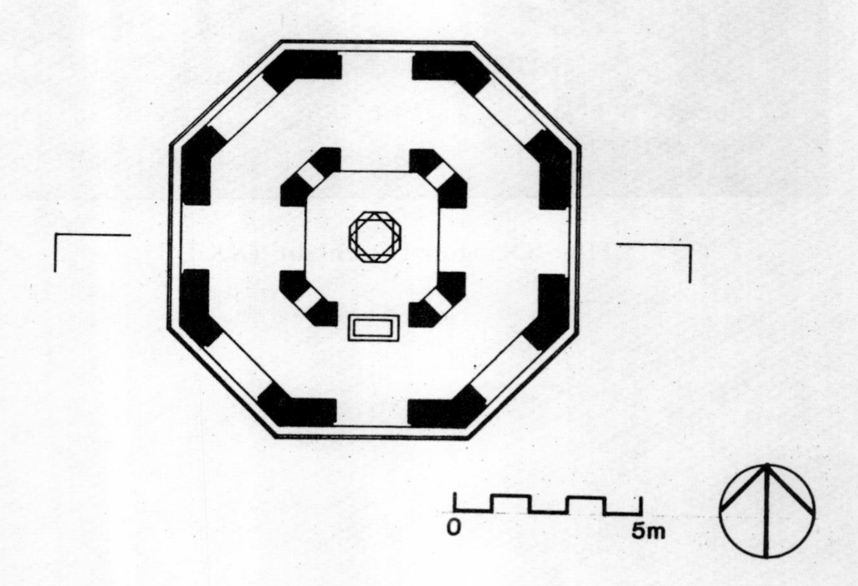

3.119 'Octagonal Fountain' (XXII/1); plan, section and elevation (VRP).

3.120 'Bhojanśāla'; view to northwest.

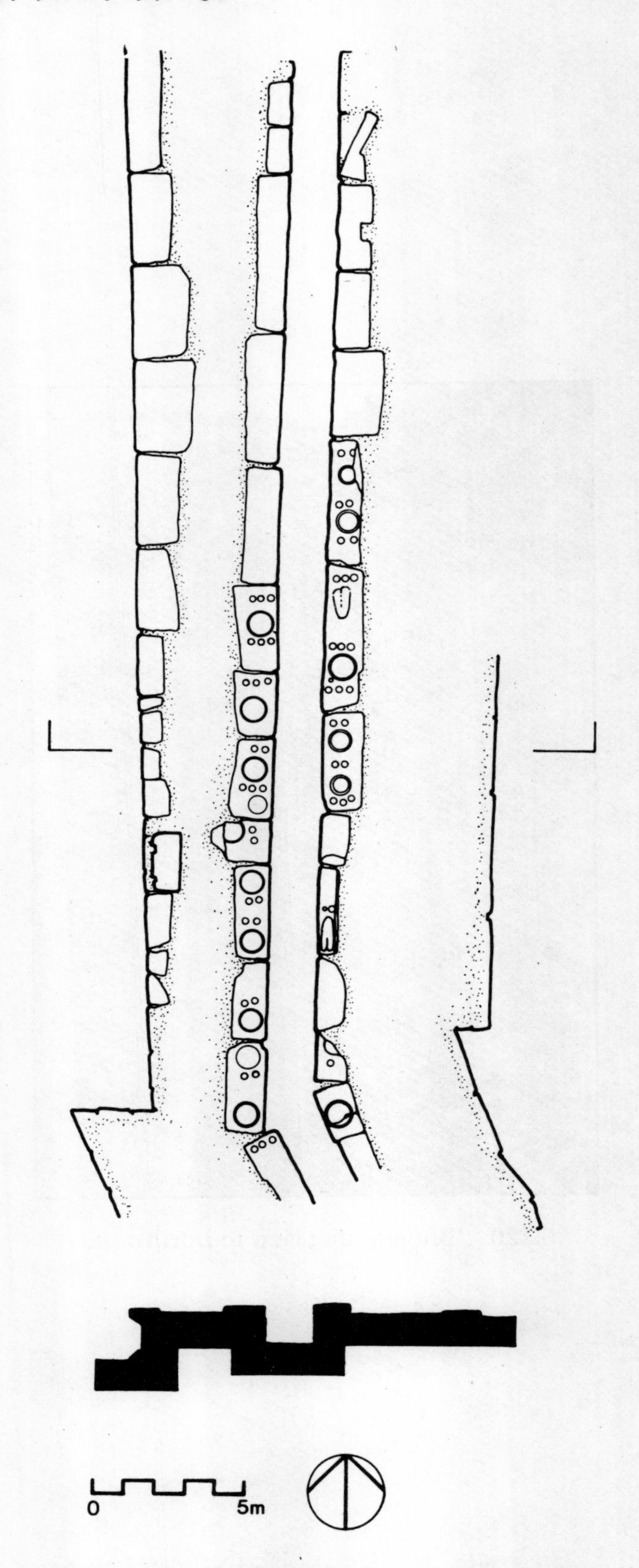

3.121 'Bhojanśāla'; plan and section (VRP).

3.122 'Bhojanśāla'; carved slabs flanking the channel.

3.123 'Bhojanśāla', close-up of carved slab.

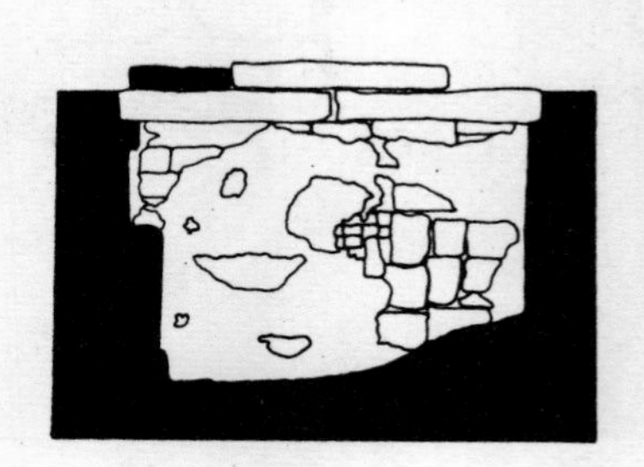

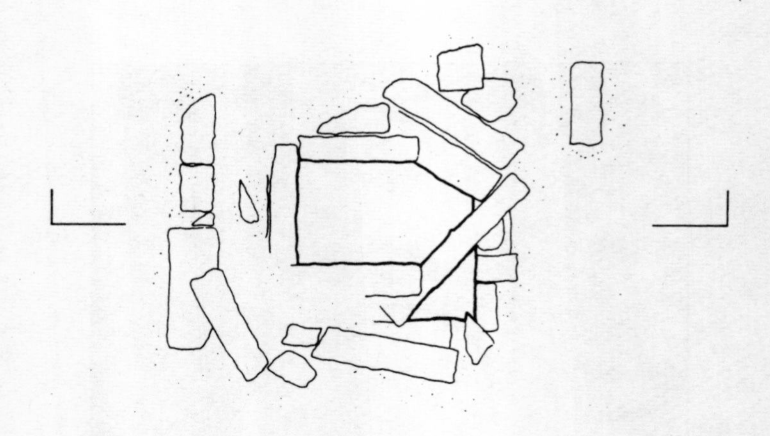

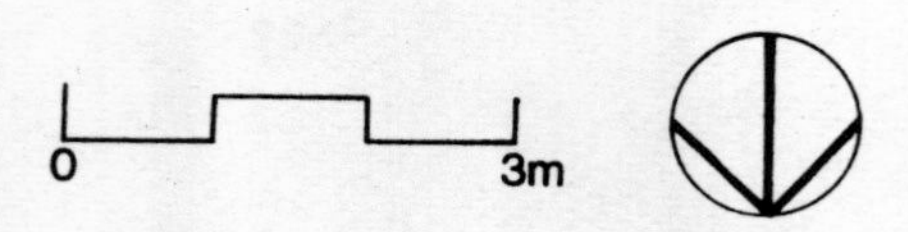

3.124 Well (Ib/9); plan and section (VRP).

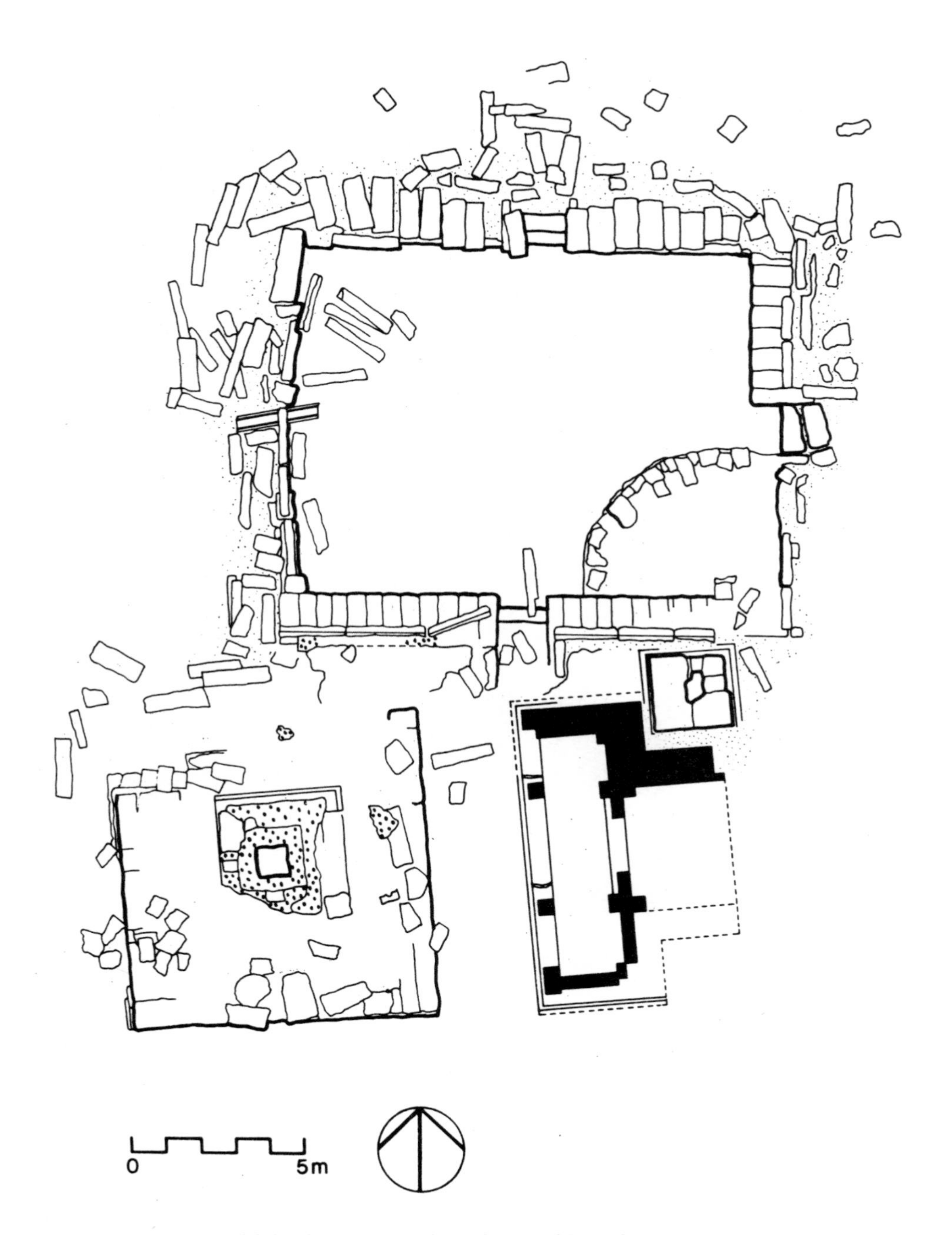

3.125 Stone-lined tank (Ic/12); plan (VRP).

3.126 Plaster-lined, stone tank (IIIa/5).

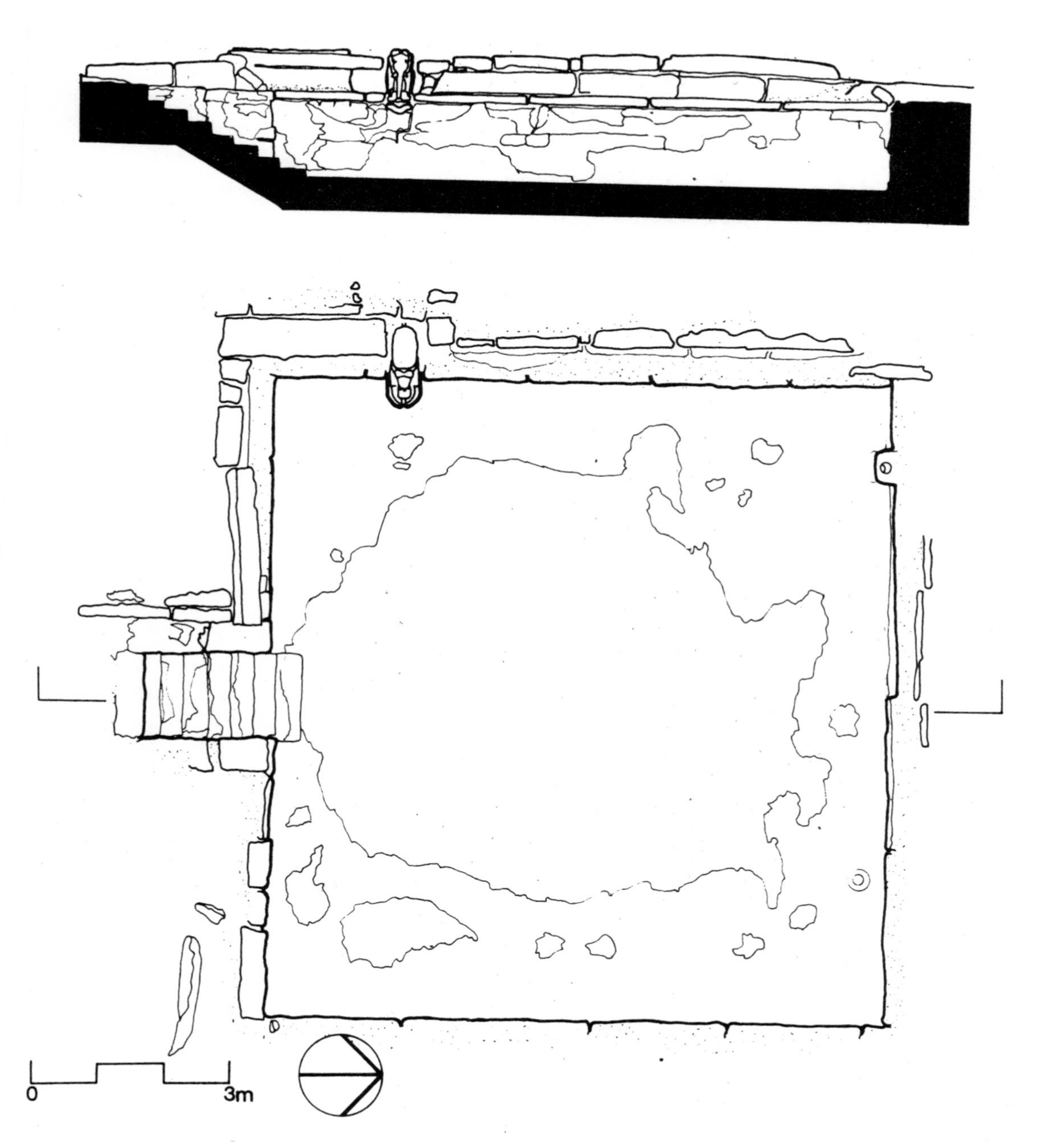

3.127 Plaster-lined, stone tank (IIIa/5); plan and section (VRP).

3.128 Monolithic, stone trough (IIIa/7).

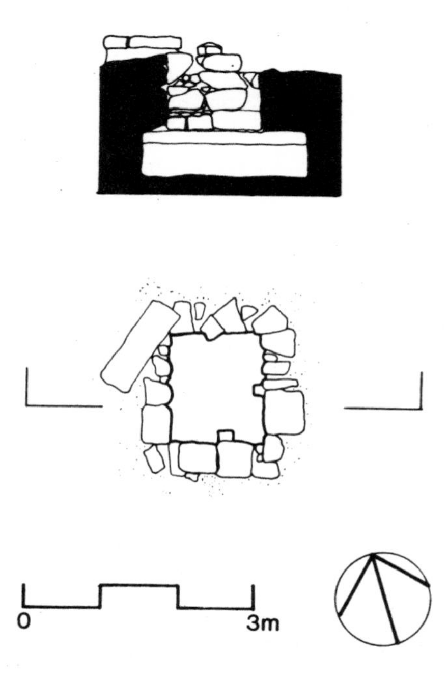

3.129 Well (V/4); plan and section (VRP).

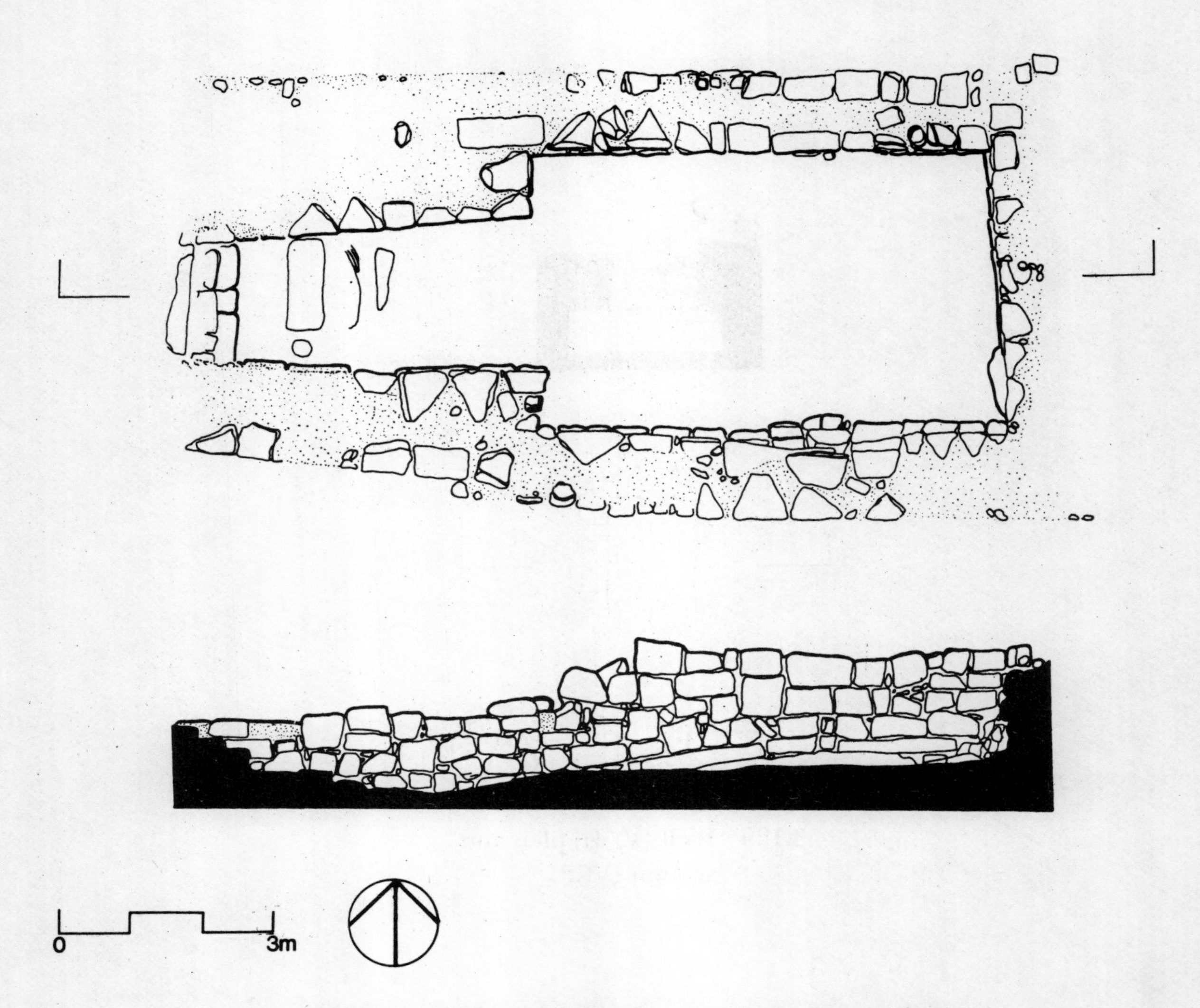

3.130 Plaster-lined tank (V/10); plan and section (VRP).

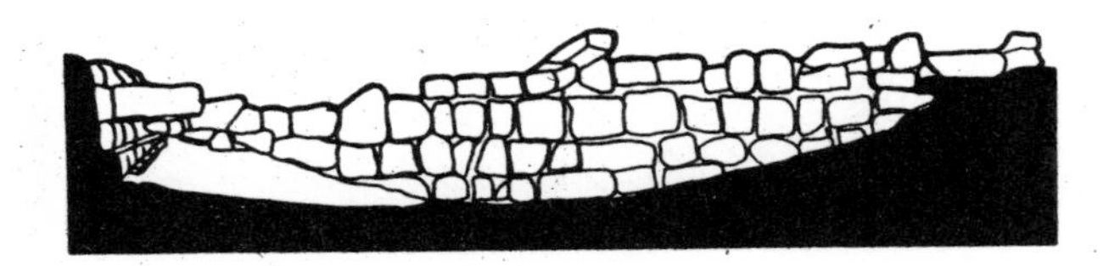

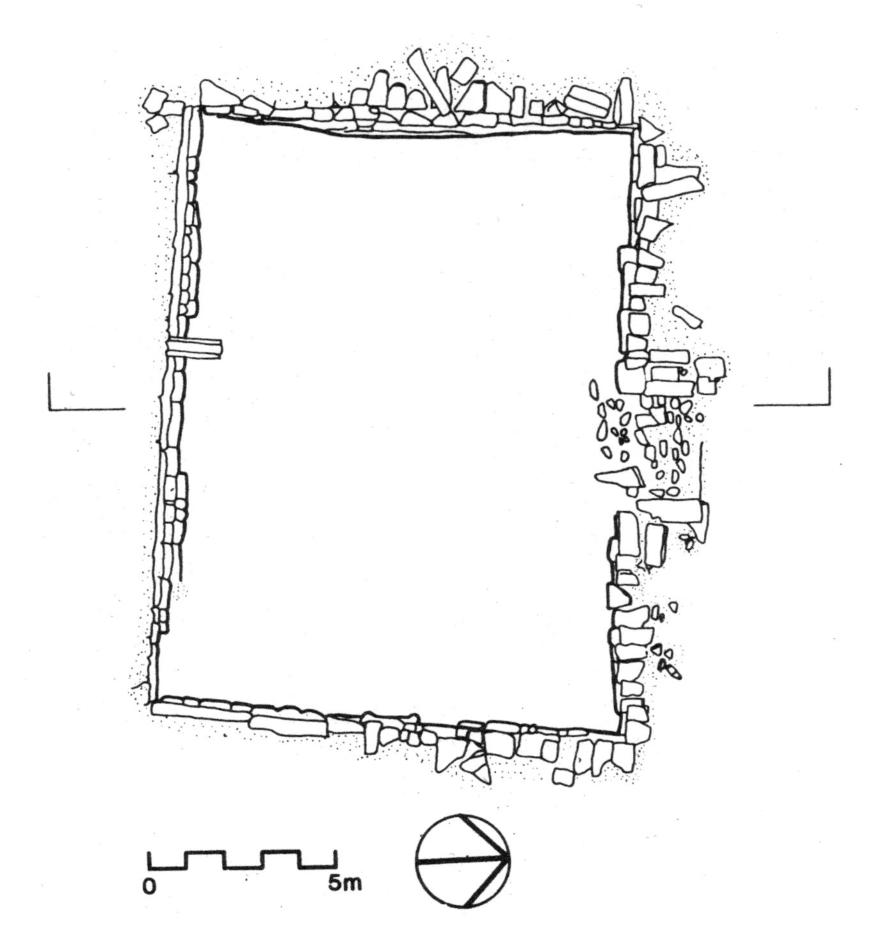

3.131 Rubble-lined tank (V/12); plan and section (VRP).

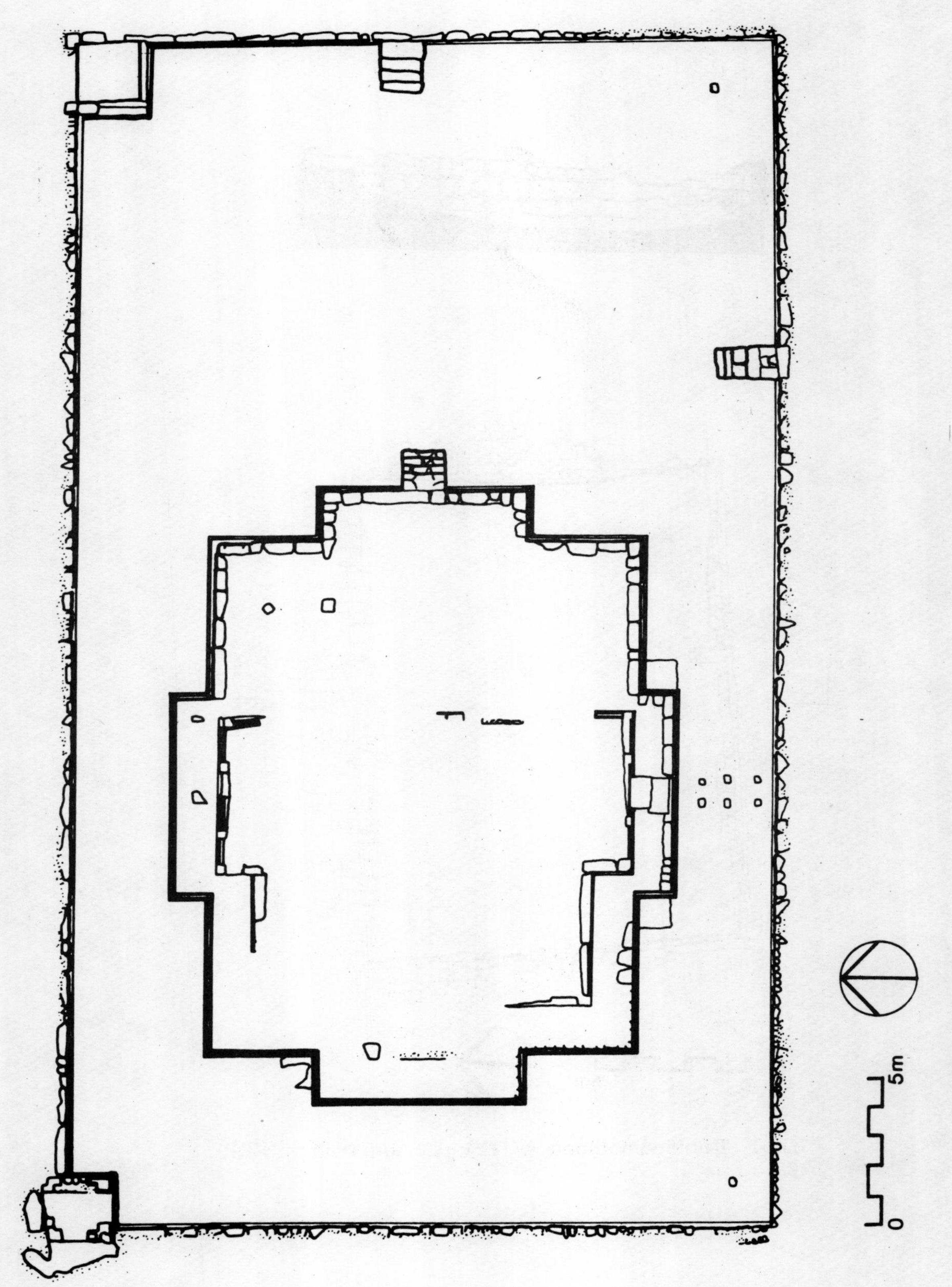

3.132 'Water Pavilion' (XIV/2); plan (VRP).

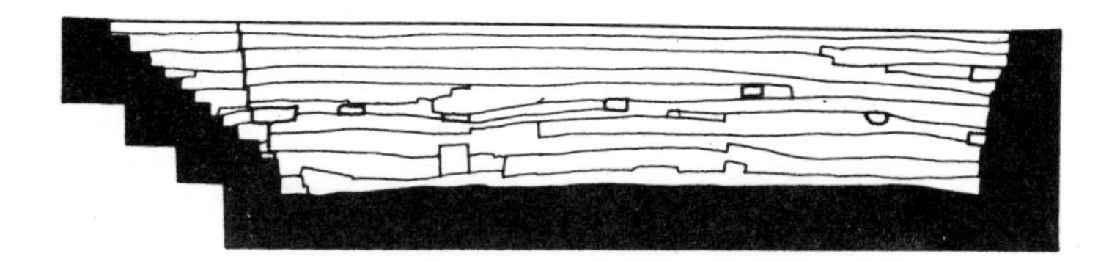

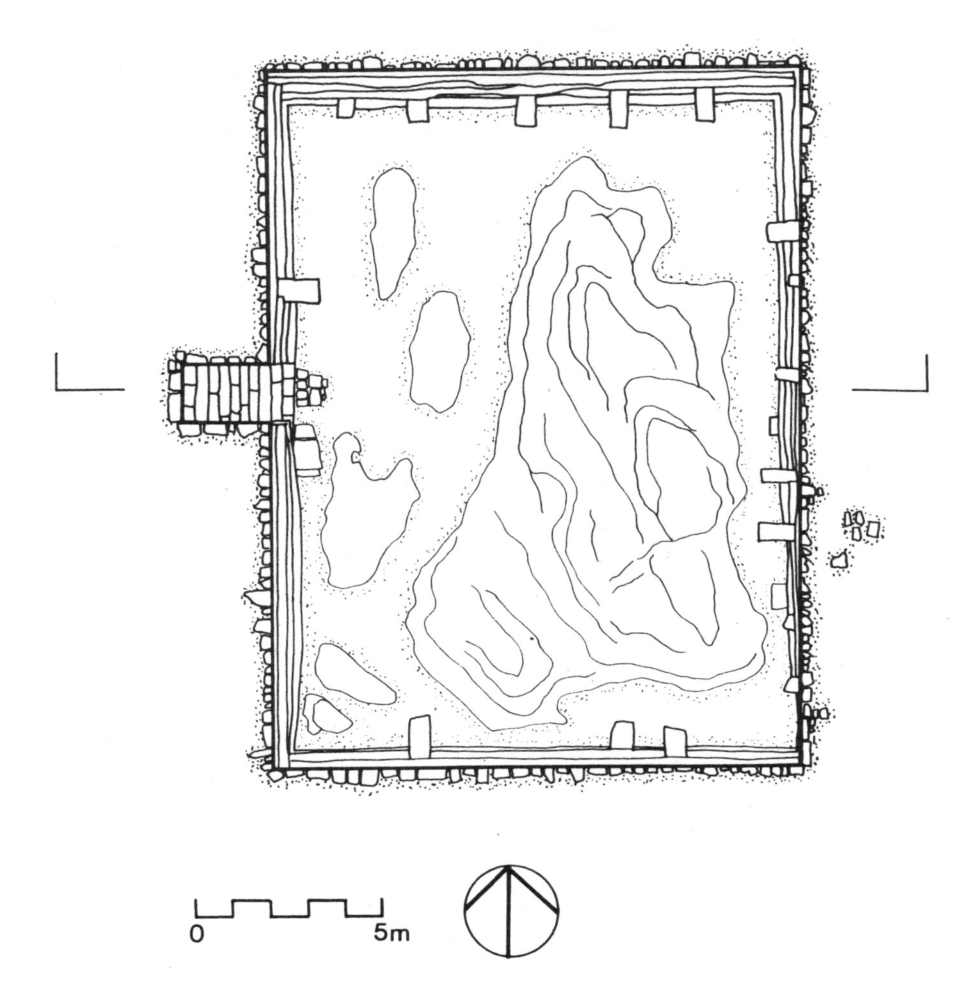

3.133 Stone-lined, rainwater collection tank (XIV/9); plan and section (VRP).

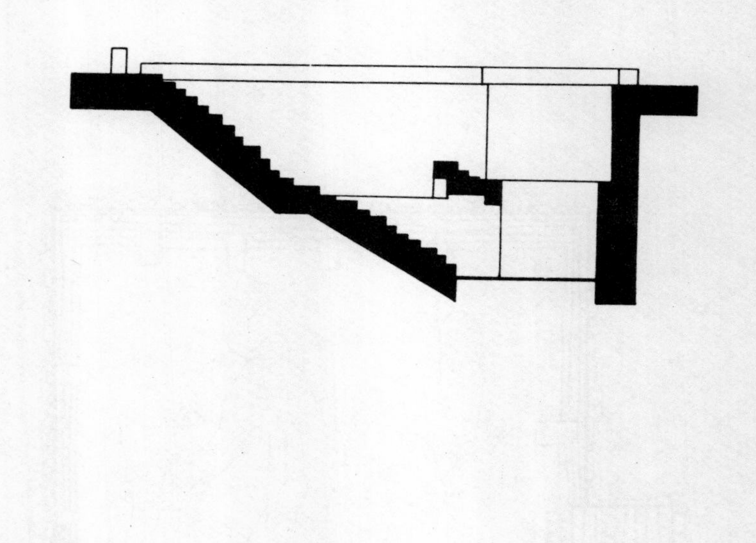

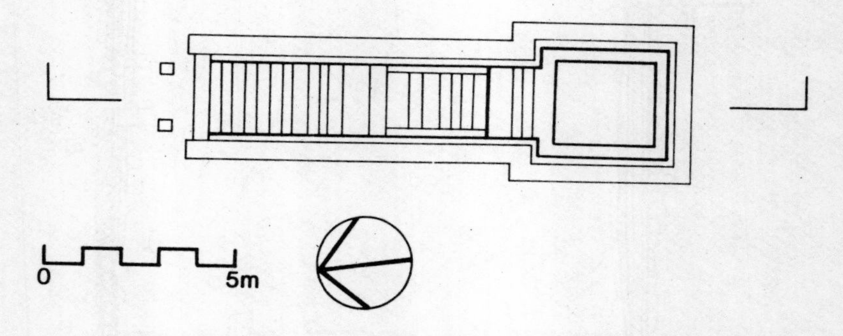

3.134 Step-well (XVa/8); plan and section (VRP).

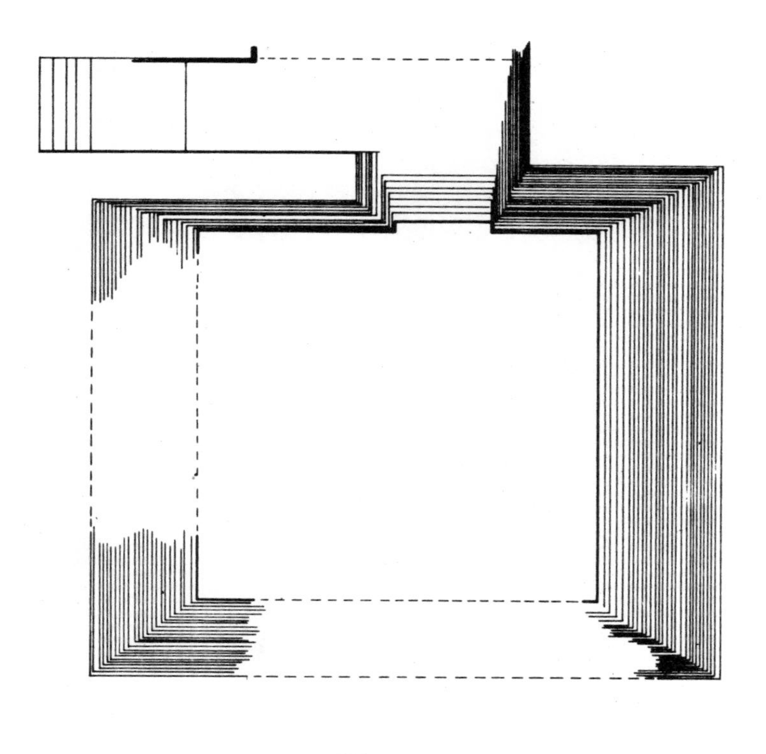

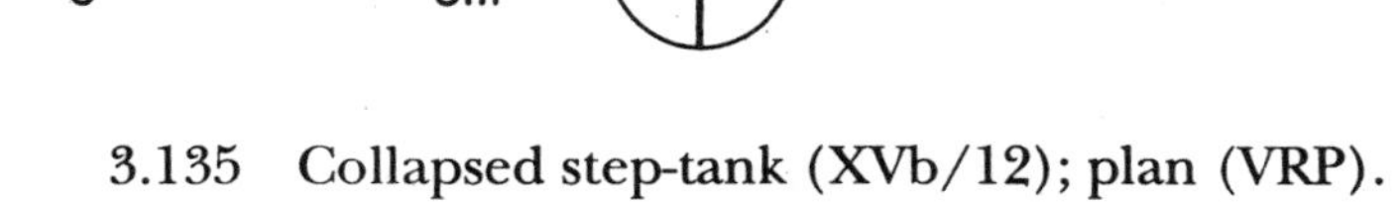

3.135 Collapsed step-tank (XVb/12); plan (VRP).

3.136 Pipeline in XVb.

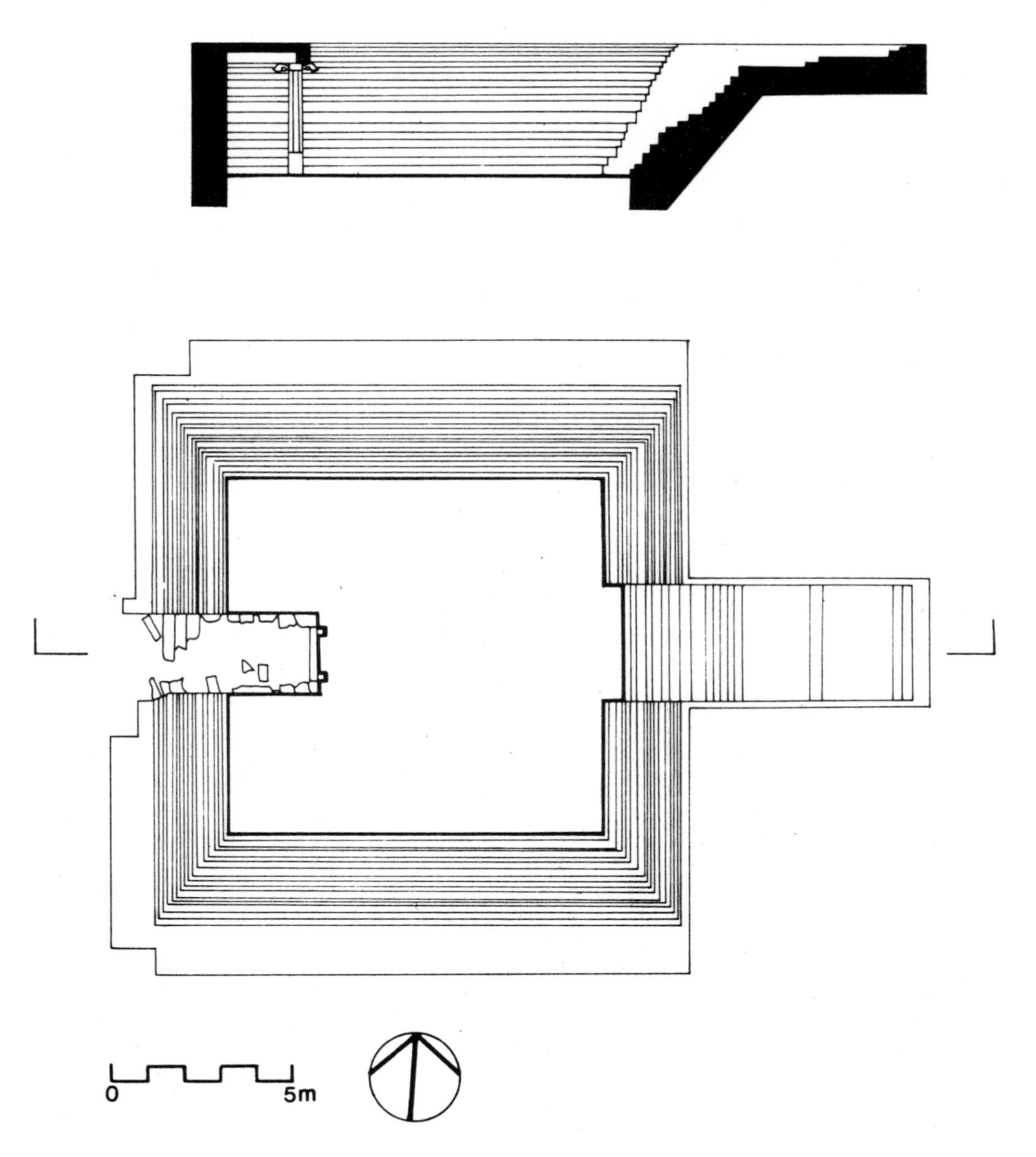

3.137 Step-tank (NQy/3); plan and section (VRP).

3.138 Hemakutam hill; first rainwater collection tank.

3.139 Hemakutam hill; second rainwater collection tank.

3.140 Hemakutam hill; third rainwater collection tank.

3.141 Hemakutam hill; fourth rainwater collection tank.

3.142 Hemakutam hill conduit; view to south.

3.143 Hemakutam hill conduit; tank at the southwest corner of the Vīrūpākṣa temple.

3.144 Hemakutam hill conduit passing through prākāra wall of temple.

3.145 Hemakutam hill conduit flowing across the northern toe of the hill.

3.146 Rainwater collection tank at the Parameśvara maṭha.

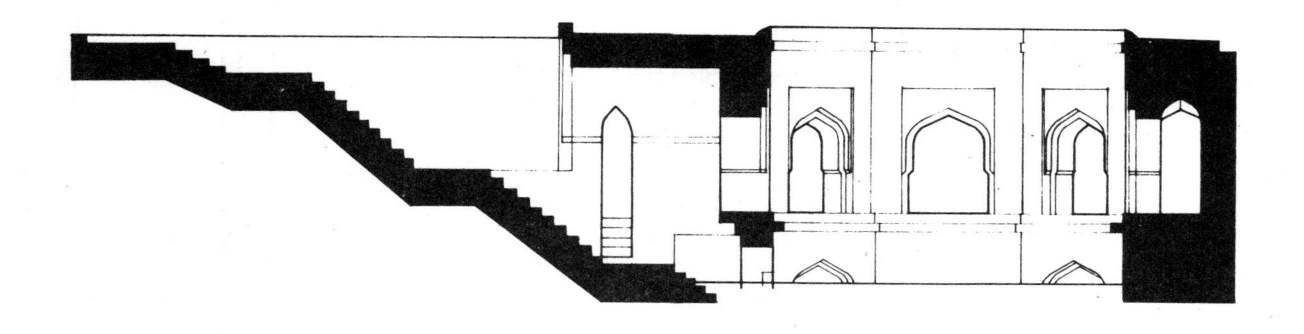

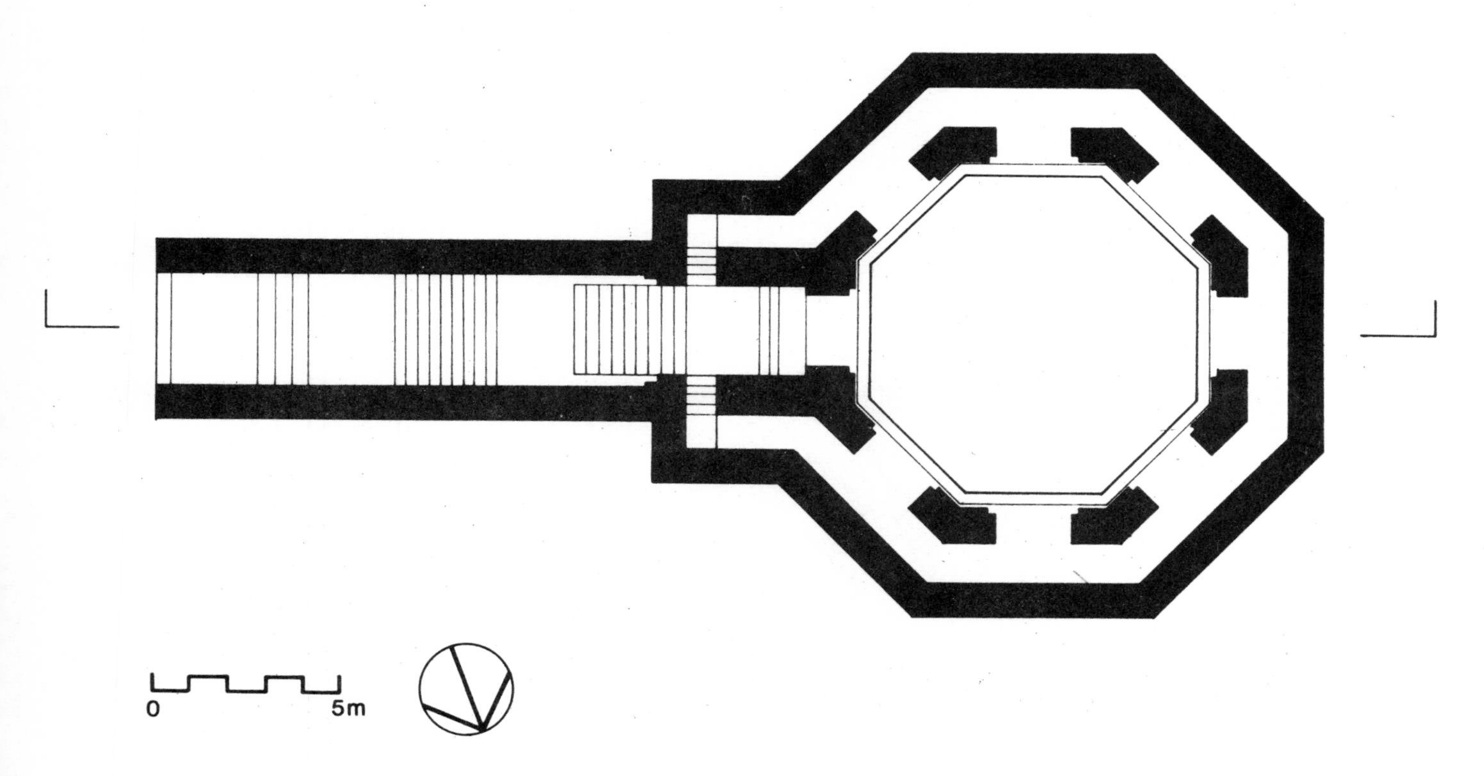

3.147 Mallappannagudi step-well; plan and section (VRP).

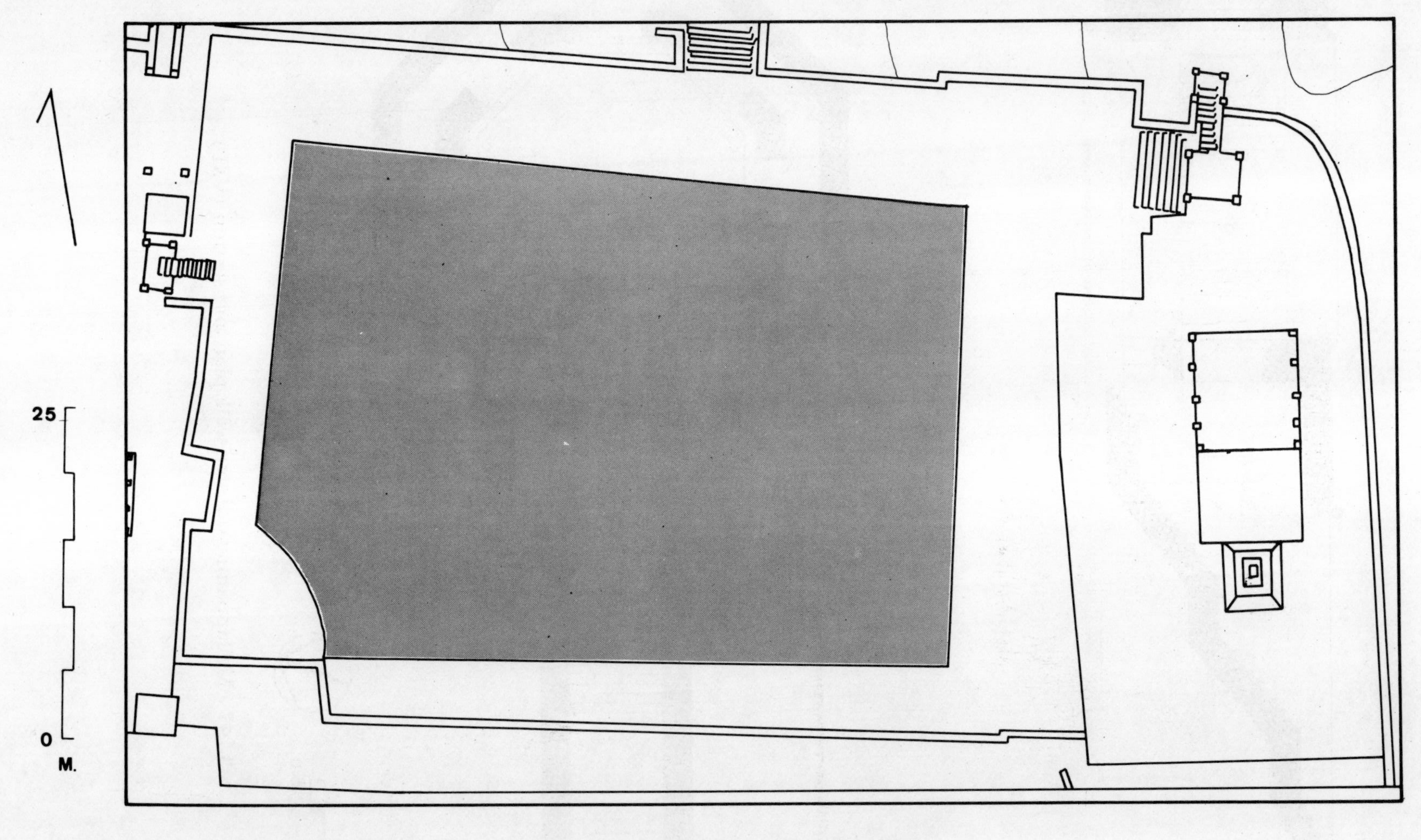

3.148 Manmatha tank; plan (after VRP).

3.149 Manmatha tank; view to northwest.

3.150 Manmatha tank; entrance pavilion.

3.151 Manmatha tank; outlet.

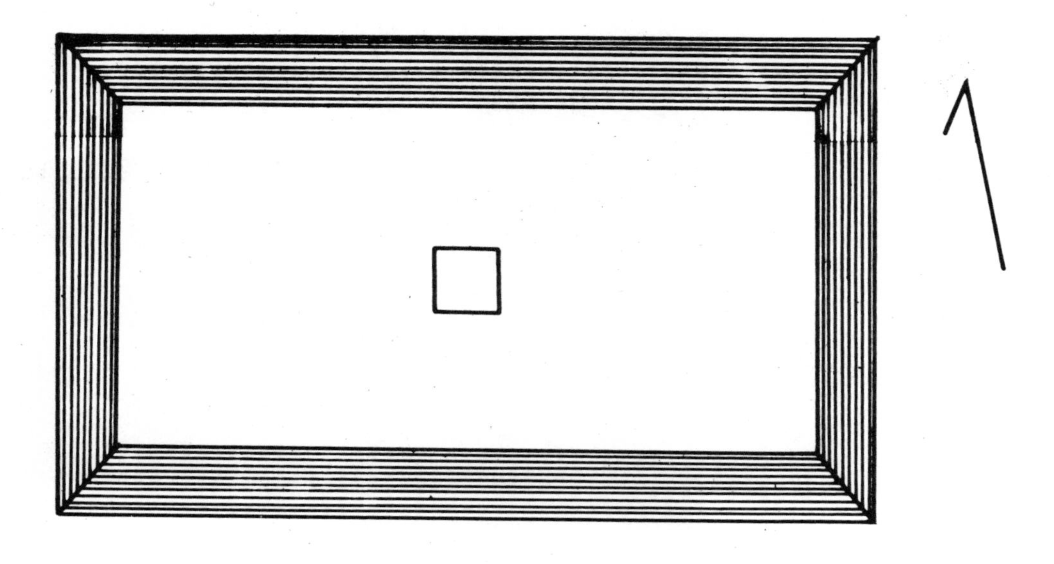

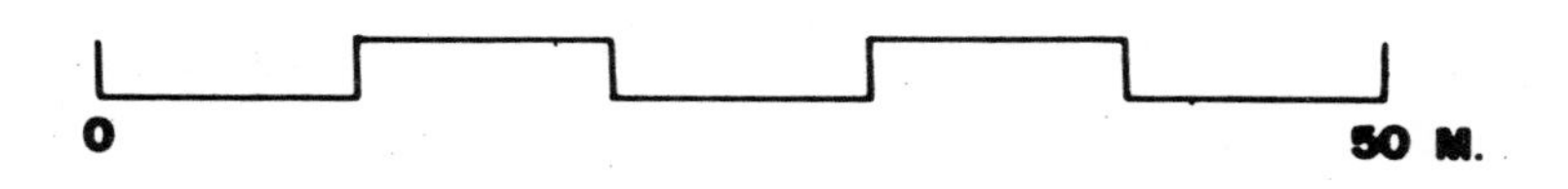

3.152 Viṭṭhala tank; plan and section.

3.153 Bālakṛṣṇa tank; view to east.

3.154 Lokapāvana tank; view to northeast.

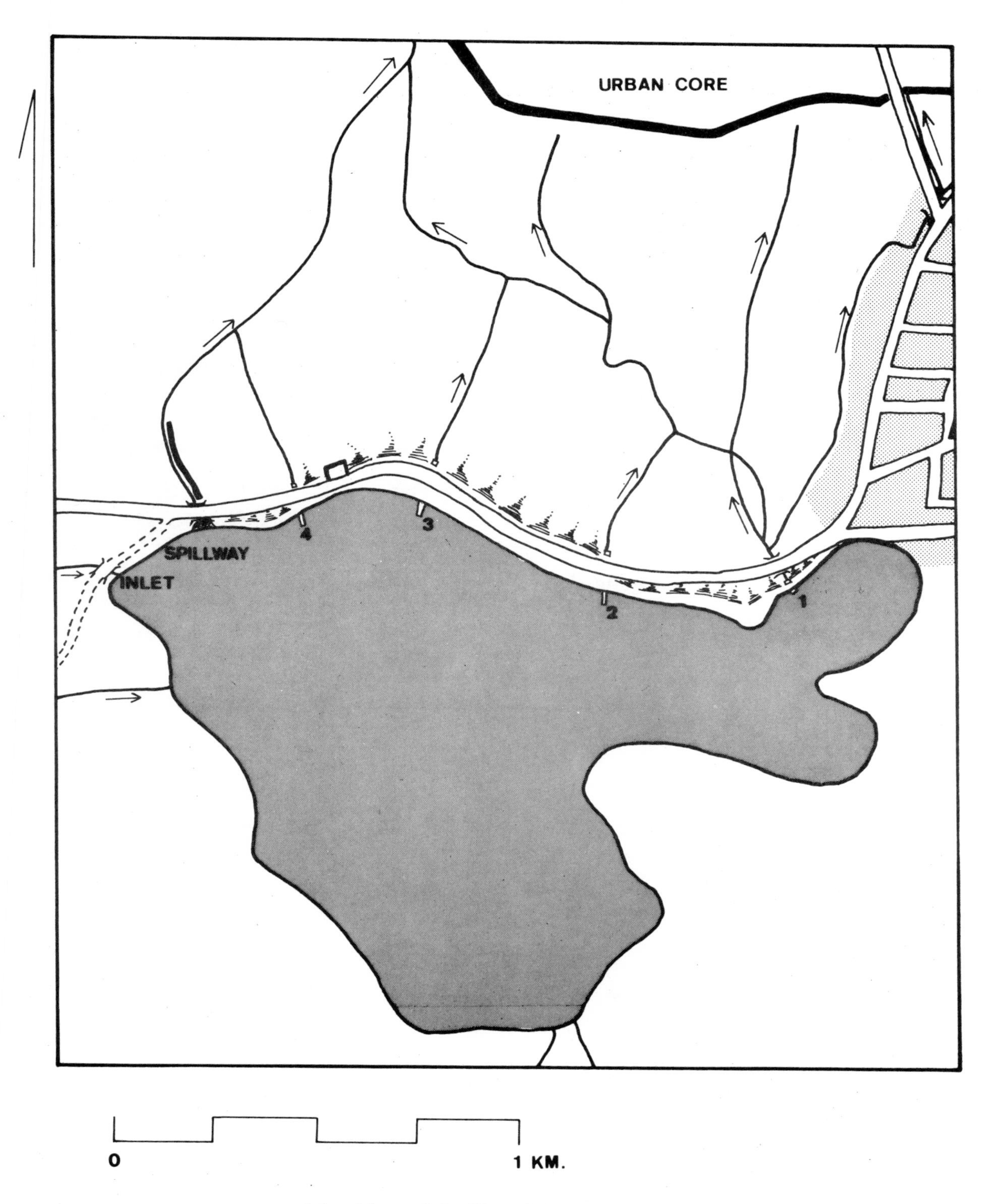

4.1 Map of the Kamalāpuram tank.

4.2 Road running along the top of the Kamalāpuram bund.

4.3 Outcrop included in the bund.

4.4 Close-up of the upstream revetting at the east end of the bund.

4.5 The upstream revetting in the centre of the bund.

4.6 Three-sided, masonry structure.

4.7 Kamalāpuram tank; first outlet.

4.8 Irrigation channel taking off from the first outlet; view to northeast.

4.9 Kamalāpuram tank; second outlet.

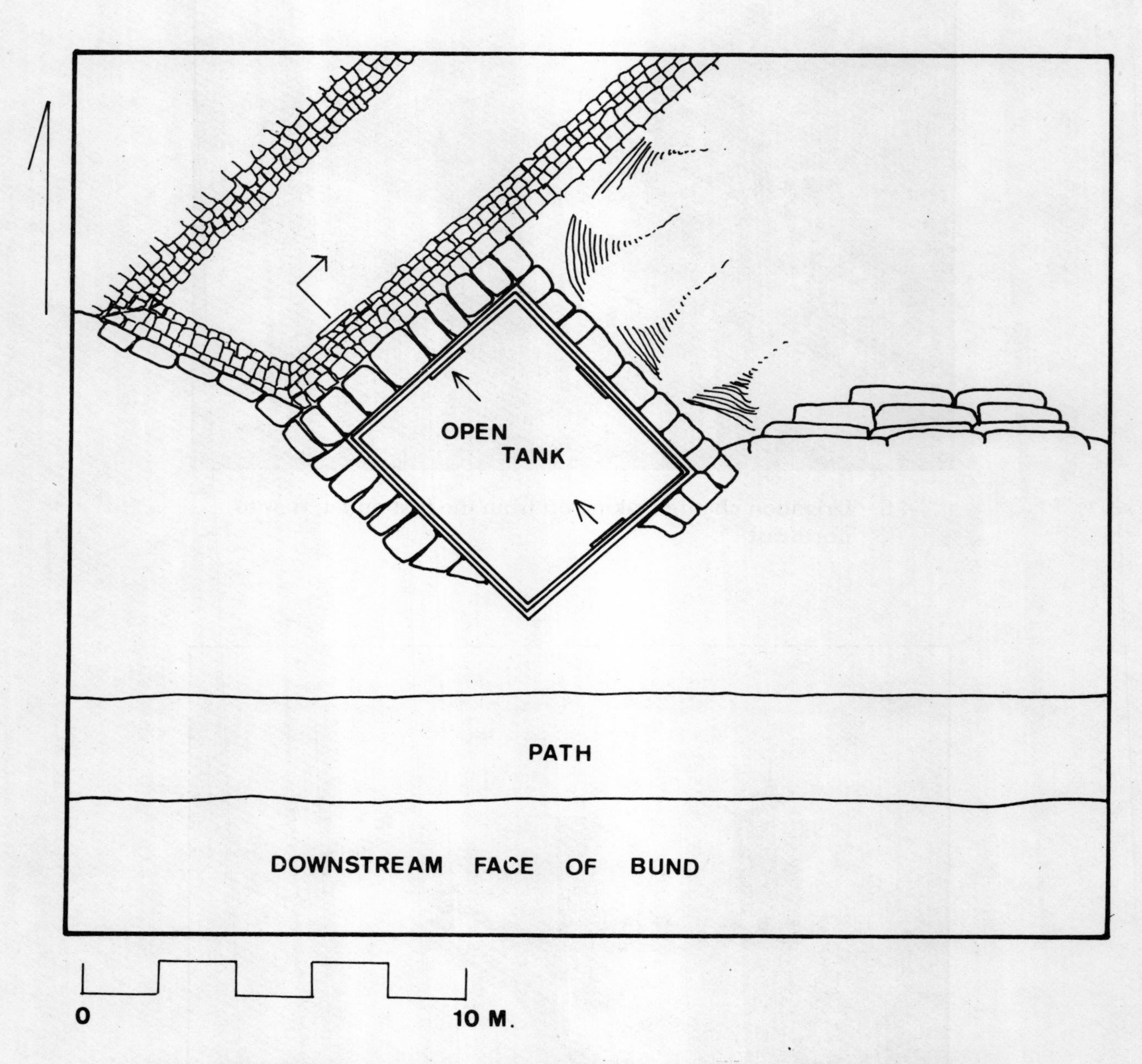

4.10 Kamalāpuram tank; second outlet, plan of the open tank on downstream face of the bund.

4.11 Kamalāpuram tank; second outlet, open tank on downstream face of the bund.

4.12 Kamalāpuram tank; second outlet, close-up of interior of the open tank on downstream face of the bund.

4.13 Kamalāpuram tank; second outlet, water passing into an irrigation channel.

4.14 Kamalāpuram tank; third outlet.

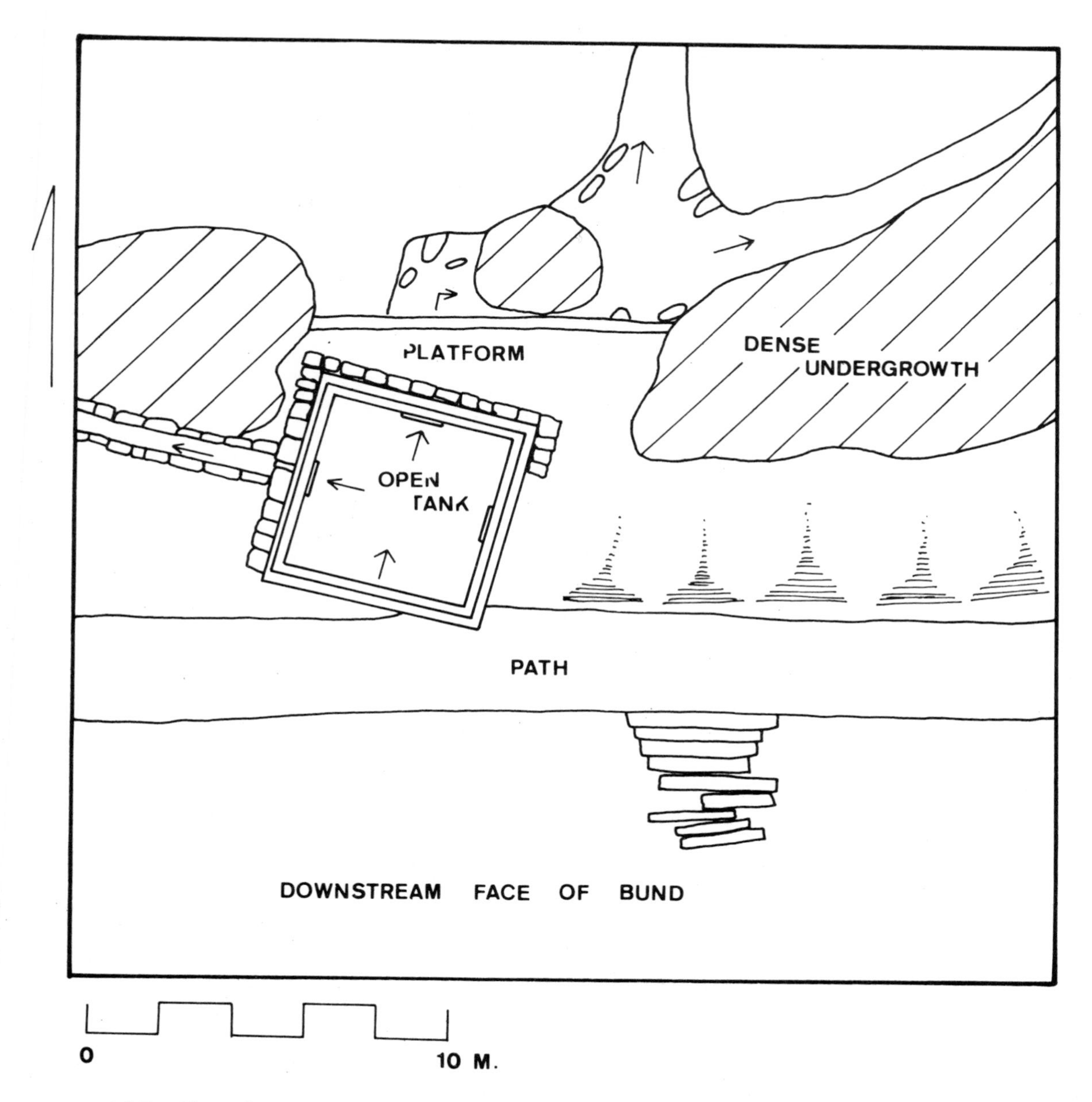

4.15 Kamalāpuram tank; third oulet, plan of the open tank on downstream face of the bund.

4.16 Kamalāpuram tank; third outlet, the open tank on downstream face of bund; overview from the top of the bund.

4.17 Kamalāpuram tank; third outlet, the open tank on downstream face of the bund.

4.18 Kamalāpuram tank; fourth outlet.

4.19 Kamalāpuram tank; fourth outlet, downstream face of the bund.

4.20 Kamalāpuram tank; buried, masonry structure to the east of the fourth outlet.

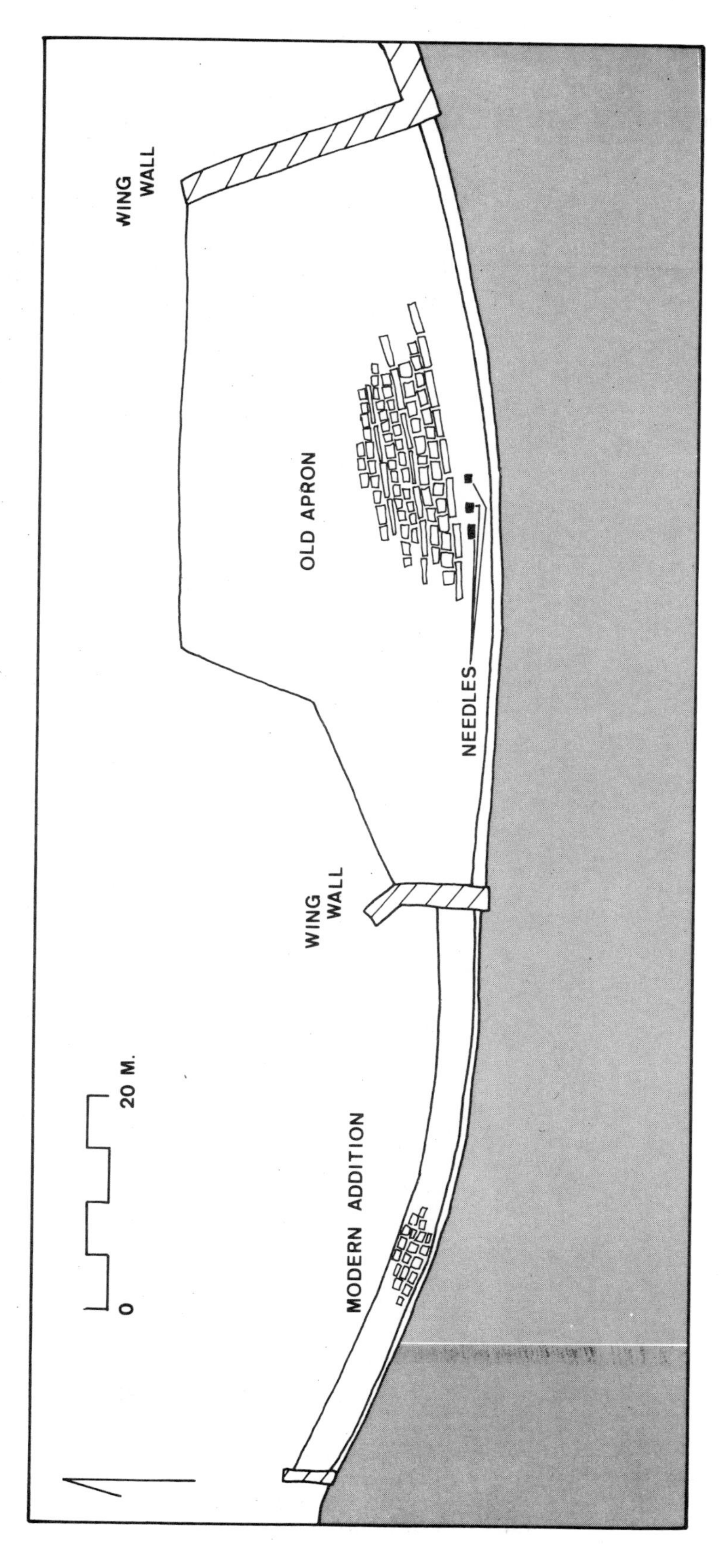

4.21 Kamalāpuram tank; plan of spillway.

4.22 Kamalāpuram tank; spillway, view to west.

4.23 Kamalāpuram tank; spillway, close-up of needles.

4.24 Kamalāpuram tank, the escape channel taking off from the spillway.

4.25 Kamalāpuram tank; fragmentary remains of a structure.

4.26 Rock-cut channel passing between the Archaeological Office and the Inspection Bungalow.

4.27 Rock-cut channel flowing parallel to the main street of Kamalāpuram.

4.28 Rock-cut channel bifurcating.

4.29 Left branch of the channel flowing towards the 'Royal Centre'.

4.30 Channel raised up on supporting piers.

4.31 Channel flowing through gateway into the 'Urban Core'.

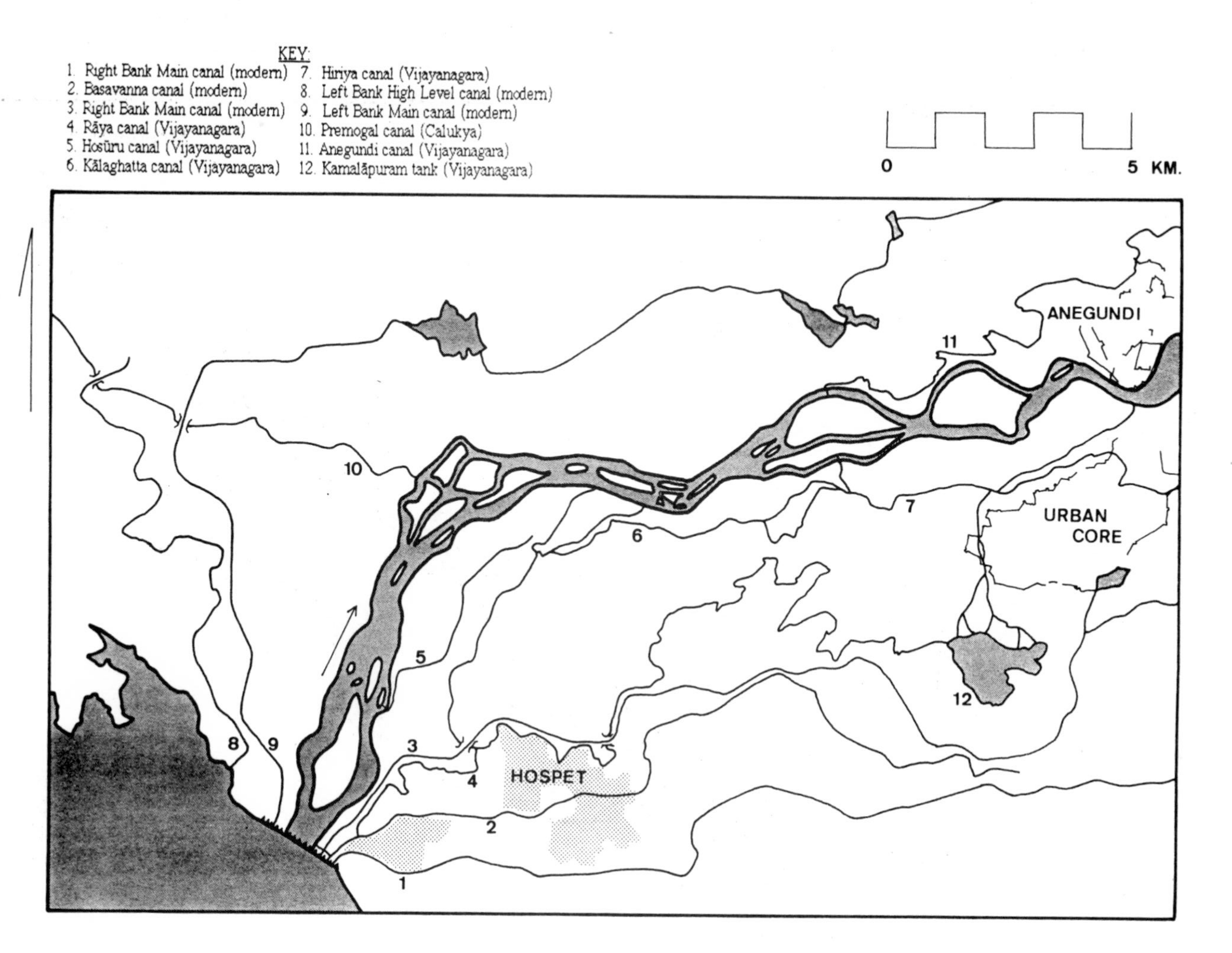

4.32 Map showing major canals after the construction of the Tungabhadrā Dam.

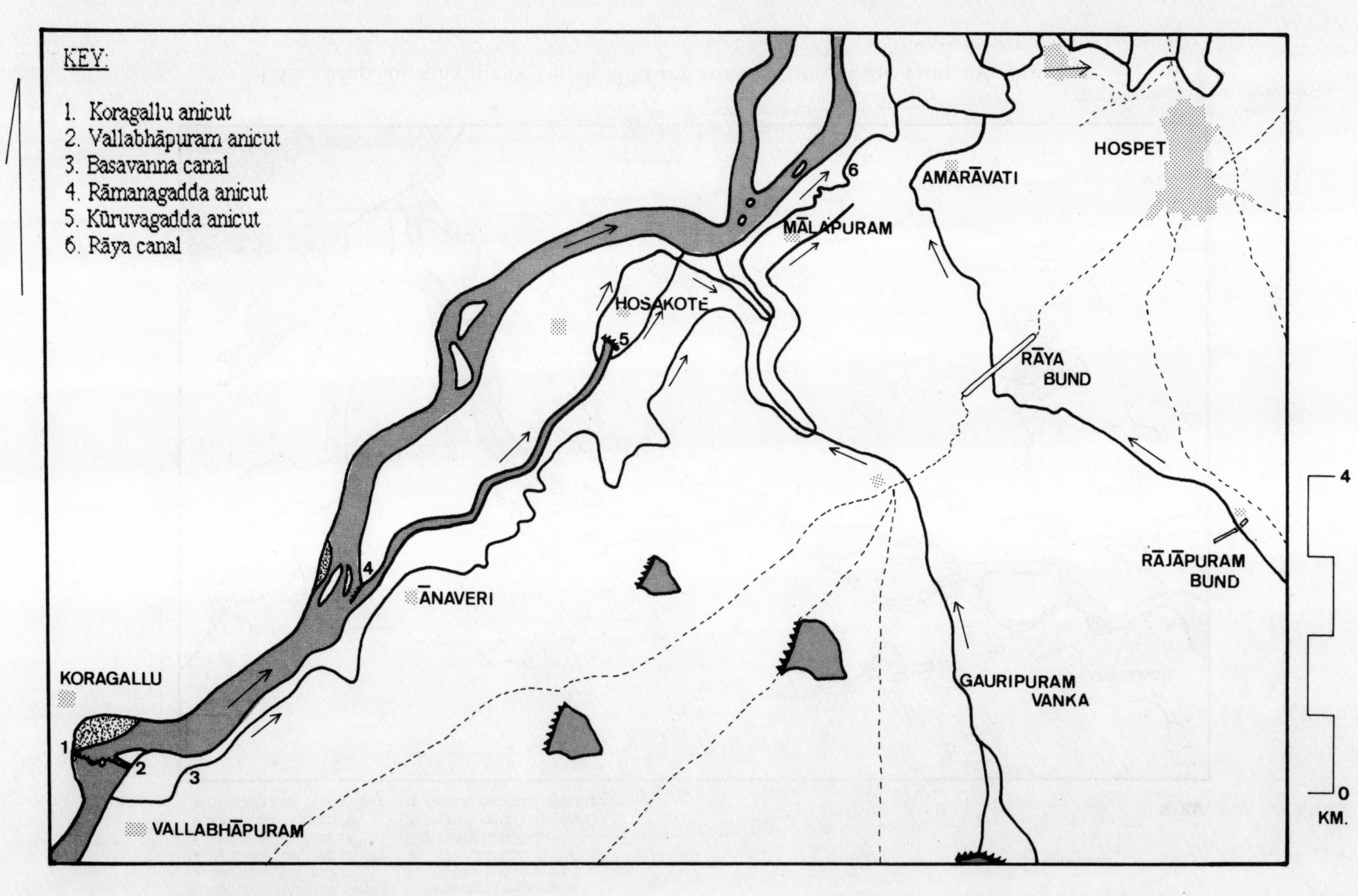

4.33 Map showing important irrigational features on the right bank of the river between Vallabhāpuram and Hospet.

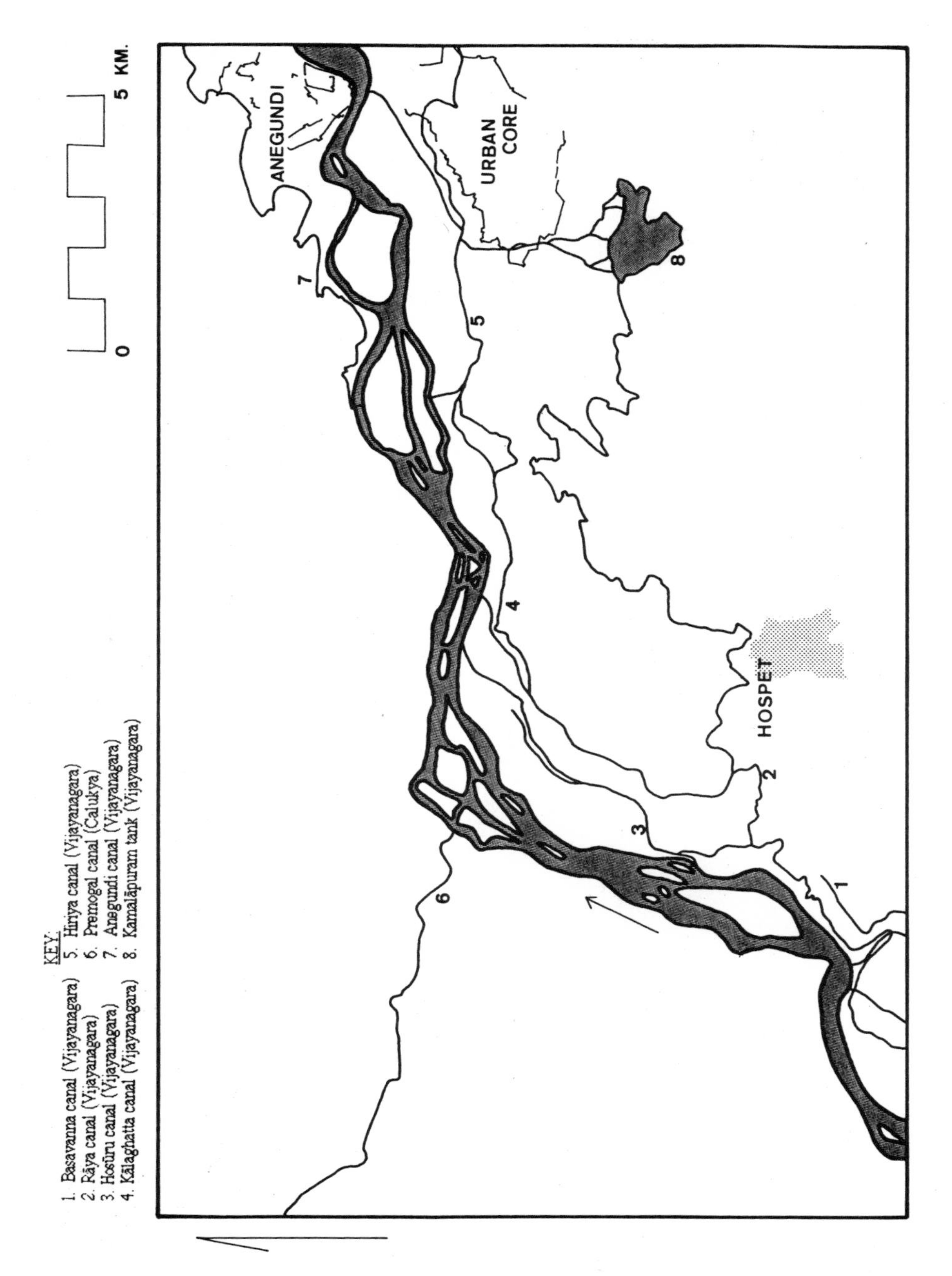

4.34 Map showing major canals before 1900.

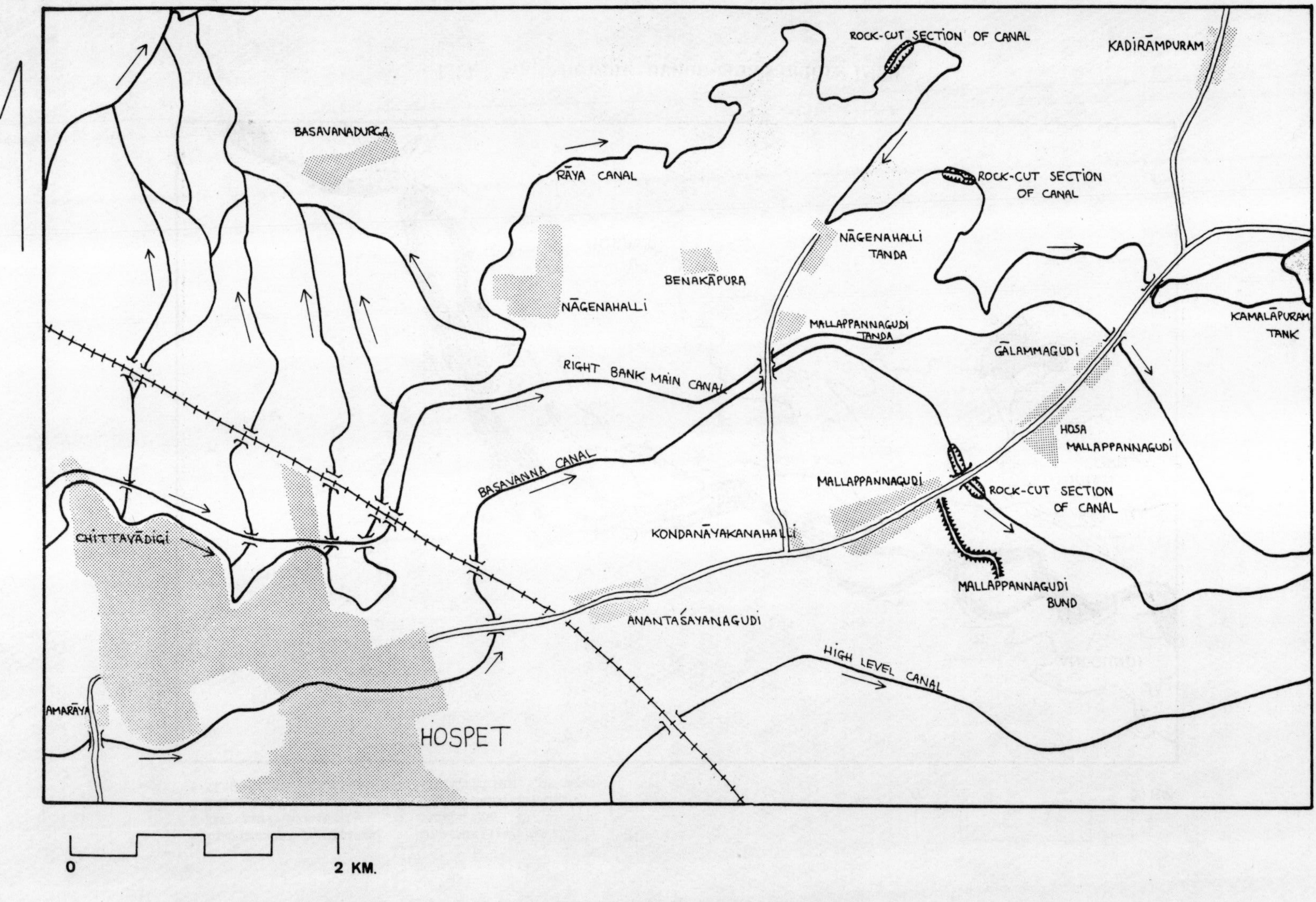

4.35 Map showing the route of the Basavanna and Rāya canals from Amarāya to Kamalāpuram.

4.36 Basavanna canal flowing east of Amarāya; view to west.

4.37 Basavanna canal flowing north after passing under the Hospet Kamalāpuram road; view to north.

4.38 Basavanna canal flowing north of Bhattarahalli; view to west.

4.39 Basavanna canal in deep excavation; view to south.

4.40 Basavanna canal; type 1 revetting.

4.41 Basavanna canal; type 2 revetting.

4.42 Rāya canal flowing close to Nāgenahalli (view to southwest).

4.43 Rāya canal flowing close to Nāgenahalli (view to northeast).

4.44 Rāya canal flowing around sheet rock.

4.45 Rāya canal flowing under the Hospet–Kamalāpuram road; view to north.

4.46 Rāya canal; rubble revetting.

4.47 Rāya canal; fragment of original block-lining.

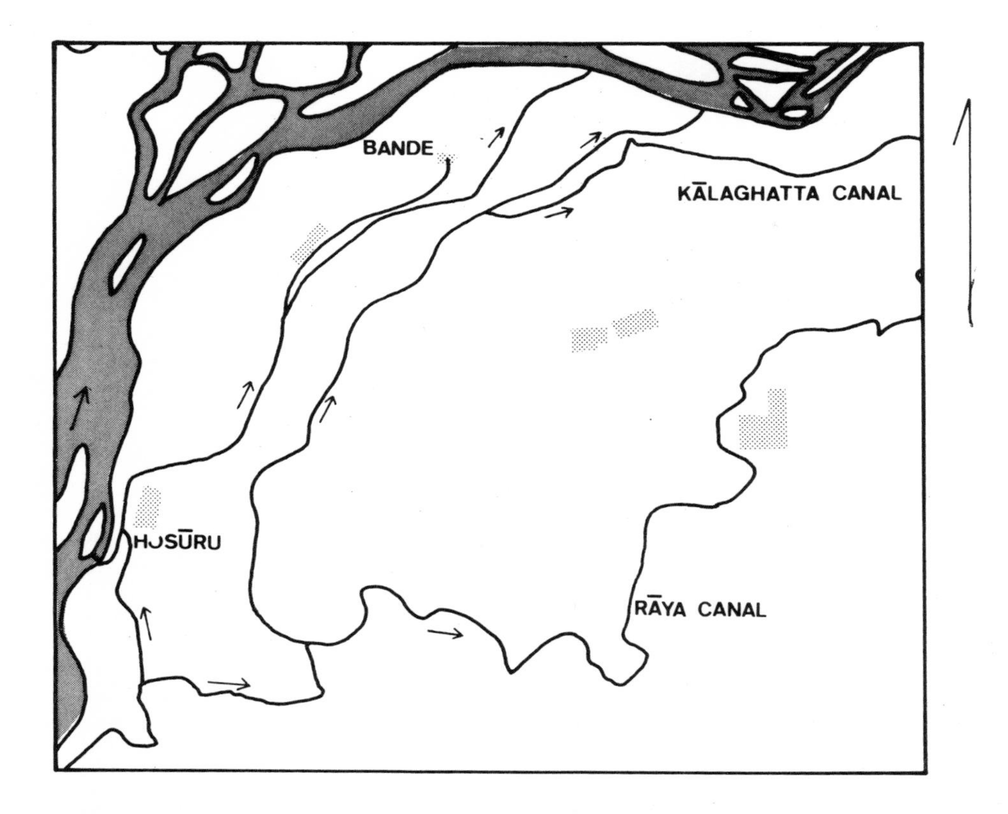

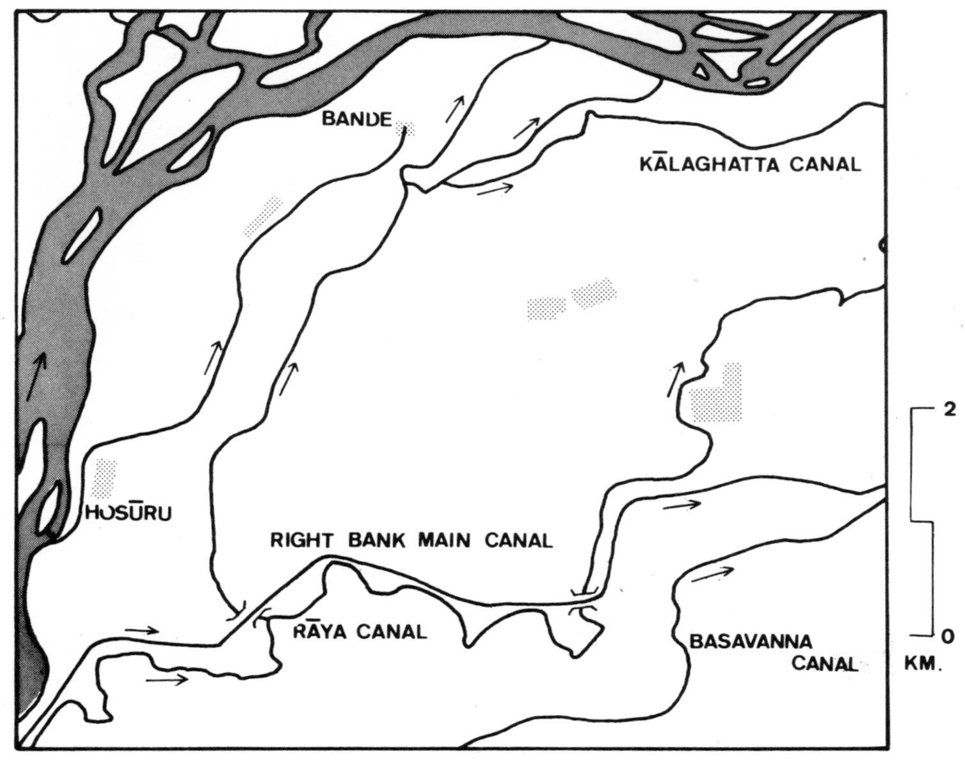

4.48 Maps showing the Hosūru canal before (above) and after (below) the construction of the Tungabhadrā Dam.

4.49 The Hosūru anicut.

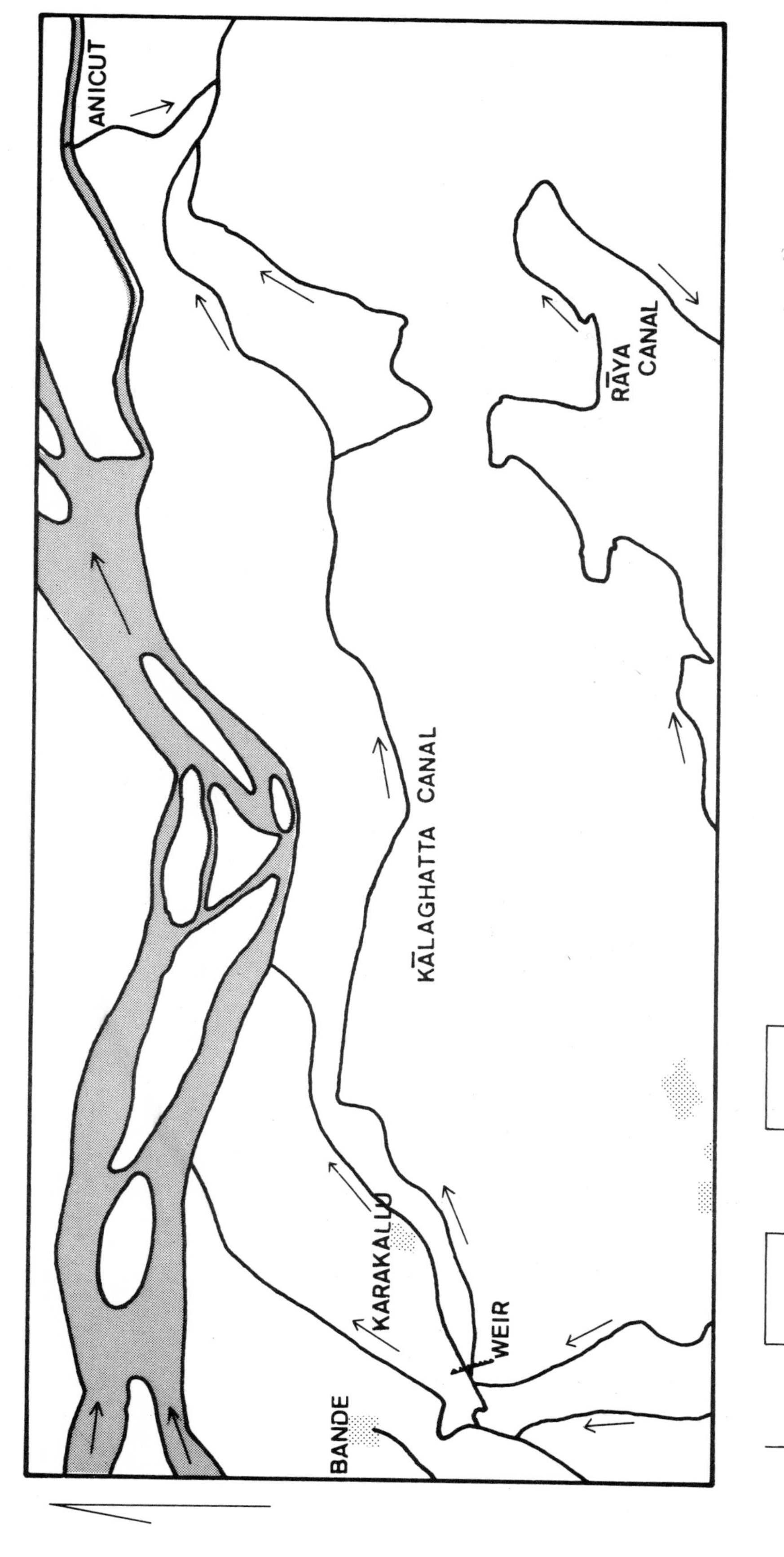

4.50 Map showing the route of the Kālaghatta canal.

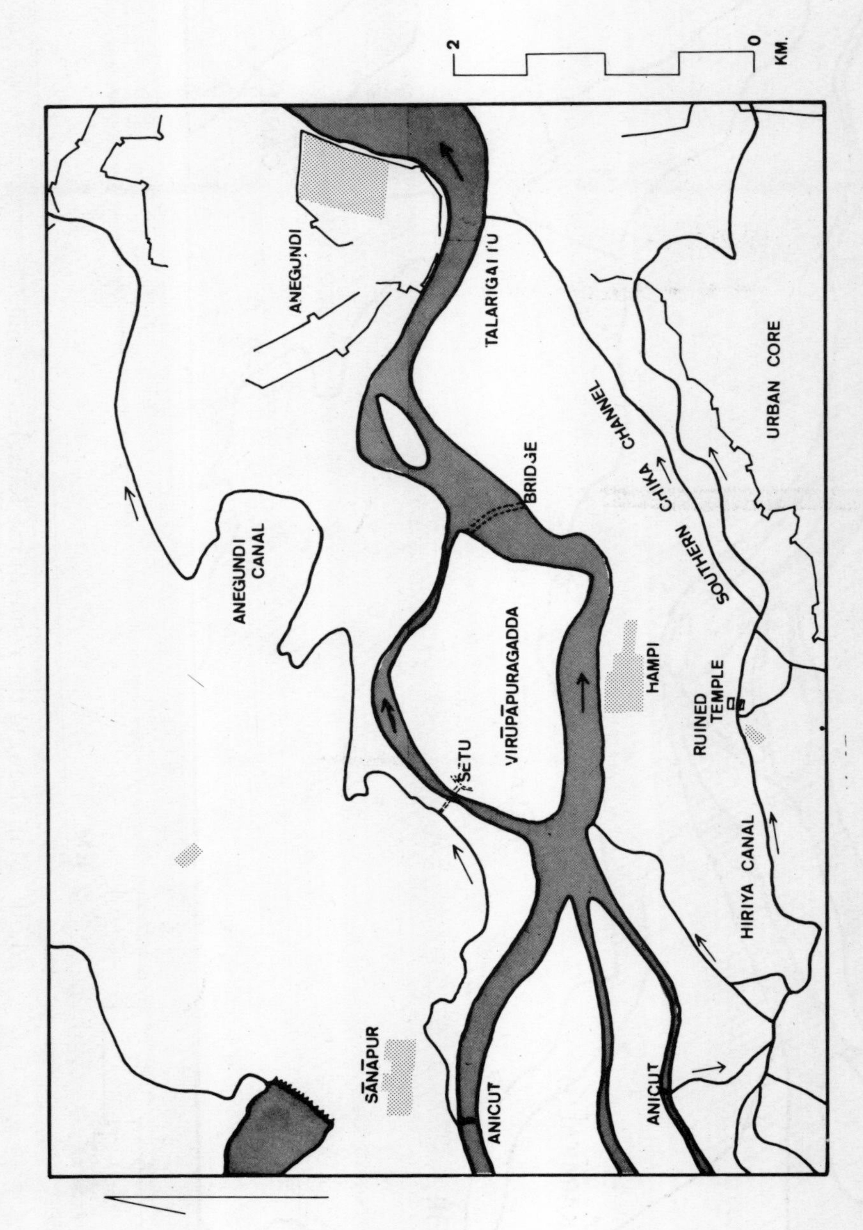

4.51 Map showing the routes of the Hiriya and Anegundi canals.

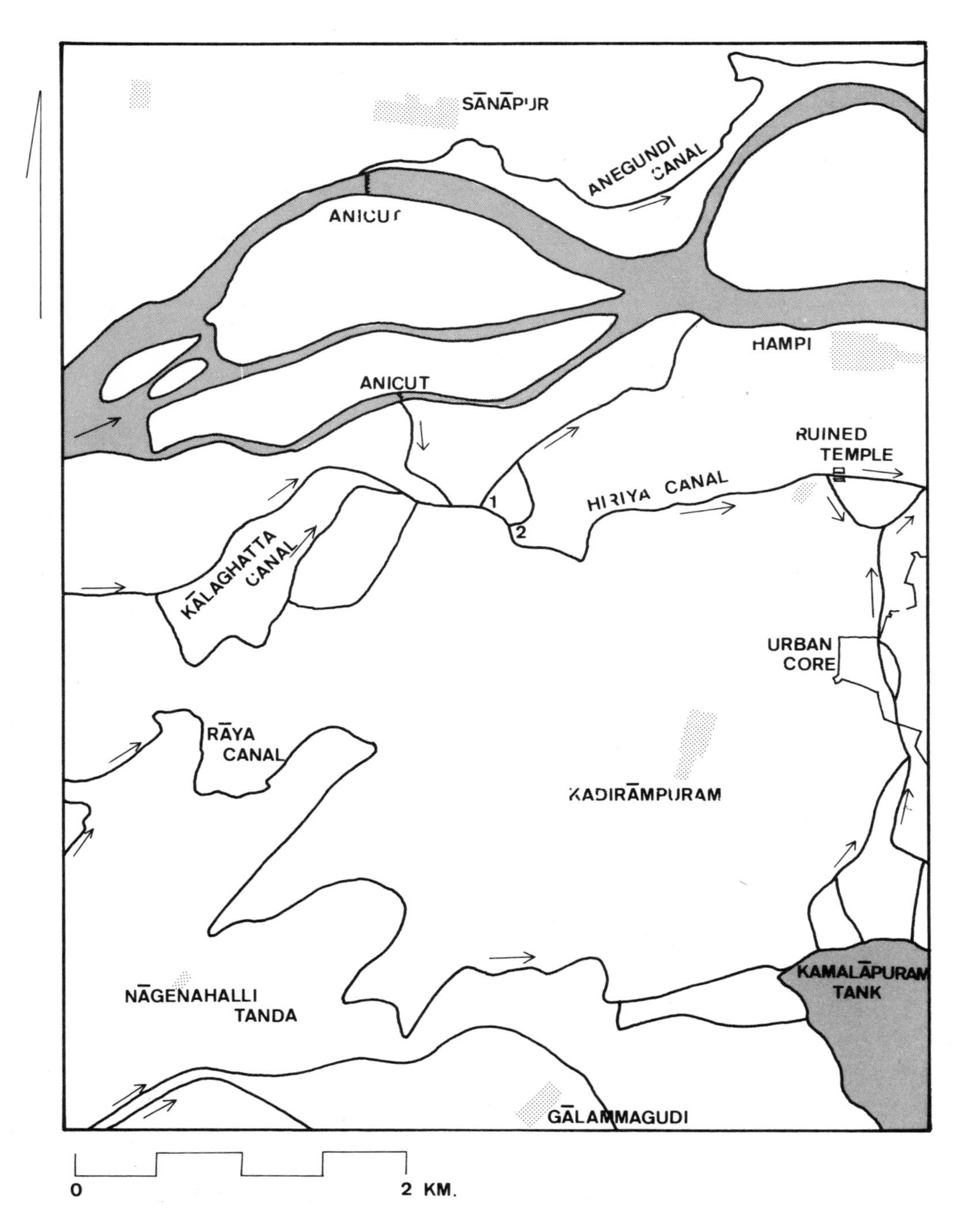

4.52 Map showing the spatial relationship between the Hiriya and the other major canals.

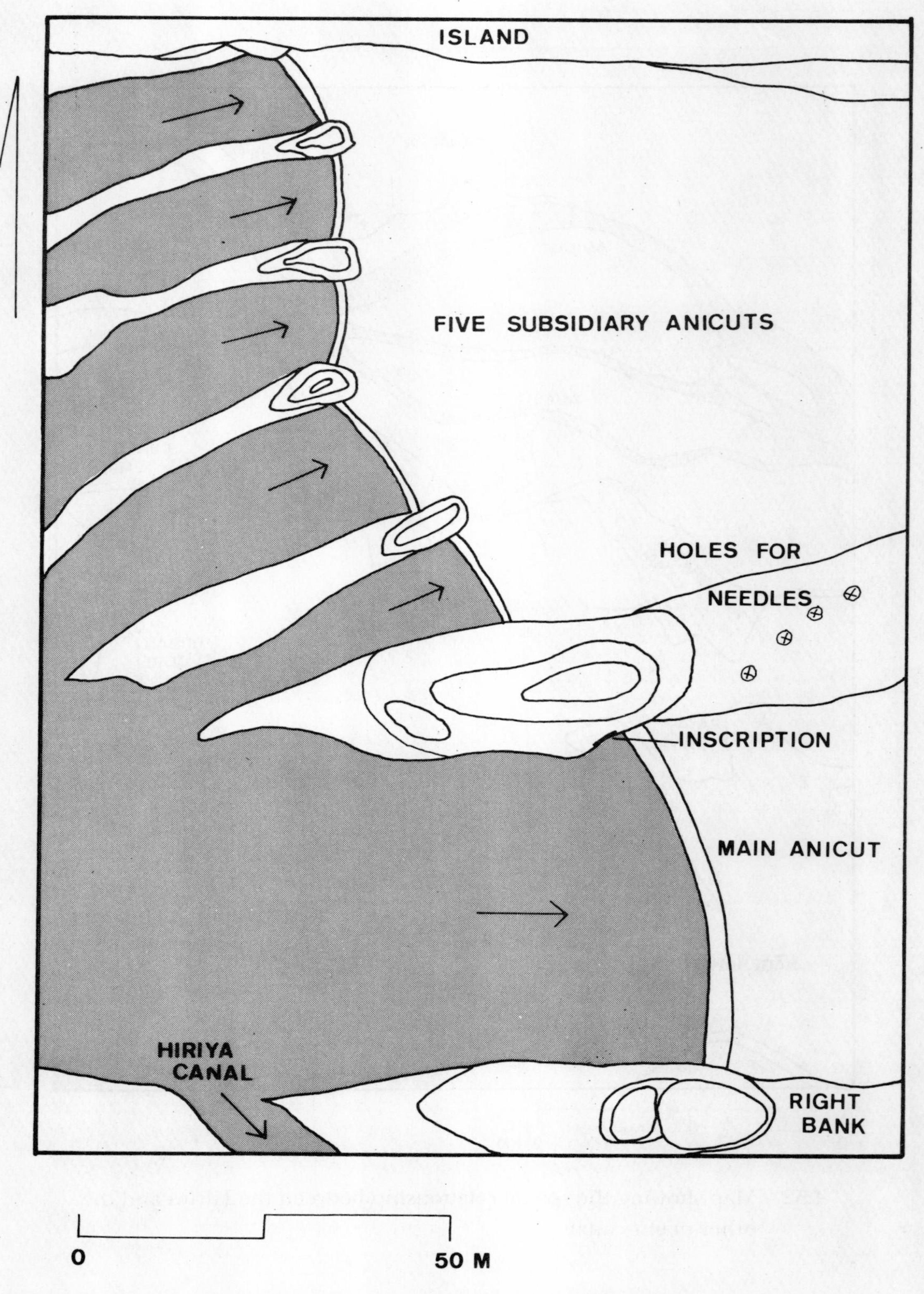

4.53 Diagrammatic plan of the Hiriya anicuts.

4.54 Hiriya anicuts; one of the five, smaller anicuts; view to northeast.

4.55 Hiriya anicuts; close-up of holes drilled in slabs.

4.56 Main Hiriya anicut; view to north.

4.57 Main Hiriya anicut; view to southeast.

4.58 Main Hiriya anicut; fragments of the original weir.

4.59 Main Hiriya anicut; inscription.

4.60 Hiriya anicuts, holes for the fitting of a temporary weir crest.

4.61 Escape weir on the Hiriya canal.

4.62 Hiriya canal; an original water feature on the first outlet, view to east.

4.63 Hiriya canal; an original water feature on the first outlet, view to north.

4.64 Hiriya canal held in embankment against an outcrop of sheet rock; view to west.

4.65 Hiriya canal flowing north of Kṛṣṇāpuram; view to west.

4.66 Hiriya canal flowing in deep excavation towards the ruined temple; view to west.

4.67 Hiriya canal flowing through the ruined temple; view to west.

4.68 Hiriya canal flowing around right side of a boulder; view to west.

4.69 Hiriya canal flowing around left side of a boulder; view to west.

4.70 Hiriya canal flowing towards the Kamalāpuram–Hampi bridge; view to west.

4.71 Hiriya canal flowing under the Kamalāpuram–Hampi bridge; view to north, showing old revetting.

4.72 Hiriya canal discharging into the river at Talarigattu.

4.73 Hiriya canal discharging into the river at Talarigattu; close-up of temporary weir.

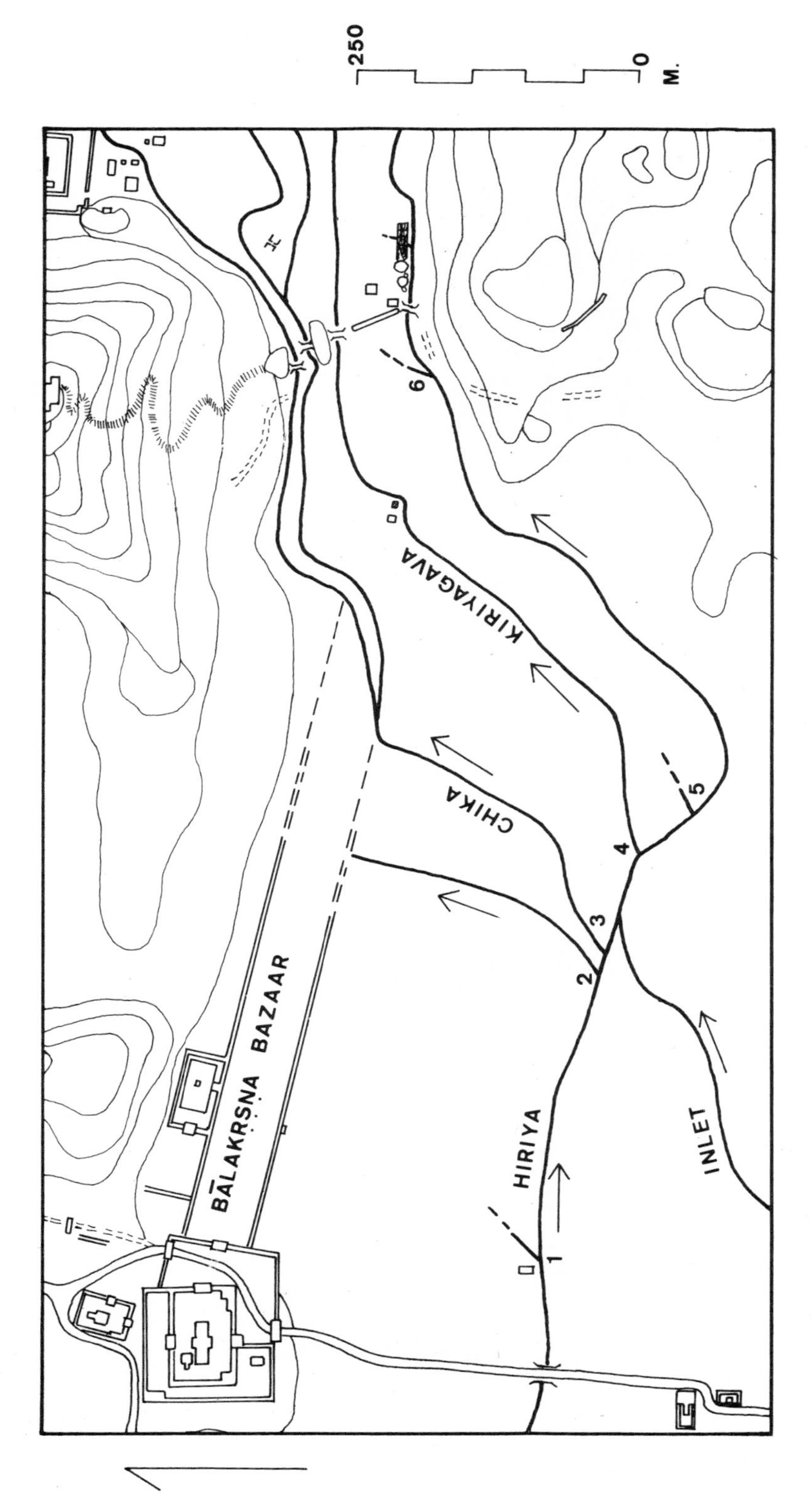

4.74 · Map of the irrigated valley from the Kamalāpuram–Hampi bridge to Matanga Parvatam (after VRP).

4.75 Panorama of the irrigated valley from the top of Matanga Parvatam; view to southwest.

4.76 Hiriya canal; the first outlet, view to west.

4.77 Hiriya canal; the fourth outlet, view to west.

4.78 Overview of the valley bund; view to southwest.

4.79 Paved road on the right bank of the Hiriya canal beneath Matanga Parvatam; view to northeast.

4.80 Right bank of canal broken away to reveal an earlier structure; view to west.

4.81 Right bank of canal broken away to reveal an earlier structure; view to south.

4.82 Bund across the valley below Matanaga Parvatam; view to northwest.

4.83 Upstream face of the valley bund; view to north.

4.84 Panorama of the exposed core of the valley bund; view to west.

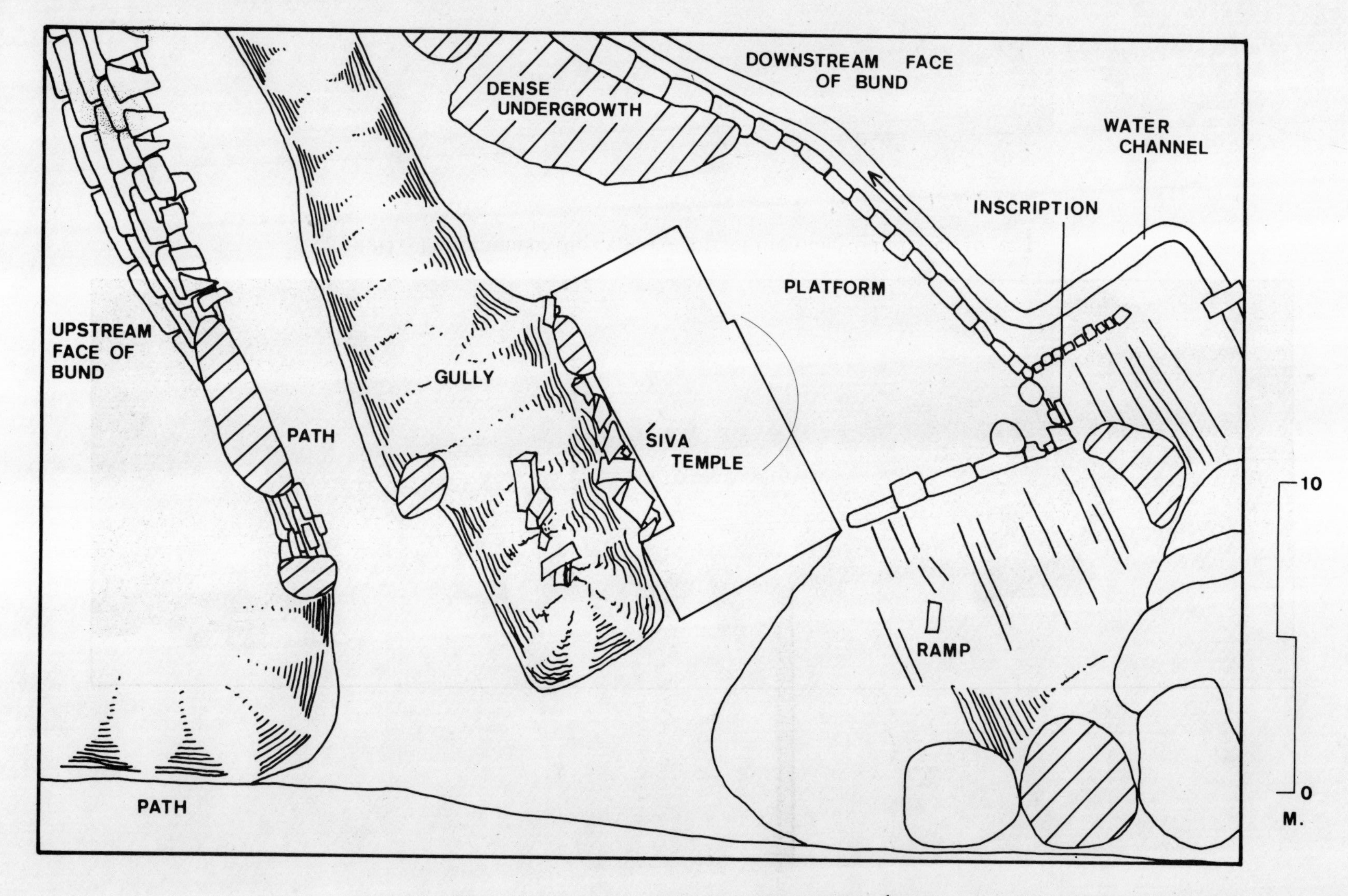

4.85 Plan of the south end of the valley and the Śiva temple.

4.86 Valley bund; close-up of the revetting on the upstream face.

4.87 Valley bund; close-up of the revetting on the downstream face.

4.88 Upstream face of the Hiriya bridge; view to east.

4.89 Downstream face of the Hiriya bridge; view to northwest.

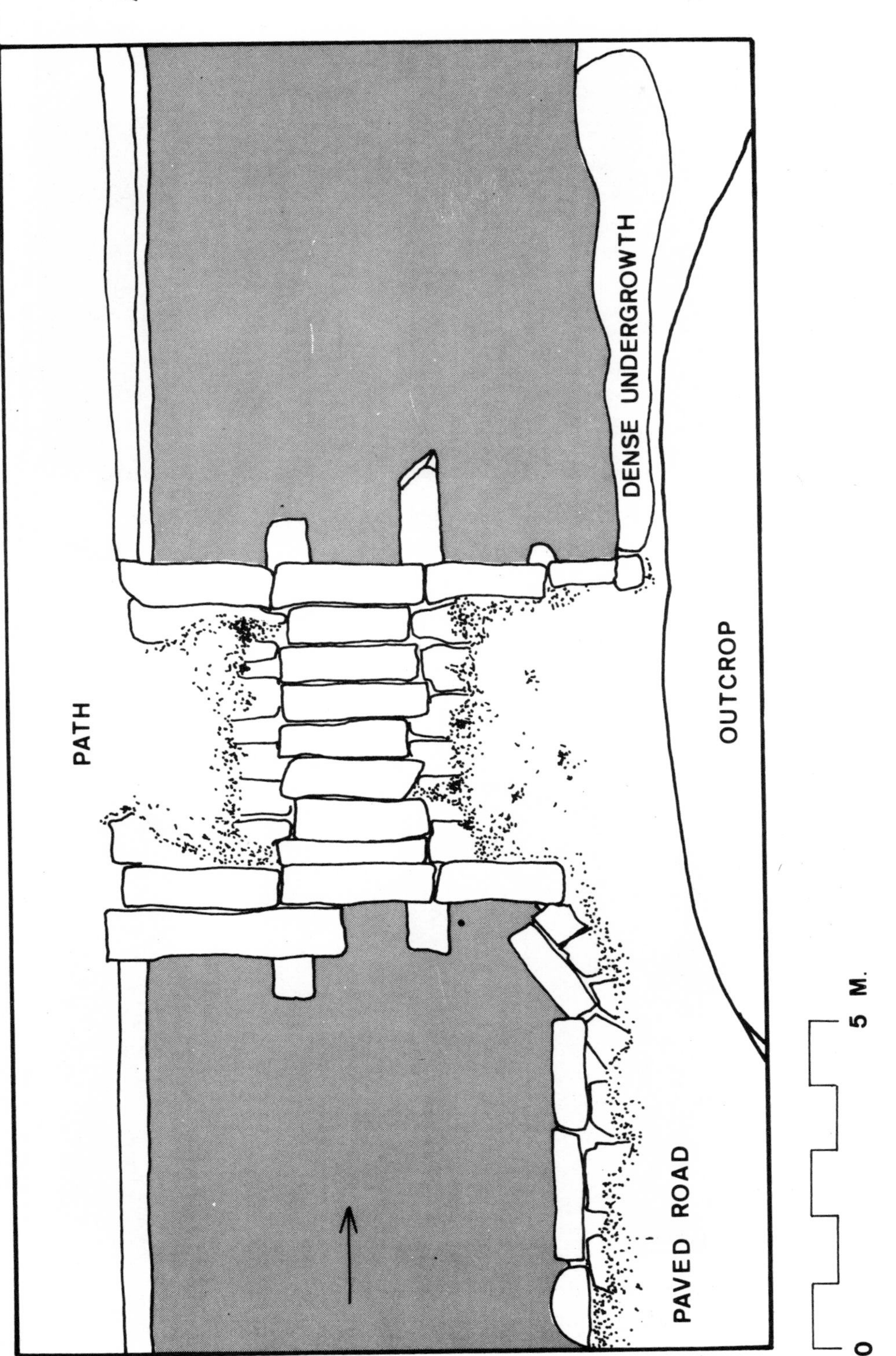

4.90 Plan of the Hiriya bridge.

4.91 Śiva temple on the valley bund; view to northwest.

4.92 Paved ramp descending from the top of the bund of the valley floor; view to northwest.

4.93 Close-up of the inscription marking the boundary of Achyutāpura.

4.94 Second bridge below Matanga Parvatam; view to northwest.

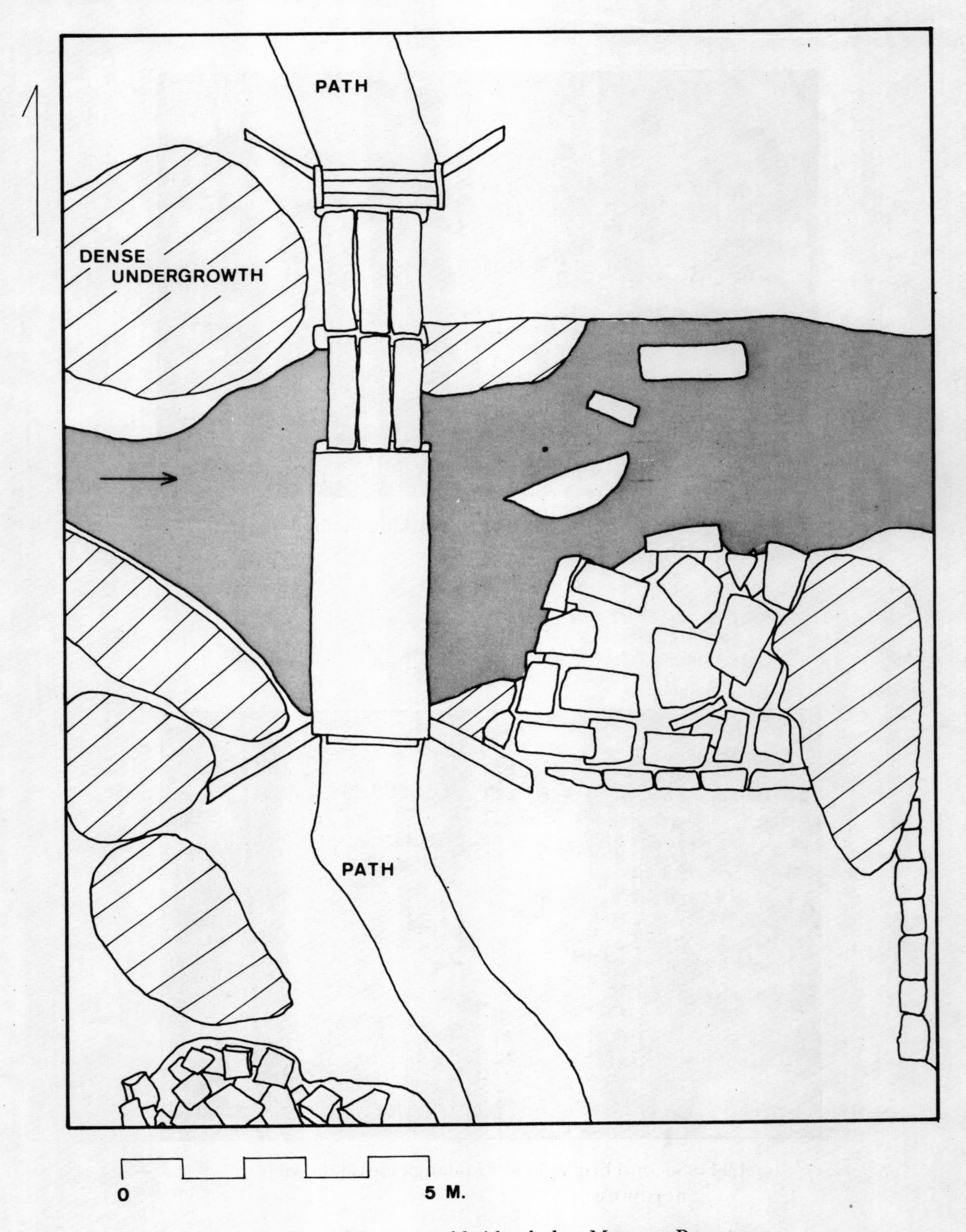

4.95 Plan of the second bridge below Matanga Parvatam.

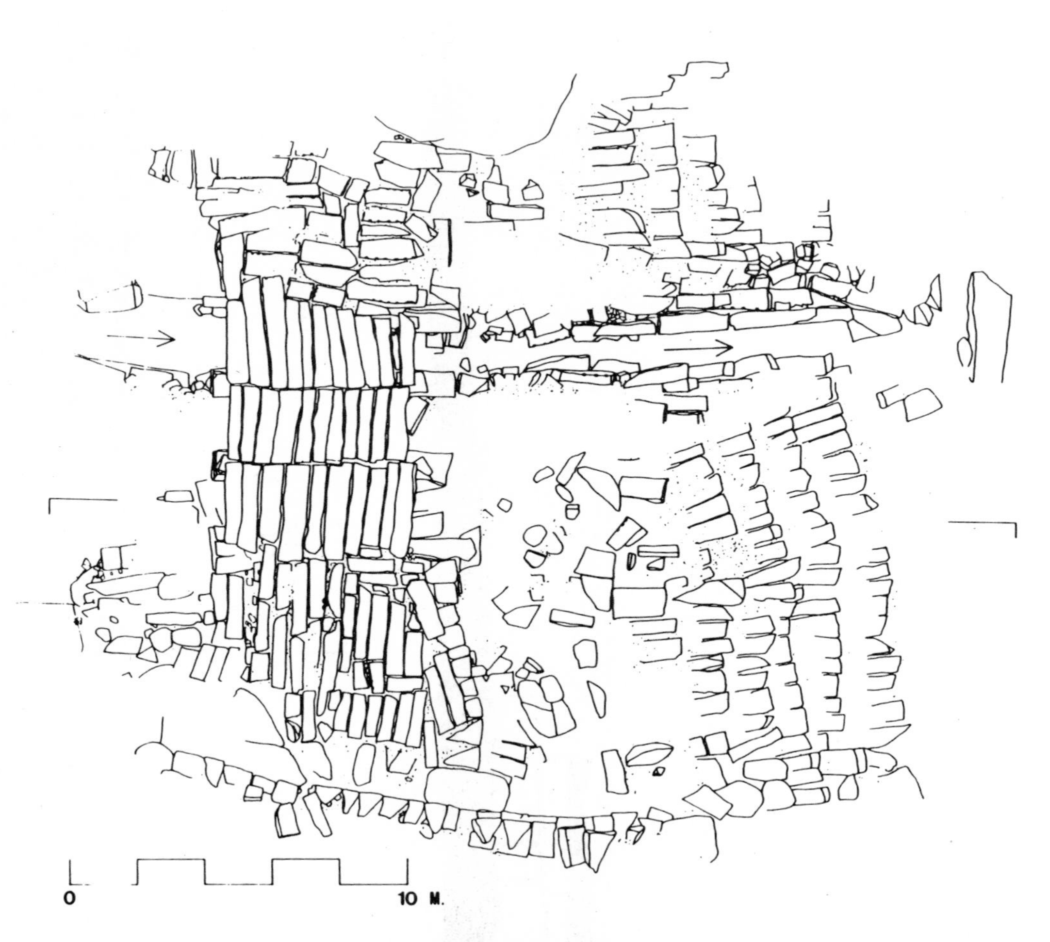

4.96 Plan of the waste weir (VRP).

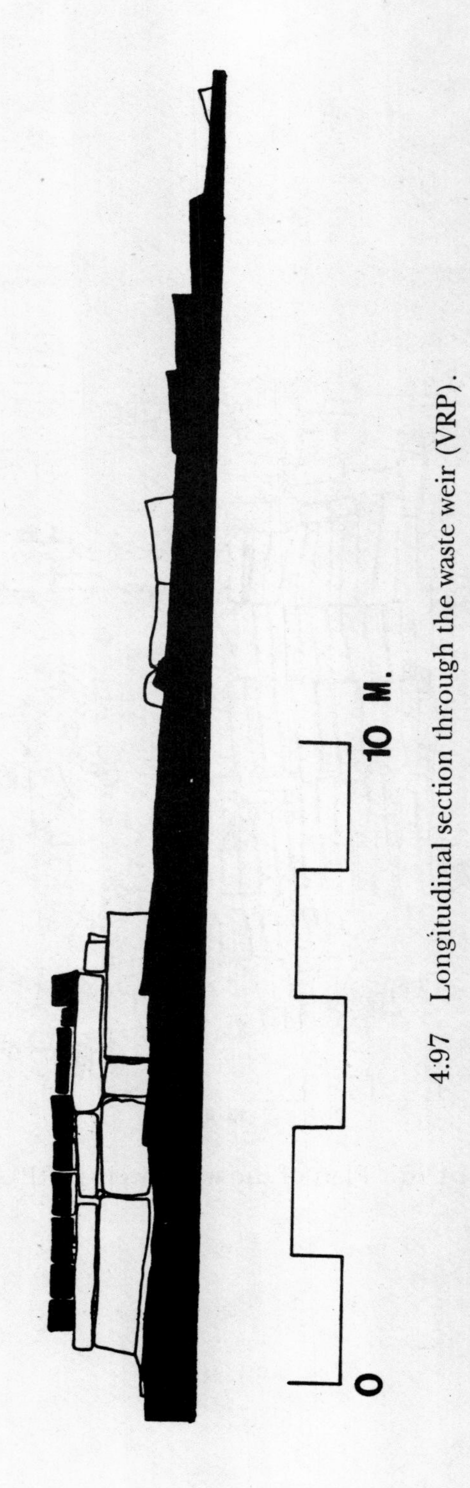

4:97 Longitudinal section through the waste weir (VRP).

4.98 Waste weir; view to northeast.

4.99 Waste weir; view to southwest.

4.100 Outlet feeding the old sluice; view to north.

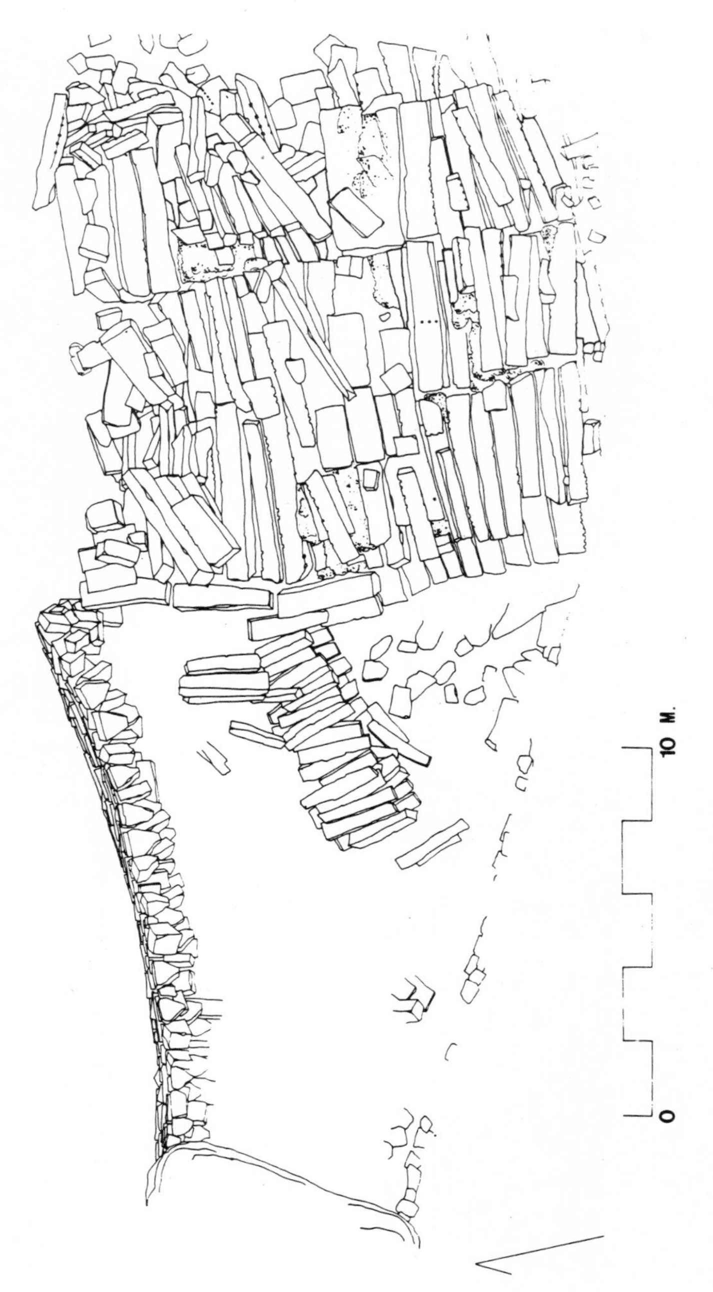

4.101 Plan of the three-bayed structure (VRP).

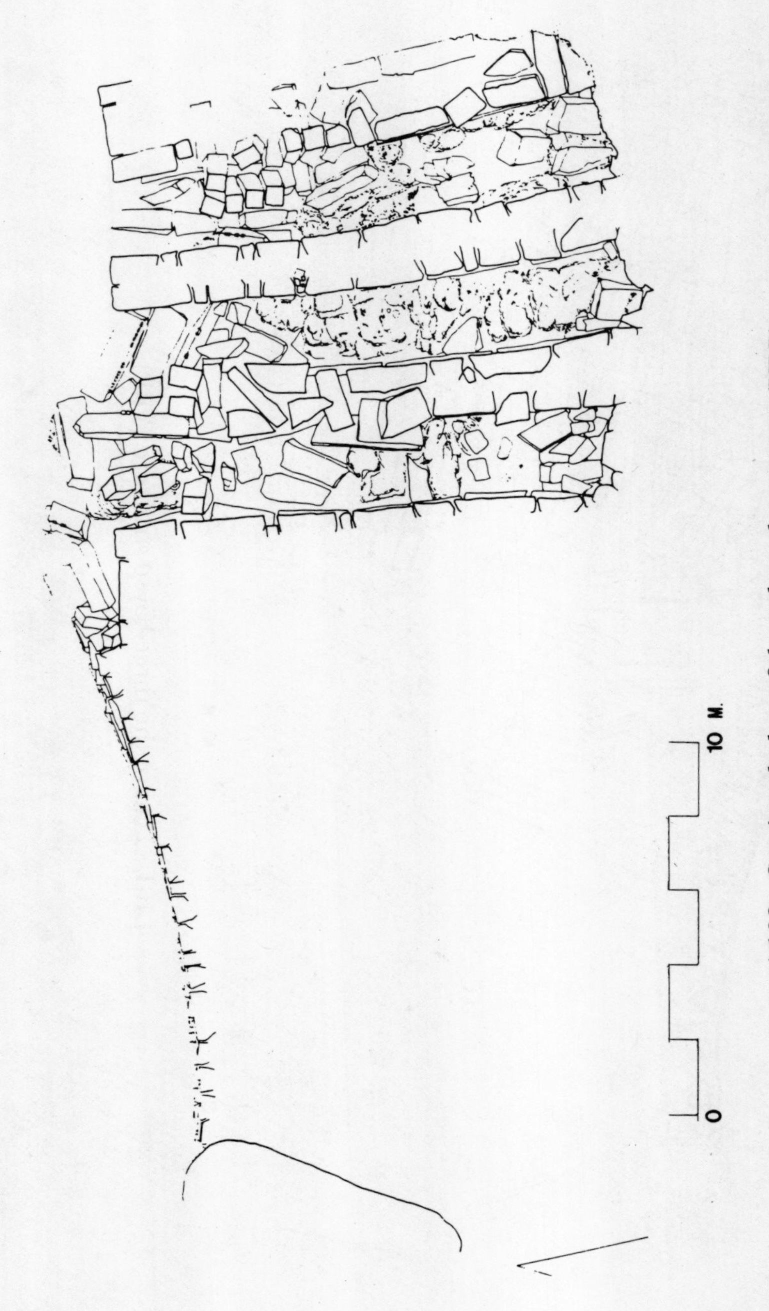

4.102 Sectional plan of three-bayed structure (VRP)

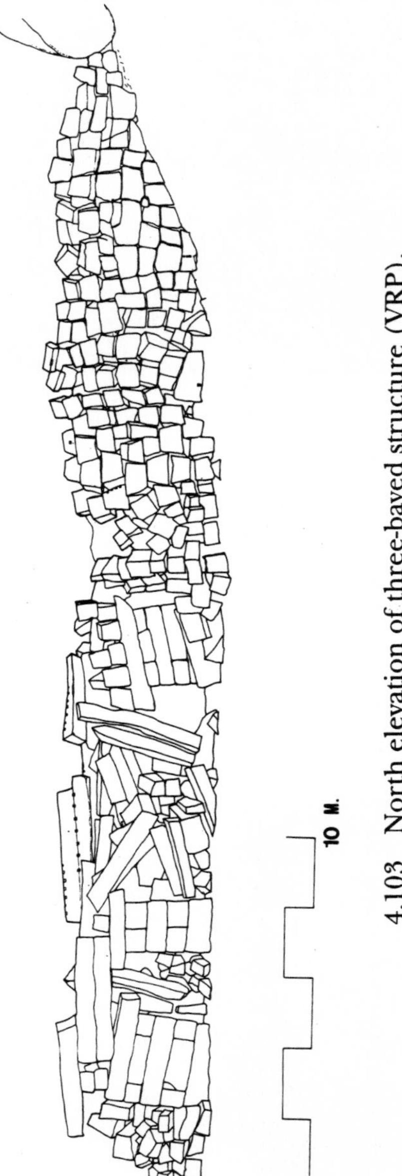

4.103 North elevation of three-bayed structure (VRP).

4.104 Three-bayed structure before clearance; view to west.

4.105 Three-bayed structure after clearance; view to west (VRP).

4.106 Ceiling beams laid on top of three-bayed structure, view to southwest (VRP).

4.107 Remains of channel running over three-bayed structure; view to east (VRP).

4.108 Hiriya canal; inscription on the inside of crevice, view to east.

4.109 Hiriya canal; close-up of inscription.

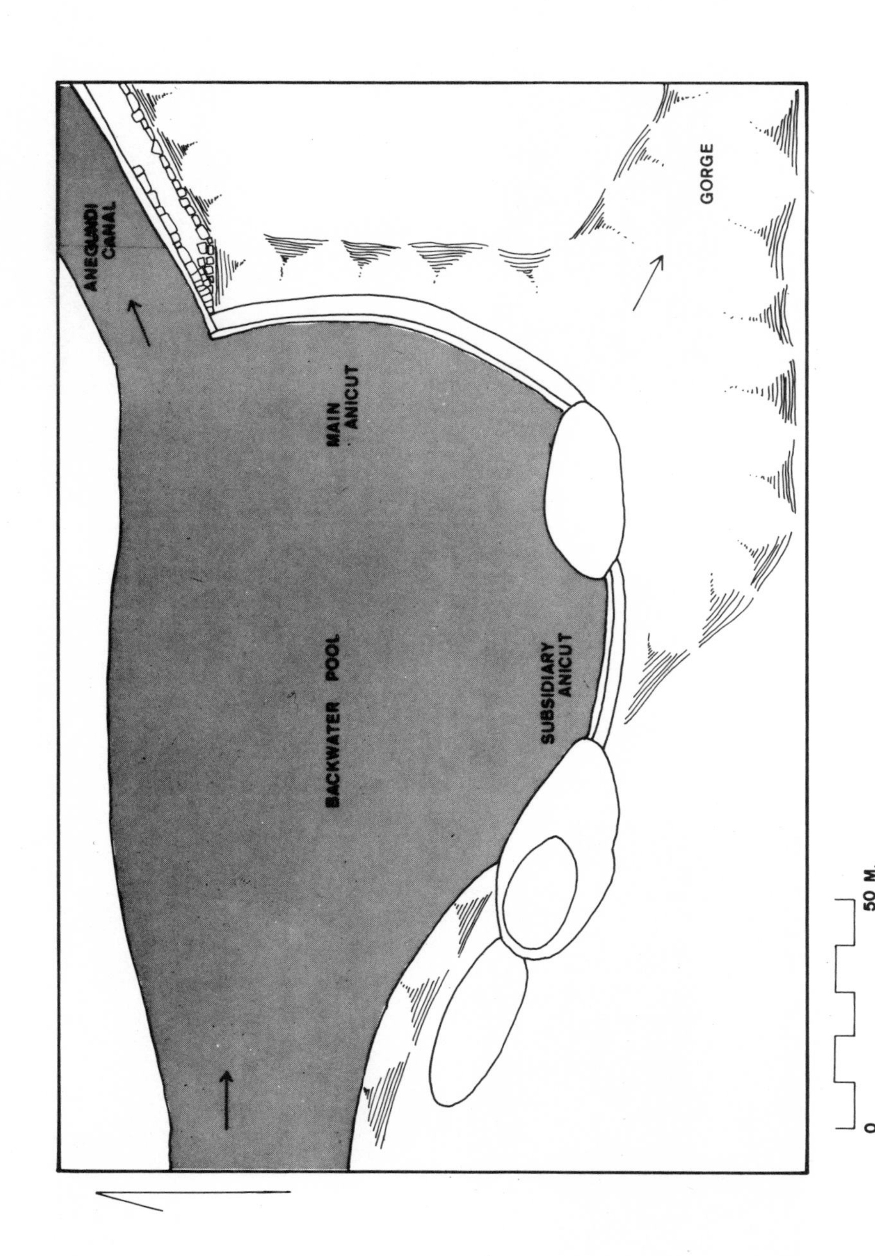

4.110 Diagrammatic plan of the Anegundi anicut.

4.111 Anegundi anicut; crest of main weir, view to north.

4.112 Anegundi anicut; downstream apron, view to north.

4.113 Anegundi anicut; vertical face of the downstream apron, view to north.

4.114 Anegundi anicut; close-up of constructional technique.

4.115 Anegundi canal flowing east; view to east.

4.116 Setu aqueduct; view to northeast.

4.117 Upstream face of the Rāya bund; view to north.

4.118 Rāya bund; first outlet, view to northwest.

4.119 Rāya bund; southeast elevation of the first outlet.

4.120 Rāya bund; second outlet, view to northwest.